Statistical Methods for Trend Detection and Analysis in the Environmental Sciences

Statistics in Practice

Statistics in Practice is an important international series of texts which provide detailed coverage of statistical concepts, methods and worked case studies in specific fields of investigation and study.

With sound motivation and many worked practical examples, the books show in down-to-earth terms how to select and use an appropriate range of statistical techniques in a particular practical field within each title's special topic area.

The books provide statistical support for professionals and research workers across a range of employment fields and research environments. Subject areas covered include medicine and pharmaceutics; industry, finance and commerce; public services; the earth and environmental sciences, and so on.

The books also provide support to students studying statistical courses applied to the above areas. The demand for graduates to be equipped for the work environment has led to such courses becoming increasingly prevalent at universities and colleges.

It is our aim to present judiciously chosen and well-written workbooks to meet everyday practical needs. Feedback of views from readers will be most valuable to monitor the success of this aim.

A complete list of titles in this series appears at the end of the volume.

Statistical Methods for Trend Detection and Analysis in the Environmental Sciences

Richard E. Chandler

Department of Statistical Science, University College London, UK

E. Marian Scott

School of Mathematics and Statistics, University of Glasgow, UK

A John Wiley & Sons, Ltd., Publication

Library of Congress Cataloging-in-Publication Data

Chandler, R. E. (Richard E.)
   Statistical methods for trend detection and analysis in the environmental sciences /
Richard Chandler, E. Marian Scott.
     p. cm. – (Statistics in practice ; 90)
   Includes bibliographical references and index.
   ISBN 978-0-470-01543-8 (hardback)
   1. Environmental sciences – Statistical methods. I. Scott, E. Marian. II. Title.
   GE45.S73C43 2011
   577.01′1 – dc22

                                                                2010051076

A catalogue record for this book is available from the British Library.

Print ISBN: 978-0-470-01543-8
ePDF ISBN: 978-1-119-99156-4
oBook ISBN: 978-1-119-99157-1
ePub ISBN: 978-1-119-99196-0

Typeset in 9/11 Times by Laserwords Private Limited, Chennai, India
Printed in Great Britian by TJ International Ltd, Padstow, Cornwall

Contents

Part II   CASE STUDIES   265

Preface

The life story of this book has been long and rather involved. It started with general conversations about the ubiquitous nature of the questions 'What is changing and by how much?' within environmental sciences, moved on to discussions about how some of the newer, potentially more powerful and almost certainly more realistic statistical approaches to answering such questions had not yet been translated over to the applied sciences communities and then ultimately reached the 'Wouldn't a book be a good idea?' stage. Many of those initial conversations were held around meetings organised by the Environmental Statistics section of the Royal Statistical Society. Momentum was maintained subsequently through a series of training courses for environmental science PhD students, loosely titled 'Quantifying the environment' and sponsored by the Natural Environment Research Council. As always, having formulated our grand and ambitious plans, the task of actually putting them into practice took much longer than we either hoped or expected. So, now that we have finally reached our objective, we owe a huge debt of thanks to all the staff at Wiley for their patience over more years than we care to admit.

Many people have contributed to the discussions above. Others have helped by responding constructively to our unsolicited requests for preprints or queries regarding technical details of their research. With apologies that we cannot mention all of these people by name, we take the opportunity here to thank all of them for their time and input. In particular, some of the work described herein has developed from past PhD projects co-supervised with Adrian Bowman at Glasgow, notably those of Andrew McMullan, Marco Giannitrapani and especially Claire Ferguson: the review chapters in Claire's PhD thesis formed the nugget for some of the sections in Chapter 4. We are also grateful to Rebecca Wilson at the University of Leicester, to Job Verkaik at KNMI and to Gavin Simpson and Don Monteith at UCL, for providing the ozone, wind speed and alkalinity data used in Part I of the book and for dealing with our queries with patience and good grace.

The book is split into two parts: the first a textbook-style introduction to the area and the second a collection of extended case studies demonstrating the practical application of modern statistical approaches to the analysis of trends in real environmental studies. These case studies have been chosen to span different environmental application areas as well as to highlight different methodologies, statistical issues and styles of analysis. To present this range and depth of material would not have been possible without the help of our many collaborators and contributors, who have been most helpful and forgiving throughout the long gestation of the book. To them we offer our warmest thanks and appreciation. We hope that the final result of this collaborative effort will appeal both to environmental scientists and to statisticians, and perhaps in a small way will encourage further interaction and engagement between the two communities.

This book includes an accompanying website. Please visit www.wiley.com/go/ trend_detection for more information.

Marian Scott and Richard Chandler
Glasgow and London

Contributing authors

Bryson C. Bates
Climate Adaptation Flagship
CSIRO Marine and Atmospheric
Research, Wembley, Western Australia
Australia

Adrian Bowman
School of Mathematics and Statistics
University of Glasgow
Glasgow G12 8QW
UK

Richard E. Chandler
Department of Statistical Science
UCL, Gower Street
London WC1E 6BT
UK

Stephen P. Charles
Climate Adaptation Flagship
CSIRO Land and Water
Wembley, Western Australia
Australia

Ludwig Fahrmeir
Department of Statistics
Ludwig-Maximilians-University Munich
Ludwigstraße 33
D-80539 München
Germany

Carla Rita Ferrari
ARPA Emilia-Romagna
Struttura Oceanografica Daphne, V le
Vespucci 2, 47042 Cesenatico (FC)
Italy

Marco Giannitrapani
Formerly of the School of
Mathematics and Statistics,
University of Glasgow.
Now at Novartis Farmaceutica S.A.,
E-08013 Barcelona, Spain

Elena N. Ieno
Highland Statistics
Suite N 226, Av Finlandia 21
CC Gran Alacant Local 9
03130 Santa Pola
Spain

Thomas Kneib
Department of Mathematics
Carl von Ossietzky University Oldenburg
D-26111 Oldenburg
Germany

Cristina Mazziotti
ARPA Emilia-Romagna
Struttura Oceanografica Daphne, V le
Vespucci 2, 47042 Cesenatico (FC)
Italy

Giuseppe Montanari
ARPA Emilia-Romagna
Struttura Oceanografica Daphne, V le
Vespucci 2, 47042 Cesenatico (FC)
Italy

Attilio Rinaldi
ARPA Emilia-Romagna
Struttura Oceanografica Daphne, V le
Vespucci 2, 47042 Cesenatico (FC)
Italy

E. Marian Scott
School of Mathematics and Statistics
University of Glasgow
Glasgow G12 8QW
UK

Ron Smith
CEH Edinburgh
Bush Estate, Penicuik
Midlothian EH26 0QB,
UK

Alain F. Zuur
Highland Statistics Ltd
6 Laverock Road
Newburgh AB41 6FN
UK

Part I

METHODOLOGY

1

Introduction

All scientific investigations are concerned with obtaining a deeper understanding of the world in which we live. Such investigations may be motivated primarily by curiosity, but more often by a recognition that such an understanding is beneficial to the wellbeing of humanity. Environmental science is an area where many would argue that the current level of knowledge pales into insignificance beside the capacity to enact rapid changes on an unprecedented scale, and therefore that it is not only beneficial but essential to understand more clearly the processes at work. Examples of current concerns include the response of climate to greenhouse gas emissions and the knock-on effects in areas such as water resources, agriculture and human health; the effects of industrial activity upon the quality of air and drinking water; the implications of intensive agricultural practices for biodiversity; and the sustainability of industrial-scale fishing operations. In all of these examples, the most compelling grounds for concern are observational: increases in global mean temperatures and various indices of extreme weather (Solomon *et al.*, 2007), increases in the incidence of respiratory diseases associated with particulate matter in the atmosphere (Anderson *et al.*, 1996; Zmirou *et al.*, 1998), declines in the numbers of farmland birds in Europe (Siriwardena *et al.*, 1998) and shrinking fish catches (Pauly *et al.* 2002), to cite just a few examples.

Broadly speaking, to develop an understanding of such phenomena there are two possible approaches. The first is to consider the fundamental processes that are believed to be operating and to build a more or less detailed model of these processes that can be used to make predictions and explore alternative scenarios. Examples of this 'process based' approach include the physical and chemical models of the atmosphere and oceans that are routinely used to provide projections of the earth's climate throughout the twenty-first century (Saunders, 1999; Solomon *et al.*, 2007).

The second approach is to analyse the available data, either to look for relationships that could explain how the system works or to test hypotheses suggested by process based considerations. 'Trend analysis' can be defined as the use of such an empirical approach

Statistical Methods for Trend Detection and Analysis in the Environmental Sciences, First Edition.
Richard E. Chandler and E. Marian Scott.
© 2011 John Wiley & Sons, Ltd. Published 2011 by John Wiley & Sons, Ltd.

to quantify and explain changes in a system over a period of time.[1] The statistical tools required to carry out a trend analysis range from the simple to the very advanced. However, the complexity of most environmental systems, often coupled with difficulties in making accurate observations, ensures that simple methods are rarely adequate for more than a preliminary inspection of the data. At best, such methods may fail to extract all of the available information (which, given the cost of obtaining much environmental data, is a waste of resources) and, at worst, they may yield misleading conclusions. To avoid these pitfalls it is therefore usually necessary to use more advanced methods, such as those described in the following chapters. Many of these have been developed relatively recently, and therefore are unlikely to be encountered in a traditional introductory statistics course for environmental scientists. However, all of them are well established in the statistical literature and have been found to be useful in a wide variety of applications. Furthermore, many of them can be implemented easily using freely available software.

Before proceeding any further, it is worth clarifying the subject matter of the book by defining what is meant by a 'trend', examining some of the questions that might lead one to carry out a trend analysis and summarising some of the difficulties and features that are commonly encountered in the analysis of environmental data. The stage will then be set for the statistics.

1.1   What is a trend?

We have already defined trend analysis as the investigation of changes in a system over a period of time. However, this is rather a loose definition. The use of quantitative methods requires a precise statement of the scientific question(s) of interest, framed in numerical terms. We therefore consider the behaviour of a system to be encapsulated by the values of some collection of numeric variables (for example the mean global temperature or the numbers of reported incidents of respiratory illness in a particular location) through time. In the simplest case, the data available for the analysis of such a system might consist of a sequence of regularly spaced observations of a single variable taken at equal time intervals: $y_1, \ldots, y_T$, say. Such data are often analysed using time series analysis techniques. In the time series literature, definitions of trend often refer to changes in the mean level of such a series. Chatfield (2003) defines trend in almost exactly these terms: '[Trend] may be loosely defined as "long-term change in the mean level". ' Kendall and Ord (1990) describe trend as 'long-term movement' – again implying a change in the mean level.

Most modern statistical methods require that an observed sequence $y_1, \ldots, y_T$ is regarded as the realised value of a corresponding sequence $Y_1, \ldots, Y_T$ of random variables. Equivalently, if all of the observations are assembled into a single column vector $\mathbf{y} = (y_1 \cdots y_T)'$ (here and elsewhere, a prime $'$ is used to denote the transpose of a vector or matrix), then $\mathbf{y}$ is considered to be the realised value of a random vector $\mathbf{Y} = (Y_1 \cdots Y_T)'$. This viewpoint, although perhaps surprising when seen for the first time, enables scientific questions to be framed, in completely unambiguous terms, as questions about the probability distribution from which the observations were drawn. Consider, for example,

---

[1] The term 'trend' is also sometimes used to describe the variation of some quantity over a spatial region. In general, the analysis of spatial variation requires different techniques to that of temporal variation: it is therefore not the primary focus of this book. However, tools for the analysis of space–time data are considered briefly in Section 6.1, and several of the contributions to Part II consider both spatial and temporal variation.

the expected values of the random variables $Y_1, \ldots, Y_T$:

$$\mathrm{E}\,(Y_t) = \mu_t, \text{ say} \qquad (t = 1, \ldots, T). \tag{1.1}$$

where μ_t can be thought of as an 'underlying' mean of the process at time t. In some sense then, the sequence $\mu_1, \ldots, \mu_T$ provides a formal mathematical representation of the notion of 'change in the mean level', and this sequence itself could be defined as the trend as in Diggle (1990, Section 1.4). If this definition of trend is accepted then, for example, the rather vague question 'Is there a trend in my data?' is equivalent to the much more precise 'Are $\mu_1, \ldots, \mu_T$ all equal?'.

Although the concept of trend as 'change in the mean level' is ubiquitous in the time series literature, it is not universal in the environmental sciences. For example, Robson (2002) distinguishes between 'trend' and 'fluctuation':

> A data series is said to show *trend* if, on average, the series is progressively increasing or decreasing. A data series shows *fluctuation* if the average of a series changes noticeably through time but not in any consistent direction.

Furthermore, environmental problems are not always focused upon the average behaviour of a process. For example, one may be interested in assessing changes in the frequency of 'extreme' events such as large floods or dangerously high air pollution episodes. To define 'trend' solely in terms of mean levels therefore seems unnecessarily prescriptive: long-term changes in any statistical properties are potentially of interest. However, the exact meaning of 'long-term' depends upon the application: a climatologist would probably regard a couple of degrees' rise in global mean temperature over a period of decades as convincing evidence of a warming trend, whereas a geologist may regard this as uninteresting by comparison with changes in the frequency and severity of glacial epochs.

On the basis of these considerations, we offer the following definition as the focus of the statistical methods to be discussed in this book:

Definition
'Trend' is long-term temporal variation in the statistical properties of a process, where 'long-term' depends on the application.

1.2   Why analyse trends?

As indicated above, the use of quantitative methods requires that the objectives of an investigation are specified precisely. In general, the choice of an appropriate analysis method will depend upon the questions of interest. It is therefore worth considering the kinds of situation in which an analysis of trend might be useful. In environmental applications, possible reasons for carrying out a trend analysis include:

(a) To describe the past behaviour of a process. For example, it may be of interest to quantify the nature and extent of changes in a region's climate, or in some wildlife population, over a period of time.

(b) To try and understand the mechanisms behind observed changes. A common goal is to determine whether, or to what extent, such changes are associated with human activity rather than 'natural' processes.

(c) To make assessments of possible future scenarios, by extrapolating past changes into the future. Such extrapolations are often used to assess the risk of future adverse events such as severe floods or species extinction, and also to justify the introduction of policies designed to reduce this risk if it is judged to be unacceptably high.

(d) To monitor the effectiveness of environmental control policies. For example, if new controls are introduced to limit industrial emissions, it may be of interest to determine the extent to which a response can be detected in river chemistry and aquatic ecosystems.

(e) To enable the analysis of systems where long-term changes serve to obscure the aspects of real interest. For example, an ecologist may be interested in studying the interactions between several species that have been steadily increasing or decreasing in abundance due to some 'external' factors; in this case, a first step in the analysis may be to identify and remove the trends so as to see more clearly the species interrelationships.

Such objectives are not unique to environmental applications, of course; however, some of them have perhaps a more central role here than in other areas.

To illustrate some of the issues that may arise in an environmental context, we now consider some simple examples that will be used for illustrative purposes throughout Part I of the book.

1.3   Some simple examples

1.3.1   Dutch wind speeds

Our first example relates to records of hourly wind speed from 13 weather stations in the Netherlands, provided by the Royal Netherlands Meteorological Institute (KNMI). The station locations are shown in the left-hand panel of Figure 1.1. Record lengths range from 41 years (at both Beek and Hoek van Holland) to 53 years (at Schiphol); all records ran until the end of 2002 and have been carefully quality controlled and standardised using procedures described in Verkaik (2000a, 2000b). During the quality control procedure, some hourly readings were identified as suspect; some are also missing. The sites with the highest proportions of missing or suspect readings are IJmuiden (3.2 %), Hoek van Holland (0.75 %), Deelen (0.05 %) and Gilze Rijen (0.04 %). None of the remaining sites has more than 25 missing or suspect hourly readings during the entire period of record.

We highlight two applications in which the study of these data may be of interest. The first relates to engineering risk assessment. Much of the Netherlands is below sea level and is protected from inundation by dikes. The main cause of dike failure is wave damage, which itself is primarily associated with a combination of storm surges and high wind speeds (Ettema, 2001). An assessment of the wind climate of the country is therefore essential for ensuring the safety of dike systems.

The second application relates to the search for renewable energy sources. Wind power is increasingly being seen as a cheap alternative to fossil fuels. However, to provide a dependable energy supply wind speeds must be high enough to generate the required output, but not so high as to necessitate shutting down the turbines for safety reasons. To determine the long-term viability of a wind power scheme in an area therefore, an analysis of wind speed would be helpful.

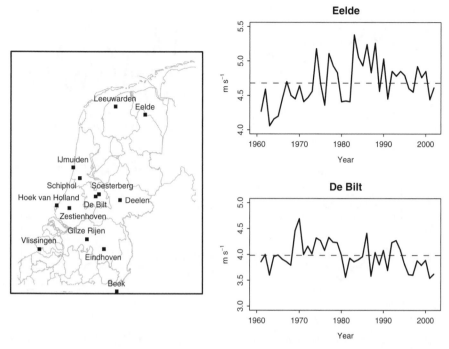

Figure 1.1   Left: locations of Dutch weather stations yielding wind speed data. Right: annual mean wind speeds for two of the stations, 1961–2002 (dashed lines indicate series means).

In both of these applications, it is clearly of interest to detect and quantify trends in wind speeds – and also, if possible, to produce useful extrapolations into the future. The objectives of a trend analysis may therefore be summarised under points (a) to (c) in Section 1.2. In an engineering risk assessment, the focus will be primarily upon the frequency of occurrence of extremely high winds; however, for wind power evaluation both high and low winds will be of interest. Motivated by such considerations, Smits *et al.* (2005) analysed the data considered here, examining trends in the annual numbers of 'events' of varying degrees of severity and reporting overall decreases in the frequency of high severity events between 1962 and 2002, although a couple of stations showed increases.

As a simple first step in examining these data, the right-hand panels of Figure 1.1 show the annual mean wind speeds at two of the stations from 1961 to 2002. The graph for De Bilt suggests a decrease over this period; however, there appears to be an upward trend at Eelde, at least from 1960 to the mid 1980s. Both of these findings are in line with the results of Smits *et al.* (2005, Figure 5), although the analysis here is slightly different since Figure 1.1 is concerned with mean wind speed whereas those authors studied trends in 'severe' events. In general, it is unwise to overinterpret the results of simple visual inspections; nonetheless, it is worth noting the possibility that trends may vary over the region.

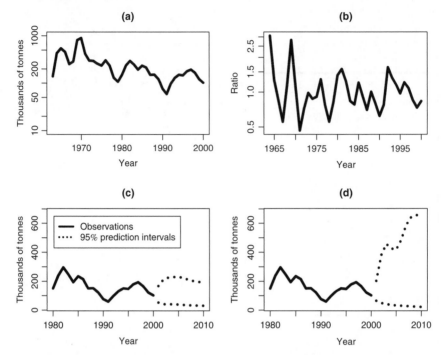

Figure 1.2   Annual North Sea haddock biomass, 1963–2000. (a) Biomass time series
(Y_t), *(b) time series of ratios* (Y_t/Y_{t-1}), *(c) observed time series 1980–2000, along with*
95% prediction intervals to 2010 obtained using model (1.2), (d) observed time series
1980–2000, along with 95% prediction intervals to 2010 obtained using model (1.3).
Logarithmic scales are used for the y axis in (a) and (b), but not in (c) and (d).

1.3.2   North Sea haddock stocks

Figure 1.2 shows some plots relating to annual estimates of spawning stock biomass for
North Sea haddock, from 1963 to 2000. These data are from DEFRA (2004). The biomass
estimates are based on mathematical models that combine information from international
catches and fishing activity, along with research vessel surveys. Over time, there has
been some variability in the quality of the data available for fish stock assessments.
Furthermore, care needs to be taken when analysing 'data' that are themselves produced
by a model. A complete analysis of the haddock stocks data would need to account both
for the variability in data quality and for the properties of the model used to produce the
data, in order to avoid overinterpreting artefacts that may be caused by either of these
features (see Section 1.4.4 below). Nonetheless, for illustrative purposes in the first part
of the book we will take these data at face value.

Biomass data are used in fisheries management to address, among other things,
questions regarding the sustainability of current levels of fishing activity. Interest
therefore lies in producing assessments of future biomass, as well as in trying to quantify
the effect of fishing activity upon stocks. It is also potentially of interest to identify
responses to past changes in the management of the fishery, since the nature of such
responses can be used to guide future management strategy. These objectives can be

summarised under points (a) to (d) in Section 1.2. We will introduce the analysis by looking at some simple future scenarios.

Figure 1.2(a) is a time series plot of the annual stock biomass time series from 1963 to 2000, in thousands of tonnes. A logarithmic scale has been used for the y axis. This is equivalent to plotting the logged biomass time series; the only difference is that the axis is labelled in the original measurement units (which are easier to interpret than their logarithms). The plot shows an apparent linear decrease over the period of record, with constant variation about this overall linear trend. This in turn suggests that the series could be represented using a linear regression model for log biomass:

$$\log Y_t = \beta_0 + \beta_1 t + \varepsilon_t, \tag{1.2}$$

where Y_t is the biomass in year t and ε_t is an 'error'. Linear regression models rely on the assumptions that the errors are drawn independently from normal distributions with mean zero and constant variance. We will see in Chapters 3 and 5 that for this particular data set, all of the assumptions seem to be satisfied except that of independence. The dependence between the errors can, however, be represented via a model of the form

$$\varepsilon_t = \phi_1 \varepsilon_{t-1} + \phi_2 \varepsilon_{t-2} + \delta_t,$$

where δ_t is a sequence of independent, normally distributed random variables with zero mean and constant variance, and ϕ_1 and ϕ_2 are extra parameters. Taking time $t = 1$ at the time of the first observation in 1963 and considering only data up to 2000, the coefficient estimates for this model are $\hat{\beta}_0 = 5.977$, $\hat{\beta}_1 = -0.034$, $\hat{\phi}_1 = 0.996$ and $\hat{\phi}_2 = -0.584$.

An alternative strategy for analysing these data is to study the proportional increase in biomass between years $t - 1$ and t:

$$R_t = Y_t / Y_{t-1}.$$

In a stable population, the ratio R_t is expected to fluctuate around the value 1. Figure 1.2(b) shows the values of these ratios from 1964 to 2000, again on a logarithmic scale. No ratio can be computed for 1963, because data for 1962 are not available. The series does indeed appear to fluctuate around 1 and, apart from a couple of points at the beginning of the series, shows fairly constant variability on the log scale. In Chapter 5, it is shown that the model

$$\log R_t = 0.493 \log R_{t-1} - 0.660 \log R_{t-2} + z_t$$

provides a good fit to the ratios. Here, (z_t) is another sequence of independent, normally distributed random variables with zero mean and constant variance. Since $\log R_t = \log(Y_t/Y_{t-1}) = \log Y_t - \log Y_{t-1}$, this model can be rewritten in terms of the actual biomass values as

$$\log Y_t = 1.493 \log Y_{t-1} - 1.153 \log Y_{t-2} + 0.660 \log Y_{t-3} + z_t. \tag{1.3}$$

In Chapter 5, we will see that models (1.2) and (1.3) both provide an excellent fit to the biomass time series; indeed, on the basis of these data it is very difficult to distinguish between them. However, the models have very different interpretations. The first asserts that over the period of record, biomass has followed a deterministic trend which is linear on the log scale. By contrast, there is no such deterministic component in the second model: here, the current year's biomass is seen as an adjustment to that of the previous year, so that any apparent pattern in the observed series is simply an accumulation of year-to-year variation.

The differences between the models are emphasised emphatically when they are used to make future projections. The bottom plots in Figure 1.2 show the last 20 years' observations, along with 95 % prediction intervals for each year from 2001 to 2010 under each model. The plots are drawn to the same scale and do not use logarithmic scaling for the y axes. This emphasises the practical differences between the models: (1.2) implies a continuing gradual decrease in the spawning stock, whereas the prediction intervals from (1.3) encompass scenarios ranging from a more rapid decrease to a population explosion. These differences should cause concern about the use of either model for extrapolation; we will see later on that such concerns are well founded.

This example illustrates nicely the difficulties of extrapolation. We have two models, which both fit very well to the data, but which yield radically different views of the future and hence carry different implications for fishery management. Of course, what is lacking in the analysis so far is any consideration of mechanisms that are known to affect the biomass – the models are purely descriptive. If a model could be built linking biomass to some index of fishing effort, for example, one might have much more faith in its predictions. This idea is pursued later in the book.

1.3.3   Alkalinity in the Round Loch of Glenhead

Since the mid 1980s it has been widely accepted that industrial emissions of, in particular, oxides of sulfur and nitrogen can be associated with the acidification of freshwater bodies and consequent damage to the associated ecosystems; see, for example, Battarbee *et al.* (1985). Recognising this, international agreements have led to substantial reductions in sulfur dioxide emissions in many countries since 1980. To monitor the response of water bodies to these emissions reductions in the UK, the United Kingdom Acid Waters Monitoring Network (UKAWMN) was set up in 1988 (Monteith and Evans, 2001). Water samples are taken at regular intervals from 11 lakes and 11 streams, representing some of the more acid sensitive freshwaters in the UK; these are then subject to chemical analysis.

Figure 1.3 shows some data from one of the UKAWMN sites, the Round Loch of Glenhead in southwest Scotland. At this site, samples are taken at roughly three-month intervals; the data in Figure 1.3 run from March 1988 to June 2002. Although there is some variation in the interobservation times, most of them are sufficiently close to three months that the series can be regarded as regularly spaced for practical purposes. The main variable of interest is alkalinity, defined as the capacity of water to neutralise an acid and measured in microequivalents of calcium carbonate per litre ($\mu eq\,l^{-1}$). The alkalinity series in Figure 1.3 shows pronounced seasonal oscillation, with a fairly clear increase from 1996 onwards. It is of interest to determine whether this is a response to declining sulfur emissions. This could, in principle, be investigated by relating the alkalinity measurements to sulfur emissions from nearby industrial areas. Unfortunately, however, emissions data are not available at a fine enough scale to support such an analysis. Instead, therefore, sulfate concentrations in the water samples have been taken as a surrogate for sulfur deposition. This introduces a further complication: as well as industrial sulfur emissions, sulfate can be introduced as a neutral salt in sea-spray, which reaches the Round Loch in small quantities since the site is only around 30 km from the coast. The influence of marine deposition at this site is confirmed by the presence of chloride ions in the water; there is no other plausible explanation for these. Assuming that the chloride is entirely marine-derived, the amount of sulfate from marine sources can be estimated according to the known ratio of sulfate to chloride in sea water (Evans, Monteith and Harriman, 2001). The remainder is taken as a surrogate for industrial sulfur deposition.

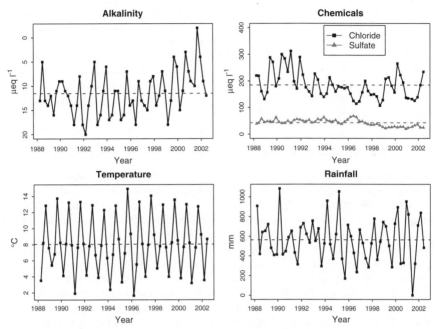

Figure 1.3   Quarterly time series of alkalinity, sulfate and chloride concentrations and meteorological variables at the Round Loch of Glenhead, 1988–2002.

The second panel of Figure 1.3 shows the quarterly time series of both chloride and nonmarine sulfate at the site. The sulfate series appears stable until about 1996, and declines thereafter; this seems consistent with the trend in alkalinity already observed, although the sulfate series shows no obvious seasonality. The chloride series in this panel is of interest because marine deposition has other effects besides the introduction of additional sulfate (Evans, Monteith and Harriman, 2001). It may therefore be worth determining whether there is any relationship between alkalinity and chloride. A quick inspection reveals that the main feature of the chloride series is a period of enhanced deposition during the early 1990s, as well as some indication of seasonality. These are both probably associated with increased rates of marine deposition when weather conditions are favourable.

Weather conditions themselves control surface water chemistry in a variety of ways: for example, weathering rates are dependent on both temperature and rainfall. The bottom two panels in Figure 1.3 show quarterly time series of temperature and rainfall from a weather station 3 km from the Round Loch, for the same period as the water chemistry data. The original data were daily; each temperature in Figure 1.3 is an average over the 91 days prior to sampling, and the corresponding rainfall is a total over the same period. Both series show pronounced seasonality. Apart from this, the temperatures appear fairly stable. The rainfall series, on the other hand, seems to show a sudden drop in the second half of 1996, with a subsequent gradual recovery to the initial level except for an unusually dry summer in 2001.

The primary motivation for this analysis is to determine whether there have been any significant changes in alkalinity at this site, and if so whether they are associated with reductions in industrial sulfur deposition. Clearly, however, the processes controlling alkalinity are complex and it is necessary to disentangle the relative contributions of weather, marine deposition and industrial deposition before we can answer the questions of primary interest. The objectives of the study can therefore be summarised under points (a), (b) and (d) in Section 1.2. The example has been chosen to illustrate a situation in which it is important to discriminate between the effects of different factors upon a quantity of interest. In Chapter 3, we show that rainfall and temperature are strongly associated with alkalinity, as is chloride; however, the existence of a relationship with sulfate is more difficult to establish.

1.3.4   Atmospheric ozone in eastern England

Ozone (O_3) is generated in the lower atmosphere as a result of reactions between hydrocarbons (generated from the burning of fossil fuels) and oxides of nitrogen (NO_x), triggered by exposure to ultraviolet radiation. Ozone is known to pose risks to human health and to be associated with increases in mortality (Anderson et al., 1996); it also damages crops and vegetation (Fowler et al., 1999). It has been identified as a greenhouse gas, with the potential to influence climate change as well as air quality (AQEG, 2007). As a result, many countries and organisations have defined targets that aim to limit atmospheric ozone concentrations for the protection of human health and vegetation. In the European Union (EU), for example, Directive 2008/50/EC of the European Parliament sets ozone targets; it also specifies that member states must collect, exchange and disseminate air quality information. Furthermore the World Health Organisation has issued its own recommendations (WHO, 2000) regarding levels of ozone exposure.

Against this background, the EU-funded GEOmon project[2] aims to collate data on various atmospheric pollutants including ozone, from a network of sites across Europe. Data from different locations have been 'harmonised' to ensure that they are comparable (Henne and Fleming, 2008). Figure 1.4 shows a specimen series from one of these sites, the Weybourne Atmospheric Observatory in eastern England. The series is at a daily timescale and runs from January 1989 to December 2008. However, the measurement station was resited in 1992, to a location around 2 km from its original position. This may have induced some inhomogeneity in the series, although the data harmonisation should have alleviated the problem to some extent. As with the haddock stocks example in Section 1.3.2, we will take these data at face value for illustrative purposes. For more details of the data and measurement station, see Penkett et al. (1999) and Fleming et al. (2006).

Each daily value in Figure 1.4 is an average of eight readings, taken at hourly intervals from 09:00 to 16:00 GMT. However, there are extended periods during which the station did not operate due to shortage of funding and many other days where one or more hourly readings are missing for a variety of reasons including instrument error, meteorological effects or contamination. In Figure 1.4 and all subsequent analyses, data have been used only for days with a full set of eight hourly readings. This is to avoid introducing artificial inhomogeneities, affecting variability in particular, by calculating daily means from different numbers of observations. In principle, such artefacts could be accounted for within a statistical analysis; however, by restricting attention here to days with complete data we aim to keep things reasonably simple. As a result, a total of 2821

---

[2] See http://www.geomon.eu.

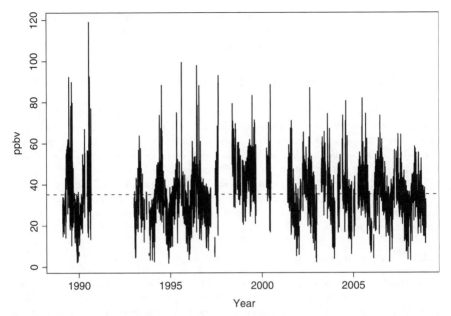

Figure 1.4   Daily ozone concentrations (parts per billion volume, ppbv) at Weybourne, Norfolk, 1989–2008. Dashed line indicates mean of available observations.

of 7305 daily readings are missing from the record as analysed here: Figure 1.4 shows the missing periods clearly. It also shows pronounced seasonality, with highest values in the summer months and an apparent decreasing trend in extreme ozone levels in particular. Overall levels seem particularly high during 1998 and 1999. However, there are many missing values during this period and it is possible that the available measurements are concentrated during the summer months when ozone levels are naturally high: thus the large proportion of missing data makes the plot difficult to interpret.

As discussed above, there is considerable interest in monitoring ozone levels across Europe, as well as in other parts of the world. With respect to Figure 1.4, questions include whether or not there are trends in the data, and if so whether these can be related to legislation that is designed to limit pollutant emissions and hence reduce ozone concentrations in the long term. These correspond to objectives (a) and (d) in Section 1.2. The mechanisms of ozone formation suggest that emissions reductions should lead in particular to declines in summer ozone peaks. Conversely, reductions in NO_x emissions are thought to create conditions that are more favourable for ozone formation during the winter months; one might therefore expect increasing ozone trends in winter. Figure 1.4 suggests that there is indeed a gradual decline in peak ozone values over time, although any trends during winter are not immediately apparent. Given that the seasonality is of particular interest, it would be useful to be able to extract a 'seasonal' component from the data so as to visualise any changes more clearly. From this perspective, other features of the data – such as irregular episodes or long-term changes in the mean level – may be seen as a nuisance that obscures some of the structure of interest. Thus, objective (e) in Section 1.2 is also relevant: if we can identify any overall trend in the data, we can then remove it and examine the seasonal cycle in more detail.

This example has been chosen primarily because it provides an opportunity to illustrate methods for the analysis of series with large proportions of missing data. This situation is not uncommon in environmental applications.

1.4   Considerations and difficulties

The examples above highlight some features that are commonly encountered in environmental studies. These include missing or irregularly spaced observations, data quality issues due to measurement difficulties and the analysis of data from a network of sites. The need to accommodate such features will often be an important consideration when determining an appropriate strategy for analysis, along with several other issues of a statistical nature. We here outline some of the issues that will be discussed in greater depth later in the book. For a perspective on wider considerations that might be relevant in any statistical investigation, the excellent review by Cox (2007) is strongly recommended.

1.4.1   Autocorrelation

Most 'standard' statistical techniques assume that the available data can be regarded, at some level, as an independent random sample from a 'population' of interest. This assumption is critical to the construction of standard hypothesis tests and confidence intervals. Of necessity, however, a trend analysis invariably relies upon the analysis of time series data and, in general, successive observations in a time series will not be independent. Consider, for example, a time series of hourly wind speeds from one of the Dutch weather stations in Figure 1.1. High wind speeds are associated with storms that may last for several hours or even days: therefore, if the wind speed at 9 a.m. today is much higher than average for the time of year, it is likely that the wind speed at 10 a.m. will also be higher than average.

Dependence between successive observations in a time series is referred to as 'autocorrelation'. In practice, autocorrelation is mostly (but not always) positive, as in the wind speed example: high and low values tend to cluster together more than would be expected in an independent sequence. This clustering can sometimes be mistaken for a trend, especially in relatively short series. A more sophisticated view is that a positively autocorrelated sequence contains less information than the same number of independent observations, so that the 'effective sample size' is reduced in the presence of positive autocorrelation. Therefore, techniques based on independent observations will tend to overestimate the precision with which quantities of interest can be estimated. To see this, consider a 'perfectly autocorrelated' sequence $y_1, \ldots, y_T$ in which y_1 is drawn from some probability distribution and then $y_2 \ldots, y_T$ are all set equal to y_1 – there is clearly only one observation's worth of information here.

There are several simple ways to account for autocorrelation in 'standard' analyses (for example by studying annual rather than hourly wind speed series because successive annual values are likely to be effectively independent). In general, however, it is preferable to use analysis methods that are specifically designed for use with autocorrelated data. Several such methods will be described in this book: the fact remains, however, that to discriminate between long-term changes and the effects of autocorrelation is often one of the most difficult tasks in a trend analysis.

1.4.2   Effect of other variables

Environmental processes do not evolve in isolation. To some extent, therefore, it is not possible to study any environmental variable without considering other aspects of the system to which it belongs. In a trend analysis, failure to do this may lead to genuine trends remaining undetected because they are 'masked' by the effects of other variables; or to uninteresting trends being identified because they are merely a response to changes in another variable. The Round Loch water chemistry study in Section 1.3.3 is a good example of this, since trends in the alkalinity of the loch could conceivably be ascribed to changes in industrial emissions, to changing weather conditions or to changes in the frequency of marine deposition events.

The need to allow for competing explanations of an observed trend is fairly obvious. The potential of other variables to obscure trends is perhaps slightly less so. Essentially, the problem is that trends usually contribute a relatively minor amount to the overall variability of a process, by comparison with other factors such as seasonality. 'Unexplained variability' is effectively synonymous with 'uncertainty' – hence, unless known sources of the variability in a process are acknowledged explicitly, it becomes difficult to make definitive statements about other aspects.

Broadly speaking, there are two ways of accounting for the effects of other variables in an analysis. The first, sometimes referred to as *normalisation*, is to adjust the variable of interest in some way beforehand. The second is to carry out inference within the framework of a statistical model that explicitly represents the possible effects of all relevant factors simultaneously. Normalisation is often used to adjust for seasonality. For example, when analysing a monthly series containing a strong seasonal cycle, a crude procedure (which is not generally to be recommended except for very preliminary purposes, for reasons that will be explained later) is to estimate the seasonal cycle by calculating the mean for each month of the year and to subtract this estimated cycle from the data prior to further analysis.

1.4.3   Lack of designed experiments

The emergence of statistical science as a serious discipline in its own right is generally credited to the work of Karl Pearson, Ronald Fisher and others in the first quarter of the twentieth century. The techniques developed at this time include procedures such as the Student t test, use of the correlation coefficient to quantify and test for association, maximum likelihood estimation and the analysis of variance. These techniques have found their way into almost all areas of quantitative investigation. It is worth bearing in mind, however, that many of them were originally intended for use in very specific and relatively simple situations. To illustrate this, consider the comparison of two groups of observations obtained under experimental conditions that are, as far as possible, identical except for a single factor of interest. Here, it is plausible to write down a mathematical model for the probability distributions generating the data and to postulate that the scientific question of interest ('Does the factor of interest have any effect on the outcome?') is directly equivalent to a question about these probability distributions (for example, 'Do they have different means?'). The most common approach to this problem is the hypothesis test: simplify the model by assuming that the two distributions have the same mean, and determine whether or not the data are consistent with this simpler model. If not, it is concluded that the factor of interest has a genuine effect upon the outcome.

The argument above is fairly standard in introductory statistics texts and courses, and underpins all hypothesis testing procedures. The basic principle is to compare a 'simple' mathematical model with an 'extended' version; evidence against the former is then regarded as evidence in support of the latter. However, this logical step is not always valid. Strictly speaking, if the data are not consistent with the simpler model then all we can conclude is that the simpler model may be incorrect. In the example above, the experiment is constructed in such a way that the most plausible alternative explanation is that the factor of interest affects the outcome. In many environmental problems, however, it is not possible to construct experiments in this way (Fisher's work on the design of agricultural experiments in the 1920s is a notable exception) and analyses are often restricted to available data that may have been collected for an entirely different purpose. Moreover, even if experiments can be designed it is not always possible to follow the protocol exactly: Manly (2001) gives examples of this, as well as a good discussion of general design issues for environmental applications. Finally, the complexity of environmental processes ensures that no model can truly capture the mechanism that generated the data – hence any 'simple' model is almost guaranteed to be incorrect!

It would be unduly pessimistic to conclude from this that hypothesis tests and other conventional statistical techniques have no place in environmental investigations. It should be clear, however, that the results of such procedures need to be interpreted carefully and thoughtfully: statistical methods (and hypothesis tests in particular) should not be regarded as a universal panacea for answering questions of interest. For example, it is not uncommon to see an analysis of trends over some region, in which a test for trend has been performed at different locations and a map has been produced showing the areas where 'significant' trends have been found. The implication of such a map, intentional or not, is that trends exist in these areas but not elsewhere. In reality, of course, it is much more likely that there are trends everywhere but that they vary over the region: if a trend is statistically insignificant at a particular location, this merely indicates that it is relatively weak.

1.4.4   Consideration of auxiliary information

All statistical analyses are designed to assist in the interpretation of data. Most 'classical' procedures operate by taking the recorded values of variables of interest, carrying out some mathematical operations on these values and producing summary information (such as regression coefficients and p-values) that can be interpreted by the analyst. By definition, the conclusions from such analyses are limited to the information contained in the values analysed. However, these values rarely tell the whole story: they are often accompanied by different types of auxiliary information. This might involve aspects of data quality (for example the knowledge that measurements have been obtained using a variety of different techniques), as well as a more or less detailed understanding of the process being studied.

A simple (nonenvironmental) example serves to illustrate the potential usefulness of auxiliary information. Suppose we wish to determine the probability that a tossed coin will show heads. The coin is tossed 10 times and six heads are obtained. On the basis of these data alone, the estimated probability of obtaining heads is 0.6; the associated standard error is 0.076, so that an approximate 95 % confidence interval for the true underlying probability is $0.6 \pm (1.96 \times 0.076) = (0.45, 0.75)$. Few people would accept that this confidence interval provides a realistic assessment of uncertainty in this example: centuries of experience, along with the geometry of the situation (a coin is effectively symmetric so there is no reason to prefer one side over the other), suggest that the probability of obtaining heads is extremely close to 0.5 for most coins. It would be extremely useful to be able to incorporate such information into an analysis.

Of course, environmental problems are much more complex than this. Nonetheless, there are few (if any) cases where the investigator has no understanding of the processes involved prior to seeing the data. In many applications, process based models (discussed at the beginning of this chapter) provide a particularly rich source of auxiliary information. Of course, these models are themselves approximations of reality, and different process based models of the same system can yield rather different outputs (see, for example, the projections of global mean temperature produced by different climate models in Meehl *et al.*, 2007). It is clear, therefore, that neither statistical models based solely on measurements nor process based models based primarily upon physics and chemistry can provide a complete solution to a given problem. In many application areas, it is increasingly being recognised that there is a need for better integration of the two modelling approaches.

In very simple terms, the main challenge in developing an integrated modelling approach is to determine the relative importance to be attached to each source of information (in the coin-tossing example, most people would attach much more importance to their prior understanding of coins than to the observations from a small number of tosses). In fact, this is the main issue to be confronted when dealing with any type of auxiliary information – for example, if some observations are known to be more accurate than others then it is natural to give them more 'weight' in an analysis. Some types of auxiliary information can be accommodated straightforwardly, for example by using an appropriate criterion to attach explicit weights to each observation according to its accuracy. In general, however, more sophisticated techniques may be required. In particular, Bayesian methods provide an appealing conceptual framework within which to represent the relationships between different sources of information, as well as the uncertainties involved.

1.4.5   The necessity of extrapolation

In Section 1.2, it was noted that one possible reason for carrying out a trend analysis is to gain some insight into the future. Of course, no observations are ever available beyond the present. We are not the first to note that it can be dangerous to stray beyond the limits of the available data – indeed, elementary statistics courses and texts invariably carry a strong warning against doing so. If insight into the future is required, however, extrapolation beyond the available data is unavoidable. It would hardly be appropriate to argue that the design of flood defences is futile because it is dangerous to assess future flood risk on the basis of past observations! What *is* appropriate, however, is to recognise the difficulty of the problem and to proceed extremely carefully when using statistical (or, indeed, any other) methods for extrapolation purposes. A particular challenge is to quantify, reliably and credibly, the uncertainty associated with an extrapolation or forecast.

Quantitative extrapolation, and assessment of the associated uncertainty, is invariably based on models. The main reason for this is that the only alternative is to make a more or less educated guess. However, there are two difficulties with model based extrapolation. Firstly, as in the haddock stocks example described in Section 1.3, there will usually be many models that provide apparently reasonable descriptions of the available data, but that have very different implications for the future; and, secondly, there is always a danger of unforeseen future changes to the system that invalidate the current model. It is almost guaranteed that any model will fail at some point in the future since, as noted previously, a model is at best an approximation to a complex system. In this context, careful modelling involves:

- Incorporating, as far as possible, knowledge of the mechanism that generated the data, particularly if there are grounds for believing that this mechanism will remain largely unchanged in the future. For example, fishing activity in the North Sea is the primary driver of trends in fish stocks there. Relationships between fishing activity and population growth can be deduced from simple principles that may be expected to hold quite generally (Haddon, 2001); confidence in extrapolations will be increased if they use these relationships.

  A potential difficulty with this approach is that to derive extrapolations for the variables of interest, it is necessary to know the future values of other variables as well (in the example above, to forecast fish stocks it will be necessary to know about future fishing activity). There are two potential solutions to this problem. The first is to exploit relationships with variables that can, in principle, be controlled – this enables a variety of scenarios to be presented (the consequences of different fishery management options could be compared, for example). The second is to exploit relationships with variables for which 'trusted' extrapolations exist. This is the basis of 'downscaling' in climatology (Wilby and Wigley, 1997), in which the aim is to produce future scenarios for variables of interest at a fine spatial scale. Here, large atmosphere–ocean general circulation models (GCMs) are often regarded as producing useful extrapolations of some variables at large spatial scales (e.g. mean global temperature), but the finer-scale structure is generally accepted as less reliable. Downscaling therefore aims to construct plausible models for the relationships between the variables of interest and the large-scale structures, and then to use the GCM outputs to drive the future scenarios.

- Making full use of auxiliary information, as discussed above, to guide the selection of an appropriate model. When building models for medium- or long-term extrapolation, it can be particularly useful to consider what is known about the long-term behaviour of a system. In this respect, environmental problems often yield more information than those in other areas such as economics. This is because many environmental systems are forced to operate within defined limits, which provide some (albeit limited) knowledge of the very long-term future. Few people would disagree, for example, that global mean temperature will remain within tens of degrees of its current value during any timescale of potential interest (which, for some environmental applications such as the safety assessment of nuclear waste repositories, can run into hundreds of thousands of years). By contrast, many of the models commonly used for forecasting and extrapolation have the property that the uncertainty in a forecast grows without bound as the forecast horizon increases. Such models may provide perfectly adequate approximations for short-term extrapolations of global mean temperature, but probably should not be used in the longer term.

- Understanding clearly the limitations of the model being used and, as far as possible, being explicit about the kinds of future event that could lead to failure of the extrapolations or their uncertainty assessments.

To some extent, the last point here is the most important: the limitations of forecasts must be recognised if the information they provide is to be used effectively. Unfortunately, in environmental science as elsewhere, the provision of information regarding the limitations of future projections is often inadequate at present. There is growing recognition that some assessment of uncertainty needs to be provided, but this on its own is not enough: it is also necessary to demonstrate that the uncertainty assessment is realistic, or at least credible. This can only be achieved by carrying out projections, with uncertainty assessments, for observations that were not used in the model-building process and then using a suitable measure to compare these observations with both the projections and the uncertainty assessment. This topic will not be treated further in this book; there is, however, an extensive literature on the subject. A good starting point, for atmospheric scientists in particular, is Jolliffe and Stephenson (2003). For its insights into the dangers of extrapolation in general, Chatfield (2000) is also worth reading.

1.5   Scope of the book

The remainder of Part I is intended as a handbook of modern statistical methods that may be useful for the analysis of environmental trends. The methods are illustrated using simple examples, chosen primarily for clarity of exposition. For a similar reason, Part I focuses mostly on situations in which there is a single series of primary interest (although one might be interested in examining the possibility that trends in this series are associated with changes in other factors). The simultaneous analysis of trends in several series (for example wind speed records from all of the Dutch weather stations shown in Figure 1.1) introduces additional complications that are not helpful at an introductory level. Chapter 6 provides pointers to some more advanced and specialised techniques, and serves as a stepping stone to Part II of the book.

Part II presents four substantial case studies, in which modern statistical methods of trend analysis have been applied to real environmental problems. The problems considered here are more involved than the simple examples presented in Part I: they serve to illustrate how the basic techniques can be used in more complicated situations, as well as to indicate some extensions that may be needed to meet the requirements of a particular application.

The book is aimed at researchers and graduate students in environmental science, as well as at statisticians. The reader is assumed to be familiar, at an operational level, with the basic statistical concepts of estimation, confidence intervals, hypothesis testing and regression. Some exposure to probability distributions and random variables (again at an operational level) would also be helpful, although this is not essential. Most introductory statistics courses will provide the required level of background knowledge. For those wishing to consolidate their statistical background, there are a number of books on the market that give an overview of statistical methods for more or less specialised environmental problems; some of these are summarised in Section 1.6 below.

Throughout Part I of the book, techniques are illustrated where possible using software written in the statistical programming environment R (R Development Core Team, 2004). R is based on the S language, which is designed for easy implementation of a wide range of advanced statistical and graphical procedures. It is rapidly becoming the computing environment of choice within the statistical research community: apart from the attractive price (it is free), this is mainly due to its flexibility and the ease of developing add-on packages to implement new methods. The Comprehensive R Archive Network (CRAN)[3] provides a means of distributing user-contributed packages that extend the basic

---

[3] http://cran.r-project.org/mirrors.html.

capabilities of the environment. In practice, this means that tested implementations of the latest statistical methods are freely and widely available. Such resources are enormously valuable: without them, it is unlikely that many applied researchers would have the time or the energy to try out many of the methods described in this book. As it is, the difficult work has all been done: all that is required is to know what methods are available, and to experiment with them!

The data sets used in the examples can all be downloaded from the website for the book,[4] as can the R scripts used in the analyses. Those unfamiliar with R may wish to use these scripts as models for their own work. Introductory texts on the S language, and on R in particular, include Dalgaard (2002), Fox (2002) and Maindonald and Braun (2003). Venables and Ripley (1999) is a compact and comprehensive reference source.

1.6   Further reading

The techniques covered in the book are largely drawn from the areas of time series analysis and regression modelling: smoothing methods are particularly important in the latter context. Although references are provided throughout the text, at this stage it may be useful to suggest other books that give an accessible overview of these areas. In addition, we indicate some texts that provide an introduction to statistical methods for environmental applications. In the latter category, Manly (2001) provides a clear and accessible overview of a wide variety of topics in environmental statistics; his discussion of the design of experiments for environmental monitoring and trend detection is particularly useful in the current context. More modern statistical developments are covered by Piegorsch and Bailer (2005); the treatment here is slightly more advanced, although still accessible to a general readership. Townend (2003) is another good general introductory text. Oceanic and atmospheric scientists may find Thiebaux (1994) and Wilks (2005) helpful.

For an introduction to time series analysis, Chatfield (2003) is one of the most popular books on the market; it contains an accessible overview of the subject, along with pointers to other relevant literature. For an accessible and detailed discussion of trends from a time series perspective, Kendall and Ord (1990) is hard to beat; unfortunately, this is now out of print, but copies should still be available in libraries. Brockwell and Davis (2002) and Diggle (1990) are also useful references.

The basics of regression analysis are covered by most introductory texts such as those cited already; for a slightly more in-depth treatment it may be necessary to consult a standard statistical text such as Rice (2006) or Wackerly, Mendenhall and Scheaffer (2007). Weisberg (2005) and Draper and Smith (1998) provide detailed accounts of regression methods, starting at an introductory level. Fox (2002) and Faraway (2005) specifically focus on the use of R for regression analyses, while Krzanowski (1998) provides a good overview that includes extensions such as generalised linear models (covered in Section 3.5 of the present volume). In a time series context, the main difficulty is the potential presence of dependence between successive observations. The econometrics literature is probably the best place to look for general information regarding regression methods in the presence of such dependence – Davidson and Mackinnon (2004) and Greene (2003) both deal with the subject. For an introduction to nonparametric regression, which is used extensively in the following chapters, Bowman and Azzalini (1997) is highly recommended: it contains many worked examples using the R environment. Another good book, at a slightly more technical level, is Simonoff (1996).

---

[4] http://www.wiley.com/go/trend_detection.

At several places in Part I, the relevant theory is most easily expressed using matrix notation. We assume familiarity with matrix multiplication, as well as the definitions and properties of matrix transposes and inverses. Readers unfamiliar with these concepts may care to consult a linear algebra text such as Poole (2006) or Cohn (1994), or one of several books that collect together the results that are most useful in statistical applications: Healy (2002) and Chapter 2 of Manly (1994) both contain excellent, compact and nontechnical summaries, whereas Searle (1982) and Schott (1997) are more detailed.

References

Anderson, H. R., de Leon, A. P., Bland, J. M., Bower, J. S. and Strachan, D. P. (1996) Air pollution and daily mortality in London: 1987–92. *British Medical Journal*, **312**, 665–669.

AQEG (2007) Air quality and climate change: a UK perspective Third Report of the UK Air Quality Expert Group. DEFRA Publications, London. Available from http://www.defra.gov.uk/environment/quality/air/airquality/publications/.

Battarbee, R. W., Flower, R. J., Stevenson, A. C. and Rippey, B. (1985) Lake acidification in Galloway: a palaeoecological test of competing hypotheses. *Nature*, **314**, 350–352.

Bowman, A. W. and Azzalini, A. (1997) *Applied Smoothing Techniques for Data Analysis – The Kernel Approach with S-Plus Illustrations*, vol. 18, Oxford Statistical Science Series. Oxford University Press, Oxford.

Brockwell, P. J. and Davis, R. A. (2002) *Introduction to Time Series and Forecasting*, 2nd edition. Springer-Verlag, New York.

Chatfield, C (2000) *Time-Series Forecasting*. Chapman and Hall/CRC, Boca Raton, Florida.

Chatfield, C. (2003) *The Analysis of Time Series – An Introduction*, 6th edition. Chapman & Hall/CRC Press, Boca Raton, Florida.

Cohn, P. M. (1994) *Elements of Linear Algebra*. Chapman and Hall/CRC Press, Boca Raton, Florida.

Cox, D. R. (2007) Applied statistics: a review. *Annals of Applied Statistics*, **1**, 1–16.

Dalgaard, P. (2002) *Introductory Statistics with* R. Springer-Verlag, New York.

Davidson, R. and Mackinnon, J. G. (2004) *Econometric Theory and Methods*. Oxford University Press, New York.

DEFRA (2004) Department for Environment, Food and Rural Affairs: e-Digest of Environmental Statistics. Available from http://www.defra.gov.uk/environment/statistics/index.htm.

Diggle, P. J. (1990) *Time Series: A Biostatistical Introduction* , Oxford University Press, Oxford.

Draper, N. R. and Smith, H. (1998) *Applied Regression Analysis*, 3rd edition. John Wiley & Sons, Inc., New York.

Ettema, J. (2001) Analysis of wind fields for wind climate assessment of the Netherlands. Technical Report, Royal Netherlands Meteorological Institute (KNMI), De Bilt, the Netherlands. Available from http://www.knmi.nl/samenw/hydra/documents/windfields/index.html.

Evans, C. D., Monteith, D. T. and Harriman, R. (2001) Long-term variability in the deposition of marine ions at west coast sites in the UK Acid Waters Monitoring Network: impacts on surface water chemistry and significance for trend determination. *Science of the Total Environment* **265**, 115–129.

Faraway, J. J. (2005) *Linear Models with R*. Chapman & Hall/CRC, Boca Raton, Florida.

Fleming, Z. L., Monks, P. S., Rickard, A. R., Bandy, B. J., Brough, N., Green, T. J., Reeves, C. E. and Penkett, S. A. (2006) Seasonal dependence of peroxy radical concentrations at a Northern hemisphere marine boundary layer site during summer and winter: evidence for radical activity in winter. *Atmospheric Chemistry and Physics*, **6**, 5415–5433.

Fowler, D., Cape, J. N., Coyle, M., Flechard, C., Kuylenstierna, J., Hicks, K., Derwent, D., Johnson, C. and Stevenson, D. (1999) The global exposure of forests to air pollutants. *Water, Air, and Soil Pollution*, **116**, 5–32.

Fox, J. (2002) *An R and S-PLUS Companion to Applied Regression*. Sage Publications, Thousand Oaks, California.

Greene, W. H. (2003) *Econometric Analysis*, 5th edition. Prentice-Hall, New Jersey.

Haddon, M. (2001) *Modelling and Quantitative Analysis in Fisheries*. Chapman & Hall/CRC, Boca Raton, Florida.

Healy, M. J. R. (2002) *Matrices for Statistics*, 2nd edition. Oxford University Press, Oxford.

Henne, S. and Fleming, Z. (2008) GEOmon harmonized data set of in-situ ground based data of O_3, NO_2, CO. Technical Report, GEOmon project. Available from http://geomon.empa.ch/index.php. Accessed 1 September 2010.

Jolliffe, I. T. and Stephenson, D. B. (eds) (2003) *Forecast Verification: a Practitioner's Guide in Atmospheric Science*. John Wiley & Sons, Ltd, Chichester.

Kendall, M. and Ord, J. (1990) *Time Series*, 3rd edition. Edward Arnold.

Krzanowski, W. J. (1998) *An Introduction to Statistical Modelling*. Arnold, London.

Maindonald, J. H. and Braun, J. (2003) *Data Analysis and Graphics Using R: An Example-Based Approach*. Cambridge University Press, Cambridge.

Manly, B. F. J. (1994) *Multivariate Statistical Methods: A Primer*, 2nd edition. Chapman & Hall/CRC, Boca Raton, Florida.

Manly, B. F. J. (2001) *Statistics for Environmental Science and Management*. Chapman and Hall/CRC, Boca Raton, Florida.

Meehl, G. A., Stocker, T. F., Collins, W. D., Friedlingstein, P., Gaye, A. T., Gregory, J. M., Kitoh, A., Knutti, R., Murphy, J. M., Noda, A., Raper, S. C. B., Watterson, I. G., Weaver, A. J. and Zhao, Z. C. (2007) Global climate projections. In *Climate Change 2007: The Physical Science Basis. Contribution of Working Group I to the Fourth Assessment Report of the Intergovernmental Panel on Climate Change* (eds. S. Solomon, D. Qin, M. Manning, Z. Chen, M. Marquis, K. Averyt, M. Tignor, and H. Miller,) Cambridge University Press Cambridge. pp. 747–845.

Monteith, D. T. and Evans, C. D. (2001) United Kingdom Acid Waters Monitoring Network Summary Report. UK Department of the Environment, Transport and the Regions.

Pauly, D., Christensen, V., Guenette, S., Pitcher, T. J., Sumaila, U. R., Walters, C. J., Watson, R. and Zeller, D. (2002) Towards sustainability in world fisheries. *Nature*, **418** (6898), 689–695.

Penkett, S. A., Plane, J. M. C., Comes, F. J., Clemitshaw, K. C. and Coe, H. (1999) The Weybourne Atmospheric Observatory. *Journal of Atmospheric Chemistry*, **33**, 107–110.

Piegorsch, W. W. and Bailer, A. J. (2005) *Analyzing Environmental Data*. John Wiley & Sons, Ltd, Chichester.

Poole, D. (2006) *Linear Algebra: A Modern Introduction*, 2nd edition. Thomson Brooks/Cole, Belmont, California.

R Development Core Team (2004) *R: A Language and Environment for Statistical Computing*. R Foundation for Statistical Computing, Vienna, Austria. ISBN 3-900051-07-0.

Rice, J. (2006) *Mathematical Statistics and Data Analysis*, 3rd edition. Duxbury Press, Belmont, California.

Robson, A. J. (2002) Evidence for trends in UK flooding. *Philosophical Transactions of the Royal Society, London*, **A360**, 1327–1343.

Saunders, M. A. (1999) Earth's future climate. *Philosophical Transactions of the Royal Society, London*, **A 357**, 3459–3480.

Schott, J. R. (1997) *Matrix Analysis for Statistics*. John Wiley & Sons, Inc., New York.

Searle, S. R. (1982) *Matrix Algebra Useful for Statistics*. John Wiley & Sons, Inc., New York.

Simonoff, J. (1996) *Smoothing Methods in Statistics*. Springer-Verlag, New York.

Siriwardena, G. M., Baillie, S. R., Buckland, S. T., Fewster, R. M., Marchant, J. H. and Wilson, J. D. (1998) Trends in the abundance of farmland birds: a quantitative comparison of smoothed Common Birds Census indices. *Journal of Applied Ecology*, **35**, 24–43.

Smits, A., Klein Tank, A. M. G and Können, G. P. (2005) Trends in storminess over the Netherlands, 1962–2002. *International Journal of Climatology*, **25**, 1331–1344.

Solomon, S., Qin, D., Manning, M., Chen, Z., Marquis, M., Averyt, K. B., Tignor, M. and HLM (eds.) (2007) *Climate Change 2007: The Physical Science Basis. Contribution of Working Group I to the Fourth Assessment Report of the IPCC*. Cambridge University Press, Cambridge.

Thiebaux, H. J. (1994) *Statistical Data Analysis for Ocean and Atmospheric Sciences*. Academic Press, New York.

Townend, J. (2003) *Practical Statistics for Environmental and Biological Scientists*. John Wiley & Sons, Ltd, Chichester.

Venables, W. N. and Ripley, B. D. (1999) *Modern Applied Statistics with S-Plus*. Springer-Verlag, New York.

Verkaik, J. (2000a) Documentatie windmetingen in Nederland (documentation on wind speed measurements in the Netherlands). Technical Report, Koninklijk Nederlands Meteorologisch Instituut. Available from http://www.knmi.nl/samenw/hydra/documents/docum0.htm (in Dutch).

Verkaik, J. (2000b) Evaluation of two gustiness models for exposure correction calculations. *Journal of Applied Meteorology*, **39**, 1613–26.

Wackerly, D., Mendenhall, W. and Scheaffer, R. (2007) *Mathematical Statistics with Applications*, 7th edition. Duxbury Press, Belmont, California.

Weisberg, S. (2005) *Applied Linear Regression*, 3rd edition. John Wiley & Sons, Inc., Hoboken, New Jersey.

WHO, (2000) Guidelines for air quality. World Health Organisation, Geneva. Available from http://whqlibdoc.who.int/hq/2000/WHO_SDE_OEH_00.02_pp1-104.pdf.

Wilby, R. L. and Wigley, T. M. L. (1997) Downscaling general circulation model output: a review of methods and limitations. *Progress in Physical Geography*, **21**, 530–548.

Wilks, D. S. (2005) *Statistical Methods in the Atmospheric Sciences*. Academic Press, London.

Zmirou, D., Schwartz, J., Saez, M., Zanobetti, A., Wojtyniak, B., Touloumi, G., Spix, C., de Leon, AP., Le Moullec, Y., Bacharova, L., Schouten, J., Ponka, A. and Katsouyanni, K. (1998) Time-series analysis of air pollution and cause-specific mortality. *Epidemiology*, **9**, 495–503.

2

Exploratory analysis

This chapter describes some simple procedures that may be used to gain a preliminary understanding of a data set, often as a prelude to more sophisticated analyses. Specifically, techniques for visualising time series data are discussed, as well as some of the 'classical' tests for trend. In addition, the key concepts of autocorrelation and smoothing are introduced – the former because of the need to account for it in subsequent analyses (see Section 1.4) and the latter because, in one form or another, smoothing techniques underlic many of the other methods discussed in this book.

2.1 Data visualisation

The first step in any statistical analysis is to look at the data. For all but the smallest data sets, this usually involves a variety of graphical displays designed to highlight specific features, for example to gain a preliminary indication of the structure of a data set. In the context of a trend analysis, the kinds of questions to be asked here are:

- Does it look as though trends are present? If so, how may they usefully be described? Do they affect just the mean of the variable of interest, or do they also affect other aspects such as variability? Do they appear the same throughout the year, or do they vary with the seasons?

- Apart from trends, do other factors appear to affect the variable of interest? For example, is seasonality present and if so how does it manifest itself? Are there potentially important relationships with other variables in the data set?

Visualisation can also be used to identify possible data quality problems, and if possible to rectify these at the outset. Common problems with environmental data include transcription errors, unflagged faults in measurement devices, changes due to new or recalibrated instrumentation, changes in measurement units (for example the change from Imperial to metric units in the UK in the 1970s), observations at or below the limit of detection (see Section 6.5) and the undocumented 'infilling' of missing data with fictional values. In the context of a trend analysis, it is particularly important to identify

Statistical Methods for Trend Detection and Analysis in the Environmental Sciences, First Edition.
Richard E. Chandler and E. Marian Scott.
© 2011 John Wiley & Sons, Ltd. Published 2011 by John Wiley & Sons, Ltd.

changes in measurement technology or practice over time, since these may lead to the identification of spurious trends unless they are accounted for properly. To identify data quality problems, some experience of the data collection process is usually helpful and, of course, the task is simplified considerably through contact with the person(s) who originally collected the data. Even if this is not possible, however, careful data visualisation can often identify features that merit further investigation. An obvious starting point is to look at outlying observations to see whether there is anything obviously wrong with them – bearing in mind, however, that an observation is not necessarily erroneous just because it is an outlier.

Finally, visualisation can be used to suggest assumptions that may plausibly be held in the initial stages of a subsequent analysis, and hence to guide the choice of analysis strategy. For example, if successive observations in a data set appear to be independent then it may be appropriate to use a 'standard' analysis technique, without modification and without needing to worry about possible autocorrelation. It should be emphasised, however, that assumptions made on the basis of a preliminary visualisation exercise should be regarded as indicative only. More detailed analysis may reveal features that were not apparent initially, so the use of preliminary analysis to suggest plausible assumptions does not eliminate the need for subsequent checking!

The remainder of this section discusses some techniques that are commonly used to address these aims.

2.1.1   Time series plots

As with the examples presented in Chapter 1, the obvious starting point in a trend analysis is usually a plot of the observations against time.

Example 2.1

Figure 2.1 presents time series plots of some of the Dutch wind speed data considered in Section 1.3.1. To show the main features of the data, they have been aggregated to an annual timescale. Figures 2.1(a) and (b) both show the annual mean wind speeds at De Bilt (see the map in Figure 1.1), from 1961 to 2002. There is little discernible structure in panel (a): here the data are displayed using a scatterplot with the vertical axis starting at zero (the smallest possible value of wind speed). In panel (b), the scatterplot is replaced by a line graph, where the vertical axis is chosen so that the data fill the plot region and the overall mean of the series is indicated. This plot clearly shows a gradual decreasing trend and is arguably much more useful for examining changes over time. In particular, it is striking that there are long sequences of observations above or below the overall mean (between 1969 and 1979, and then after 1994). Such features form the basis for some of the tests for trend to be discussed in Section 2.4.

It is also worth comparing Figure 2.1(b) with the corresponding plot in the bottom right-hand panel of Figure 1.1. Both plots show exactly the same data and have identical axis ranges. The only difference is the aspect ratio – Figure 2.1(b) is proportionally much wider. However, this is enough to give a rather different visual impression – the trend in Figure 1.1 looks dramatic, whereas that in Figure 2.1(b) appears rather modest!

The final plot in Figure 2.1 shows the annual mean wind speeds at IJmuiden, again over the 1961–2002 period. At this site, around 3 % of the daily values are missing (see Section 1.3.1.). To ensure that all annual means are comparable, therefore, they have been computed only for years with at least 335 available daily observations (i.e. years with

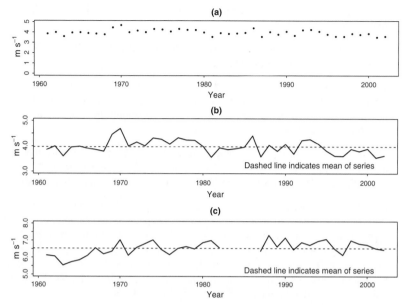

Figure 2.1   Time series plots of mean annual wind speeds at De Bilt (a, b) and IJmuiden (c), 1962–2002. See text for details.

at most a month's worth of missing data). The resulting missing values are indicated by breaks in the line graph in Figure 2.1(c); it is interesting that they are all in consecutive years. It could be argued that to visualise any trends, it would be better to connect the two parts of the graph with a straight line over the missing period. However, this would give the entirely unjustified impression that the missing values all lie in between the neighbouring points. Indeed, in this particular application it is conceivable that some daily values are missing because they were too high to be measured safely or accurately, in which case we may suspect that years with several missing daily values had high annual mean wind speeds. Another possibility is that this station experienced operational difficulties during the mid 1980s, in which case there may be grounds for checking carefully the quality of the data from this period. Of course, it would be possible in principle to infer something about the missing daily values here, for example using records from neighbouring stations. However, for present purposes it suffices to note that by inserting breaks in a line graph (which is the default behaviour in R), the reader can make their own judgement about the importance of the missing observations.   ■

This example illustrates that even with very simple data sets, it pays to think carefully about the preparation and interpretation of exploratory graphics. The following guidelines may be useful:

- When displaying observations made at equal time intervals, line graphs usually reveal structure more clearly than scatterplots. Breaks in a line should be used to indicate missing observations.

- For observations made at irregular intervals, line graphs have some undesirable features. For example, if there is a large gap between two successive observations then a line graph will give a strong visual impression of a linear trend over this period. For this reason, scatterplots may be preferable for displaying irregularly spaced series.

- To aid the interpretation of a time series plot, it is useful to add a horizontal line showing the mean of the available data.

- The time resolution of a plot should be chosen so that the main features of interest stand out clearly. This might, for example, involve the aggregation of daily or monthly data to a yearly level (an alternative is to smooth the data, as discussed in Section 2.2 below).

- The interpretation of a plot can be influenced by the choice of scale for the vertical axis. For exploratory analysis it is advisable to plot over the range of values of scientific interest. This provides an instant visual context for any apparent structure.

- As a guide to subsequent analysis, it can be worth exploring nonlinear transformations of the vertical axis in a time series plot. On page 8, for example, Figure 1.2(a) shows that the trend in haddock stocks can be regarded as linear on a logarithmic scale. Many software packages, R included, support the use of logarithmic axis scaling. To explore other transformations, however, it is usually necessary to transform the data themselves prior to plotting (this has already been discussed, in Section 1.3.2, as an alternative means of producing Figure 1.2(a)).

- Almost all environmental data have some error associated with them. Sources of error include the measurement process itself (for example when taking chemical readings in a stream), as well as sampling variation (examples include the use of wildlife census data to estimate total population numbers, and the use of data from several weather stations to estimate a spatial average). If the magnitude of the error is non-negligible and varies appreciably between observations, some measure of the precision of each observation should be indicated clearly on a time series plot. If estimated standard errors are available, the precision can be indicated by adding points two standard errors above and below each observation to yield approximate 95 % uncertainty intervals. Otherwise, or if the calculation of standard errors is not straightforward, some other measure should be indicated – for example the size of sample used to obtain an estimate of population numbers.

Further general guidance on the graphical presentation of data can be found in the excellent book by Cleveland (1994).

2.1.2   Boxplots

Time series plots are conceptually simple, but they do not fulfil all of the previously described aims of an exploratory analysis. For example, it is not easy to see structure other than in the mean of a series, and important detailed information (regarding outlying observations, for example) may be lost if a long series is aggregated to show the main features of interest more clearly. The boxplot is an alternative graphical technique that remedies some of these deficiencies.

Boxplots, such as those in Figure 2.2, are designed to facilitate comparisons between the distributions of observations falling in different groups – for example different months of the year, different years or different spatial locations. For each group of observations, the 'box' extends from the first to the third quartile of the data; the height of the box thus represents the interquartile range (IQR). The median is indicated in each box by a horizontal line. The dashed 'whiskers' extend in each direction to the most extreme data point that is at most 1.5 IQRs away from the box; observations more extreme than this are marked individually.[1] The idea is that the whiskers represent the main body of each distribution and observations falling beyond the whiskers are possible outliers; the construction is designed to ensure this interpretation for a wide range of distributional shapes. For example, if the data values in each group are drawn from normal distributions then the box heights will be around 1.35 standard deviations and only 0.7 % of observations are expected to fall outside the whiskers. If data values are drawn from exponential distributions, the lower whiskers will usually extend down to the smallest data value and around 5 % of observations will fall above the upper whiskers.

Example 2.2

Figure 2.2(a) shows the distributions of daily wind speeds at De Bilt, by month. There is a clear seasonal pattern in the medians of the distributions, but even more so in the locations of the upper whiskers – it seems that seasonality is present and is more pronounced in the upper tails of the wind speed distributions. Notice also that the heights of the boxes decrease in the summer months (May to August). This suggests a seasonal pattern in the variability of wind speed: the higher the average wind speed in a month, the higher the variability. This may need to be accounted for in subsequent analyses. The shapes of all the distributions seem very similar – the winter boxplots look like

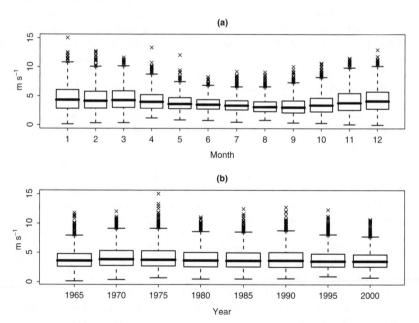

Figure 2.2   Boxplots showing distributions of daily wind speeds at De Bilt between 1961 and 2002, (a) by month, (b) by five year period (excluding 1961 and 1962).

---

[1] The exact implementation of boxplots differs slightly between software packages, particularly regarding the extent of the 'whiskers'. The description here is of the default behaviour in R version 2.8.1.

'stretched' versions of the summer ones. There are a few outlying observations in each month, but this is to be expected given the large sample size (with 40 years of daily data, the boxplot for each month represents around 1200 values); certainly, no observation appears wildly implausible.

Figure 2.2(b) shows the distributions of daily wind speeds in consecutive five year periods: 1963 to 1967, 1968 to 1972 and so on. Data for 1961 and 1962 have been omitted altogether, to ensure that each group contains roughly the same number of observations (otherwise, groups with fewer observations will tend to show fewer outlying values, which in turn could give a misleading impression of a trend in the numbers of 'extreme' observations). The most striking feature of this figure is the similarity of the distributions in each five year period. In particular, it is only just possible to see that the median lines reflect the trend in mean wind speed seen in Figure 2.1(b). This is because of the difference in vertical axis scales; that in Figure 2.2(b) is necessarily extended to accommodate the more variable daily data. The upper whiskers of the boxes broadly follow the pattern in the medians, indicating that, for this site at least, there are no obvious differences between trends in the mean and extremes of the distribution. Above the upper whiskers, the variability of outlying values is such that it is unwise to try and interpret apparent patterns.

In series with a strong seasonal component, it is possible that the seasonality may obscure any trends that are present. For example, the upper tails of all the distributions in Figure 2.2(b) are dominated by winter values, so these boxplots do not say much about trends in 'extreme' summer wind speeds. To examine this, it would be necessary to account in some way for seasonal variability before producing the five year boxplots. This could be achieved by producing boxplots for each month separately or by 'deseasonalising' the values in some way prior to plotting. A simple way to remove seasonality is by computing the difference or ratio between each value and the corresponding monthly mean. More sophisticated methods are discussed in Section 2.3. In this particular case, deseasonalising the data makes virtually no difference to the result, which is therefore not shown.

The conclusion from this exercise is that any trends in annual mean wind speeds are small by comparison with the daily variability. For the purpose of engineering risk assessment (see Section 1.3.1), this is a preliminary indication that, in this location at least, the trends are relatively unimportant. In determining the viability of wind power schemes, however, the trend in annual means may be more significant.   ■

In summary, by allowing a visual comparison of entire distributions, boxplots provide a powerful supplement to time series plots in an exploratory analysis. They are particularly useful for assessing changes in the variability and shape of distributions and for identifying possible outliers (for example, Figure 2.2 shows that the highest recorded wind speed at De Bilt is $15 \, \mathrm{m \, s^{-1}}$, and that it occurred during January between 1973 and 1977). However, relatively weak trends may be less apparent from boxplots than from time series plots. Care is also required with the interpretation of boxplots when many observations are missing – particularly when there is a suspicion that missing values do not occur at random. This may occur, for example, if especially high values are difficult to record or if access to a monitoring site becomes difficult due to adverse conditions related to the variable of interest.

2.1.3   The autocorrelation function

In Section 1.4, it was noted that dependence between successive observations in a time series may create problems for the construction of hypothesis tests and confidence intervals using standard methods. Such serial dependence is usually referred to as 'autocorrelation'; in this section we look at visual methods that are designed to indicate whether this is present.

In general, one might expect that the association between pairs of observations will depend on the time separation between them. Pairs of wind speed measurements made on successive days will tend to be more similar than pairs made a week apart, for example. It is therefore useful to examine the strength of association at a variety of time separations.

The most common statistical tool for measuring association between two quantities is the (Pearson) correlation coefficient, defined for pairs of observations: $(x_1, y_1), (x_2, y_2), \ldots, (x_n, y_n)$ as

$$r = \frac{\sum_{i=1}^{n} (x_i - \overline{x}) (y_i - \overline{y})}{\sqrt{\sum_{i=1}^{n} (x_i - \overline{x})^2 \sum_{i=1}^{n} (y_i - \overline{y})^2}},$$

where $\overline{x}$ and $\overline{y}$ are the sample means of the x and y values respectively. To measure association in a regularly spaced time series, one could therefore compute the correlation between pairs of observations at various separations. For example, given a time series $y_1, y_2, \ldots, y_T$, the association between observations separated by a single time unit could be measured by considering the successive pairs $(y_1, y_2), (y_2, y_3), \ldots, (y_{T-1}, y_T)$ as bivariate data and calculating the correlation between them in the usual way. This would be equivalent to calculating the correlation between the original series and a shifted or *lagged* version, obtained by moving each observation backwards or forwards by one time unit (hence the term 'autocorrelation' – it is the correlation between a series and itself). Note that shifting the series backwards or forwards would yield the same result.

However, time series autocorrelation and 'ordinary' correlation differ in several important respects. Most of these are rather subtle, but an obvious one is that some observations are used twice in the time series case – y_2 appears in both the first and second pairs above, for example. Because of these differences, autocorrelations are not usually calculated in the same way as ordinary correlations. For a regularly spaced time series with no missing data, the *sample autocorrelation coefficient at lag* $k \geq 0$ is usually defined as

$$r(k) = \frac{\sum_{t=k+1}^{T} (y_t - \overline{y}) (y_{t-k} - \overline{y})}{\sum_{t=1}^{T} (y_t - \overline{y})^2}, \tag{2.1}$$

where $\overline{y} = T^{-1} \sum_{t=1}^{T} y_t$ is the sample mean of all the observations. This can be interpreted as a correlation between the original series and a version shifted back k time units. For $k < 0$, corresponding to a shift forwards rather than backwards, we define $r(k) = r(-k)$. Putting $k = 0$ in (2.1) yields $r(0) = 1$. The collection $\{r(k)\}$ of all sample autocorrelation coefficients is called the *sample autocorrelation function* (ACF). For lags $k \geq 0$, a plot of $r(k)$ against k is called a *correlogram*.

The autocorrelation coefficient $r(k)$ can also be written as

$$r(k) = c(k)/c(0) \quad \text{where} \quad c(k) = T^{-1} \sum_{t=k+1}^{T} (y_t - \overline{y}) (y_{t-k} - \overline{y}).$$

The quantities $\{c(k)\}$ are called the *sample autocovariances* and stand in the same relation to the autocorrelations as do covariances to correlations in standard settings.

Note that $\bar{y}$ is used throughout (2.1) whenever a sample mean is required – this contrasts with the usual correlation coefficient, in which separate means are computed for each variable. The reason for using $\bar{y}$ throughout will be discussed below. Note also that the numerator and denominator of (2.1) have different numbers of terms–there are T contributions to the sum in the denominator, but only $T - k$ in the numerator. In view of this, it may seem natural to multiply $r(k)$ by $T/(T - k)$ to ensure that the numerator and denominator are in some sense comparable. However, this is not usually done. The reasons for this are rather technical and will not be discussed here; details can be found in Priestley (1981, Section 5.3). It suffices to note that the ratio $T/(T - k)$ will be approximately 1 if k is small relative to T; hence it is best to calculate sample autocorrelations only for relatively small values of k. Most software packages automatically determine an appropriate range of lags to display when generating a correlogram. It is also worth noting that good packages do not use formula (2.1) to calculate sample autocorrelations: an alternative based on the fast Fourier transform (FFT) is usually used instead (Chatfield, 2003, Section 7.5). Numerically, the results are identical, but the FFT is much cheaper computationally and enables the ACF to be computed rapidly for large data sets.

Of course, since the ACF is computed from a sample of data, the coefficients will not be exactly zero even in the absence of underlying correlation. To aid interpretation of a correlogram, it is therefore common to add horizontal lines showing the magnitude of coefficients that should be considered 'significantly' different from zero. It is standard practice to plot these lines at levels $\pm 1.96/\sqrt{T}$, where T is the length of the series. These lines define approximate 95 % confidence limits for individual coefficients, under the assumption that the observations are an uncorrelated sequence of values drawn from probability distributions with a common mean and variance (Chatfield, 2003, Section 4.1). If this assumption holds, at lags greater than zero around 95 % of the coefficients should lie between the lines.

Formula (2.1) assumes that there are no missing observations. If values are missing from a regular time series, it is not entirely clear how best to proceed. One option is to replace (2.1) with the corresponding expression in which terms containing missing values are dropped from each sum. For example, if the ninth observation y_9 is missing, terms corresponding to $t = 9$ and $t = k + 9$ would be dropped from the numerator and the term corresponding to $t = 9$ would be dropped from the denominator. For reasons that are again rather technical, this solution is not ideal and by default R (along with most other software packages) does not allow missing values when computing an ACF. This default behaviour can be changed, however, in which case R omits terms as just described. Providing the proportion of missing observations is not too large, the result will usually be perfectly adequate for an exploratory analysis. Otherwise, an alternative is to use methods that are appropriate for assessing dependence in irregularly spaced series; see Section 2.1.4.

As with the ordinary correlation coefficient, care is required in the interpretation of a sample ACF. Correlation is a measure of the strength of linear association between two quantities: it is well known that such association can arise either because the two quantities are intrinsically related or because they are both influenced by other factors. It is unfortunate that one such factor may itself be an underlying trend! To illustrate this, consider a series that increases at a fairly constant rate over time. In this case, the sample mean $\bar{y}$ will correspond to the level roughly halfway through the observation period. Most pairs of observations separated by k time units will fall on the same side of this halfway level if k is small relative to the length of the series, and hence will tend to be either both below or both above $\bar{y}$. This leads to positive autocorrelation at lag k. The essence of this argument applies equally in the presence of more complicated trend structures. The net result is that it can be difficult to discriminate, on the basis of a correlogram, between 'genuine' autocorrelation and trends. The ambiguity can often be resolved by considering the scientific context: a useful approach is to ascribe to autocorrelation any association at small time lags that cannot be explained by long-term variation (recall the definition of trend at the end of Section 1.1). The interpretation can be simplified greatly by considering the ACF in conjunction with other plots of the data. An example is given below.

A further point in relation to the sample ACF is that lack of correlation does not imply lack of association; it is possible for two quantities to be related to each other in a nonlinear way and to be uncorrelated. A plot of y_t against y_{t-1} can be a useful means of revealing such nonlinear association.

Example 2.3

Correlograms for the De Bilt wind speed series are presented in Figure 2.3. Panel (a) is for the time series of monthly mean wind speeds. Note that $r(0) = 1$. The clear oscillating pattern is an example of association due to factors that have not been accounted for in the analysis – it is due to the seasonality already observed in Figure 2.2(a). Hence this plot does not reveal anything new: it will be more useful to examine the autocorrelation after removing the seasonal cycle.

As noted previously, a simple way to remove seasonality is to compute the difference or ratio between each value and the corresponding monthly mean. The ratio has been

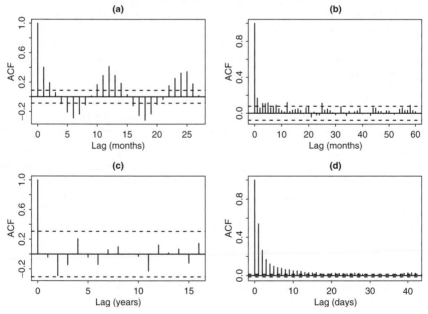

Figure 2.3   Wind speed data from De Bilt, 1961 to 2002: correlograms for (a) monthly means, (b) deseasonalised monthly means, (c) annual means, (d) deseasonalised daily values. The horizontal dashed lines on each plot are approximate 95 % confidence limits, computed under the assumption that there is no underlying trend or autocorrelation.

used here, since it is more appropriate for series that cannot take negative values and in which the variability increases with the mean (see Figure 2.2(a)). Figure 2.3(b) shows the correlogram for the resulting deseasonalised series, for lags up to five years. The coefficients decrease slowly from a value of around 0.2 at lag 1, and almost all are positive. This suggests weak association between observations at fairly large time separations, which in turn can be interpreted as a slight tendency for wind speeds to be above or below average for extended periods. Whether or not this is ascribed to 'trend' or to 'autocorrelation' depends largely on whether variation over this kind of timescale is regarded as 'long-term'.

Figure 2.3(c) shows the sample ACF for annual, rather than monthly, mean wind speeds. In contrast to the monthly ACFs, there is little structure here. This suggests that the plot may be dominated by sampling variation, and hence that the 'underlying' autocorrelation is effectively zero. Some support for this can be found by observing that all 16 coefficients lie within the approximate 95 % confidence limits shown on the plot. In the absence of any other information, this would suggest that the annual means show neither trends nor autocorrelation. This apparently contradicts both the time series plot in Figure 2.1(b) and the monthly ACF in Figure 2.3(b). An explanation for this is that the annual autocorrelations are rather imprecise since they are estimated from just 40 observations. A pragmatic interpretation of Figure 2.3(c) is that autocorrelation is unlikely to be a major problem in the analysis of annual means.

Finally, Figure 2.3(d) shows the sample ACF for deseasonalised daily wind speed data (here, seasonality has been removed by computing the ratio between each day's value and the mean of all values for that calendar day). This is rather more interesting than any of the previous plots, showing a rapid decay to a lag of around 12 days and levelling off thereafter at a very small, but nonetheless positive, level. As with the monthly plots, the persistence of positive coefficients at large lags can be attributed to trends – or at least to extended periods of above- and below-average wind speed. The behaviour at lags below 12, however, is rather different and suggests another mechanism at work. In fact, dependence at short timescales is not unexpected: it arises because in mid-latitudes, individual weather systems (and the associated winds) usually affect any given location for several successive days (McIlveen, 1992, Section 11.2). In the current context, events on the scale of individual weather systems are of no scientific interest; it is therefore convenient to interpret this short-scale dependence as 'genuine' autocorrelation.   ■

So far, we have discussed the sample autocorrelation function as a purely descriptive tool for assessing the strength of serial dependence in a series. In many ways, its interpretation can be simplified by introducing the concept of a statistical model. In Section 1.1, we saw that most modern statistical methods regard a sequence $y_1, \ldots, y_T$ of observations as the realised value of a corresponding sequence $Y_1, \ldots, Y_T$ of random variables. Such a sequence is said to be *stationary* if the following three conditions are satisfied:

(a) The random variables all have the same expectation: $E(Y_1) = \cdots = E(Y_T)$ $= \mu$, say.

(b) The random variables all have the same variance: $Var(Y_1) = \cdots = Var(Y_T)$ $= \sigma^2$, say.

(c) For any time t and lag k, the theoretical correlation between Y_t and Y_{t+k}, defined as

$$\text{Corr}(Y_t, Y_{t-k}) = \frac{1}{\sigma^2} E\left[(Y_t - \mu)(Y_{t-k} - \mu)\right] = \rho(k), \text{say}, \qquad (2.2)$$

does not depend on t.

In simple terms, the first two of these conditions state that there is no trend in the mean or the variance of the distributions, and the third states that the dependence structure does not change through time. For our purposes it is not necessary to draw the distinction, often found in mathematical texts, between sequences with these three properties (usually described in such texts as 'weakly stationary' or 'second-order stationary') and more restricted classes of stationary sequence. In Chapter 5, we will discover the surprising fact that the realised values of a stationary sequence can have trend-like behaviour. For the moment, however, to keep things simple it may be helpful to think of stationarity as 'absence of trend'. The simplest example of a stationary sequence is obtained when the random variables all have zero mean and constant variance, and are uncorrelated: the result is a *white noise sequence*.

In the context of stationary sequences, the sample autocorrelation function $\{r(k)\}$ is in fact an estimator of the underlying theoretical function $\{\rho(k)\}$ defined in Equation (2.2). This explains why $\overline{y}$ is used throughout Equation (2.1) whenever a sample mean is required – it is the natural estimator of μ in (2.2). Moreover, in some applications it can be useful to interpret the sample autocorrelation function by comparing its behaviour with that of $\{\rho(k)\}$ for particular classes of statistical models. This idea will be explored further in Chapter 5.

2.1.4   Irregularly spaced data – the variogram

The autocorrelation function can be a useful means of quantifying serial dependence in a regularly spaced time series with few missing observations. In some situations, practical considerations inevitably result in minor departures from regularity, as with the water chemistry data in Section 1.3.3. The resulting series can usually be treated as though it is regular, and the autocorrelation function used to summarise dependence. Such an approach is not appropriate, however, when observations are made at highly irregular intervals, and in this case alternative techniques are required.

To estimate a correlation, multiple pairs of observations are required. One of the problems with irregularly spaced data is the lack of such multiple pairs, since each pair of observations is separated by a different lag. One way to circumvent this is, for each lag of interest, to compute correlations between all pairs of observations with approximately the right separation. To our knowledge there is, however, no standard procedure equivalent to Equation (2.1) for use in this case; neither are there any simple methods for determining whether the resulting correlations differ significantly from zero. Note that there is nothing wrong in principle with the idea of an underlying lag-dependent correlation; the difficulty lies purely in estimating it.

As an alternative to the use of correlations to measure dependence, one can consider the squared differences between pairs of observations. This idea has been used extensively in geostatistics, where it is often necessary to measure dependence between observations made at an irregular set of spatial locations. In the current context it is helpful to think of irregularly spaced observations as samples from a process that is defined at all points in time. Conceptually, this is slightly different from the framework of the previous section, where we only considered a discrete collection of time points $1, \ldots, T$. To emphasise this difference, we use a slightly different notation and denote by $y(t)$ an observation made at time t, which is considered to be the realised value of a corresponding random variable $Y(t)$. Denote the expected value of $Y(t)$ by $\mu(t)$. Consider two particular time points, t and $t + h$, and consider the expected squared difference between observations made at these time points:

$$E\left[(Y(t+h) - Y(t))^2\right]$$
$$= E\left[\left(\{Y(t+h) - \mu(t+h)\} - \{Y(t) - \mu(t)\} + \{\mu(t+h) - \mu(t)\}\right)^2\right],$$

the second line being obtained from the first by adding both $\mu(t)$ and $\mu(t+h)$ and then subtracting them again. Suppose now that the process is stationary. In this case, $\mu(t+h) - \mu(t) = 0$ so that we have

$$E\left[(Y(t+h) - Y(t))^2\right] = E\left[\left(\{Y(t+h) - \mu(t+h)\} - \{Y(t) - \mu(t)\}\right)^2\right]$$
$$= E\left[(Y(t+h) - \mu(t+h))^2\right] + E\left[(Y(t) - \mu(t))^2\right]$$
$$-2E\left[(Y(t) - \mu(t))(Y(t+h) - \mu(t+h))\right].$$

The first two terms here are simply the variances of $Y(t+h)$ and $Y(t)$, which are equal (to σ^2, say) if the process is stationary. The final expectation is the covariance between them, which can alternatively be written as $\sigma^2 \rho(h)$, where $\rho(h)$ is the autocorrelation between $Y(t+h)$ and $Y(t)$ (which does not depend on t since the process is stationary). We therefore have

$$E\left[(Y(t+h) - Y(t))^2\right] = 2\sigma^2 [1 - \rho(h)] = 2v(h), \quad \text{say}. \tag{2.3}$$

The function $v(h) = \frac{1}{2}E\left[(Y(t+h) - Y(t))^2\right]$, which depends on the lag h but not upon the time t, is called the *(semi-)variogram* and, from Equation (2.3), clearly carries the same information as the autocorrelation function for a stationary process. Although we have considered only stationary processes in the development above, in fact it can be shown that the variogram is defined for a slightly wider class of processes, called *intrinsic processes* (Webster and Oliver, 2001, Section 4.5). If a process is not stationary, the autocorrelation function cannot be defined; it has therefore been argued in the geostatistical literature that the variogram is a more useful summary of dependence. However, for stationary processes, the relationship with the autocorrelation function in Equation (2.3) provides a useful guide to interpretation. It shows, for example, that if autocorrelation is high then the variogram will be close to zero; conversely, at large lags, where one would expect very little autocorrelation, the variogram will be approximately σ^2.

How does this help to measure dependence for irregularly spaced data? The answer is that from observations $y(t)$ and $y(t+h)$, an obvious estimate of the variogram is

$$\hat{v}(h) = (y(t+h) - y(t))^2. \tag{2.4}$$

The estimate is not very accurate, of course, but it can be computed for every pair of points in a data set. A sample of N observations will yield $N(N-1)/2$ pairs of points, with separations $h_1, \ldots, h_{N(N-1)/2}$, say. Application of Equation (2.4) to each pair of points will produce estimates $\hat{v}(h_1), \ldots, \hat{v}(h_{N(N-1)/2})$. A plot of the estimates against the separations is sometimes called a *variogram cloud* (Webster and Oliver, 2001, Chapter 5). It is usually difficult to interpret such a plot directly, because the individual estimates tend to be extremely variable; moreover, if N is large then the variogram cloud may simply contain too many points to see any structure. One way round this, which is often adopted in the applied geostatistics literature, is to average the values of $\hat{v}(h)$ over groups of pairs with similar lags. A disadvantage of this approach is that the results can depend on the choice of grouping. An alternative is to fit a smooth curve through the variogram cloud, using similar methods to those discussed in Chapter 4 of this book. The latter approach is implemented in the R command plot.Variogram, in the nlme library supplied with the standard download (Pinheiro *et al*., 2008), as well as via the variog command in the add-on geoR library (Ribeiro and Diggle, 2001).

Example 2.4

Consider the ozone concentration data introduced in Section 1.3.4. Although this series is nominally at a regular daily resolution, the high number of missing observations, and their irregular pattern of occurrence, suggests that standard methods for the analysis of regularly spaced series may not be appropriate here. Instead of computing a sample ACF, we therefore use variograms to visualise the temporal dependence structure of the data.

Figure 2.4(a) shows the sample variogram of the daily ozone series. This was produced in R using the variog command in the geoR library. Each point here represents the average squared difference between all pairs of values separated by the corresponding time lag; since the fundamental time interval here is daily and the number of observations is large, this can be regarded as a straightforward and unambiguous way to smooth the variogram cloud. The individual points in the cloud are not shown, because many of them are so large as to obscure the overall structure.

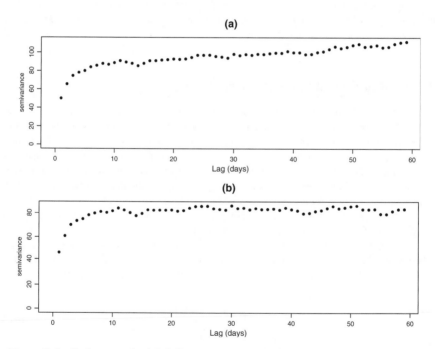

Figure 2.4   Variograms for (a) daily ozone series, (b) deasonalised daily ozone series.

Equation (2.3) shows that for a stationary process the shape of the variogram is a scaled mirror image of the ACF: whereas the ACF typically decays towards zero with increasing lag, the variogram increases to an upper limit (known as the *sill* in the geo-statistics literature) representing the variance of the process. Figure 2.4(a) does not show this behaviour, however: rather, after an initial steep rise to a lag of around 10 days, the variogram continues to increase gradually up to the maximum plotted lag of around 2 months. In variogram terms, this is the equivalent of slow decay in the sample ACF and suggests the presence of a systematic structure that cannot be attributed to short-term temporal dependence. Given the clear seasonality in Figure 1.4, this is unsurprising. For a clearer picture of short-term dependence, it is therefore helpful to remove the seasonal cycle and then recompute the variogram. Figure 2.4(b) shows the result: here, the seasonal cycle has been estimated by calculating the mean ozone concentration for each day of the year (the 20 year record makes this feasible despite the large number of missing values), and this estimated cycle has been subtracted from each year's data prior to calculating the variogram. For a 'genuinely' irregular series, of course, it would not be possible to estimate the seasonal cycle in this way. In this case one possibility would be to define a 'window' of dates around each calendar day and to estimate the seasonal cycle using averages of all observations within these overlapping windows. Another possibility would be to fit a preliminary regression model representing the seasonal cycle, for example using sine and cosine terms (see Section 3.2), and to consider the residuals from such a model as a deseasonalised series.

Figure 2.4(b) shows that deseasonalising the ozone series has eliminated the gradual increase of the variogram for lags above 10: the initial rapid increase remains, however. This suggests that there is autocorrelation in the series at lags up to around 10 days. For lags above this, the variogram fluctuates about a constant value of around 80, representing the variance of the deseasonalised process.   ∎

2.1.5   Relationships between variables

The analyses described so far are appropriate for visualising the structure of a single series. In practice, it is often of interest to assess the relationships between different series, for example to determine how some quantity x affects a response y. Given data $\{(x_t, y_t) : t = 1, \ldots, T\}$, a natural starting point is to display the pairs $\{(x_t, y_t)\}$ using a scatterplot. Although this shows the relationship between contemporary observations, it does not tell the whole story. For example, if y does not respond to changes in x until ℓ time units later, it would be more appropriate to produce a scatterplot of pairs $\{(x_t, y_{t+\ell})\}$. In reality, however, inter-series relationships are rarely as simple as this, and even if they are this simple, the value of ℓ is usually unknown.

For regularly spaced series with no missing data, a concise way of summarising relationships between series is via the *sample cross-correlation function* (CCF). The idea is very similar to the ACF: the cross-correlation coefficient at lag k is defined as

$$r_{xy}(k) = \frac{\sum_t (x_{t-k} - \overline{x})(y_t - \overline{y})}{\sqrt{\sum_{t=1}^{T} (x_t - \overline{x})^2 \sum_{t=1}^{T} (y_t - \overline{y})^2}}, \qquad (2.5)$$

where the sum in the numerator is computed over all t for which both y_t and x_{t-k} are available; $r_{xy}(k)$ is a measure of association between values of y and values of x that occurred k time units previously. The CCF is the collection $\{r_{xy}(k)\}$ of all sample cross-correlation coefficients. In the simple case when y responds to x after a delay of ℓ time units, ideally the CCF would show a prominent spike at lag ℓ. As with the autocorrelations, the cross-correlations can be written in terms of cross-covariances:

$$r_{xy}(k) = c_{xy}(k) / \sqrt{c_{xx}(0)c_{yy}(0)}, \quad \text{where} \quad c_{xy}(k) = T^{-1} \sum_t (x_{t-k} - \overline{x})(y_t - \overline{y}),$$

and $c_{xx}(0)$, $c_{yy}(0)$ are the lag zero autocovariances (i.e. the variances) of the two series.

In contrast to the ACF, $r_{xy}(k)$ is not usually equal to $r_{xy}(-k)$: the former measures association between y and a previous value of x, whereas the latter measures association between y and a subsequent value of x. Therefore the CCF must be calculated and plotted over both positive and negative lags.[2] In the absence of other complicating factors, nonzero coefficients at positive lags suggest that y responds to changes in x; on the other hand, nonzero coefficients at negative lags suggest that x responds to changes in y. Again, 95 % limits are usually added to plots of the CCF to assist in determining whether individual coefficients are significantly different from zero.

Unfortunately the CCF, like the ACF, is influenced by the effects of trends and other factors. Moreover, spurious nonzero coefficients can arise if both the x and y series are themselves autocorrelated (Brockwell and Davis, 2002, Section 7.3; see also Walther, 1997, for a compelling demonstration). This makes interpretation difficult – if there are apparently significant cross-correlations between two series that both exhibit significant autocorrelation, either of the following mechanisms could be responsible:

- There is a genuine relationship between y and x, which causes autocorrelation in x to be propagated through into y.

- x and y have nothing to do with each other; the significant cross-correlations are spurious artefacts of the autocorrelation in both series.

Of course, the scientific context may dictate that one of these explanations is more plausible than the other. Otherwise, one way to resolve the dilemma is to remove the autocorrelation from one or both series prior to calculating the CCF. This is called *pre-whitening*, and will be discussed further in Chapter 5. In the meantime, perhaps the best advice is to exercise caution in the interpretation of cross-correlations, particularly if autocorrelation is present in both series.

Example 2.5

Figure 2.5 shows correlation functions for the deseasonalised quarterly alkalinity series from the Round Loch of Glenhead (see Section 1.3.3). Seasonality has been removed by subtracting the mean for each quarter from the corresponding observations. Figure 2.5(a) is the sample ACF for the deseasonalised series. The positive coefficients up to a lag of around three years are probably due, at least in part, to the trend in alkalinity that is evident from Figure 1.3.

For illustrative purposes, Figure 2.5(b) shows the CCF of the deseasonalised alkalinity series with a lagged version of itself. Here, the series (x_t) is defined as $x_t = y_{t+1}$, so that (y_t) follows (x_t) exactly and we might expect to find $r_{xy}(1) = 1$. In fact, its value is 0.998: it is not exactly equal to 1 because, according to (2.5), it is computed from $T - 1$ pairs of observations, whereas the sample means $\overline{x}$ and $\overline{y}$ are each computed using T observations. The nonzero cross-correlations at other lags are due to autocorrelation within the alkalinity series. Notice that several of them are nominally significant according to the confidence

---

[2] It is true, however, that $r_{xy}(-k) = r_{yx}(k)$.

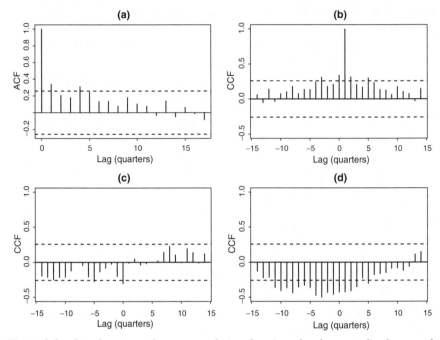

Figure 2.5   Correlogram and cross-correlation functions for deseasonalised quarterly alkalinity series from the Round Loch of Glenhead, 1988 to 2002: (a) correlogram for deseasonalised alkalinities, (b) cross-correlation with a lagged version of the same series, (c) cross-correlation with quarterly chloride series, (d) cross-correlation with quarterly series of nonmarine sulfate.

limits on the plots: a naïve interpretation would be that there is association between the series over a range of lags. This illustrates the potential difficulties in interpreting a CCF.

Figures 2.5(c) and 2.5(d) show the CCFs for deseasonalised alkalinity with chloride (Cl) and nonmarine sulfate (xSO$_4$) respectively. The Cl and xSO$_4$ series have not been deseasonalised here; this is because ultimately we want to assess directly their relationship with alkalinity, and seasonal adjustment would hinder such a direct interpretation. Seasonal adjustment of the alkalinity series, however, can be regarded as a crude way of adjusting for other factors prior to investigating the relationships of interest.

The chloride alkalinity CCF shows a small and apparently significant negative spike at lag zero, indicating the possibility of a weak simultaneous relationship between the two quantities. The remaining structure in the plot can safely be ignored, on the basis of lessons learned from Figure 2.5(b). Interpretation of the sulfate alkalinity CCF is less straightforward, with apparently significant correlations extending over a range of lags. This can to some extent be explained by the fact that the ACF of the sulfate time series (not shown here) exhibits very slow decay, indicating the presence of either substantial autocorrelation or trend. Since both the alkalinity and sulfate series appear autocorrelated, and there is no clear structure in their cross-correlation function, there is little point in trying to draw meaningful conclusions from Figure 2.5(d). In particular, it would be unwise to pay too much attention to the fact that the maximal cross-correlation appears to be at lags -3 and -4. Interpreted naïvely, this would suggest that sulfate responds to changes in alkalinity on timescales of 9 months to a year; however, it is difficult to imagine a mechanism that would have this effect, and it is much more likely that the result reflects sampling variability and the presence of other structures within both series.   ■

The example shows that CCFs may be useful when relationships between series are strong enough, but that otherwise the results may be ambiguous. To proceed further in such situations, it can be helpful to adopt a model based approach. We return to this in Chapter 3.

2.2   Simple smoothing

Time series plots are useful for visualising structure in a series providing the number of observations is not too large. However, it can be difficult to interpret such a plot if a series is long and highly variable. One way to deal with this is to aggregate the data to a coarser (e.g. annual) timescale prior to plotting. An alternative is to smooth them so as to display any trends more clearly. This section introduces some of the ideas involved. These will be elaborated later, particularly in Chapter 4.

2.2.1   Moving averages

The simplest way to create a smoothed series is to calculate a weighted average of observations in some neighbourhood of each time point. Given data $y_1, \ldots, y_T$, such a *moving average* takes the form

$$\tilde{y}_t = \sum_{j=-k}^{k} w_j y_{t+j} \tag{2.6}$$

for a collection of weights $\{w_j\}$ satisfying

$$w_j = w_{-j}(j = 1, \ldots, k) \quad \text{and} \quad \sum_{j=-k}^{k} w_j = 1. \tag{2.7}$$

A moving average is an example of a *linear filter*. Other examples are discussed in Section 2.3.

The 'symmetry' restriction $w_j = w_{-j}$ is not strictly necessary. However, it seems intuitively reasonable and the use of asymmetric weights can lead to difficulties in interpretation – for example, the smoothed series may be shifted in time relative to the original. In this chapter we will therefore focus almost exclusively upon symmetric filters. In this case, there must be an odd number of weights in the filter, since the sum in (2.6) involves $2k + 1$ terms.

In its simplest form, smoothing using (2.6) is very similar to aggregation. The main difference is that smoothing produces a value at every time point whereas aggregation does not: smoothing a monthly series produces another monthly series, but aggregation produces a series at a coarser resolution.

Clearly, in general the smoothness of the filtered series ($\tilde{y}_t$) depends on the value of k in (2.6) – larger values of k will yield smoother results. Notice also that to calculate

$\tilde{y}_t$, the values of y_{t-k} to y_{t+k} are required. A consequence of this is that $\tilde{y}_t$ cannot be calculated for $t < k + 1$ or for $t > T - k$. Methods for dealing with this are discussed in Section 2.3.3 below.

2.2.2 Local polynomial fitting

Figure 2.6 illustrates another possible approach to smoothing. The solid line here represents a time series of observations, showing a clear downward trend. Consider smoothing the series at time $t = 4$; this could be achieved by calculating a weighted average of neighbouring observations as in (2.6). Another idea is to fit a straight line to the neighbouring observations and to take the smoothed value $\tilde{y}_4$ from the fitted line. In Figure 2.6, the dashed line has been fitted to the group of five marked observations using least squares, and the fitted value $\tilde{y}_4$ is marked as a solid box. The fitted value seems more consistent with the overall pattern than the original observation y_4 (solid circle).

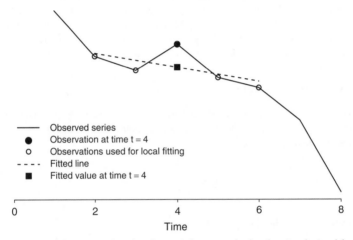

Figure 2.6 Smoothing using local polynomials: smoothed value $\tilde{y}_4$ derived by fitting a straight line to observations $y_2, \ldots, y_6$.

This idea is not restricted to fitting straight lines. One could equally consider using a quadratic, cubic or other polynomial fitted to the neighbouring observations. The general procedure is called *local polynomial smoothing*. It requires the choice of both the number of neighbouring points and the degree of the polynomial to fit (linear, quadratic, cubic or higher order).

Intuitively, it seems that local polynomial smoothing may be able to track an underlying trend more effectively than a moving average of the form (2.6), because the fitted polynomials can adapt to patterns in the data. In Figure 2.6, for example, the overall structure between times 2 and 6 seems reasonably well represented by a straight line. In fact, however, this apparent advantage is illusory. It can be shown (Kendall and Ord, 1990, Chapter 3) that the fitted values from local polynomials are actually weighted averages of the neighbouring observations. Indeed, the fitted values from a local linear fit as in Figure 2.6 are nothing more than simple averages of the neighbouring observations! This is a special case of a more general result: for regularly spaced series, (2.6) yields identical results for weights obtained from local polynomials of degrees $2p$ and $2p + 1$. A simple moving average can be regarded as the fitted value from a polynomial of degree zero (i.e. a constant), which according to this result is the same as that from a linear fit. Similarly, local quadratic and cubic fits will yield identical results. In general, higher degree polynomials yield more complicated patterns of weights. For example, local polynomial smoothing using quadratic or cubic polynomials fitted to groups of 15 observations is equivalent to filtering using (2.6), with weights $w_j = w_j^*/1105$, where w_j^* is defined as follows:

$$
\begin{array}{ccccccccc}
j & 0 & \pm1 & \pm2 & \pm3 & \pm4 & \pm5 & \pm6 & \pm7 \\
w_j^* & 167 & 162 & 147 & 122 & 87 & 42 & -13 & -78
\end{array}
\tag{2.8}
$$

Notice the presence of negative weights in this table – the $\{w_j\}$ in (2.6) are not restricted to be positive.[3]

Since local polynomial smoothing can be regarded as a special case of filtering using moving averages, it appears somewhat redundant. However, it has the important advantage of interpretability: it often makes sense to think of a trend as 'approximately linear' or 'approximately quadratic' over short time periods, and this provides some insight into exactly what the procedure is doing. By contrast, a collection of weights such as those in (2.8) is difficult to interpret out of context. A more compelling advantage is that the idea of fitting polynomials can be exploited to yield fairly accurate estimates of trend near the ends of a series; this is discussed in more detail later.

2.2.3   Further considerations

So far, we have assumed that the observations are equally spaced in time. For unequally spaced time series, the basic concepts are unchanged, but some modifications are required to cope with the irregular nature of the data. These are discussed in Chapter 4.

The implication behind the moving average (2.6) is that interest lies in the mean of the studied process. If this is not the case, it is usually possible to compute aggregated summaries of the data that reflect the aspects of interest, and then to apply smoothing techniques to these aggregated summaries. For example, to study trends in extreme daily wind speeds a first step might be to compute a time series of annual maximum wind speeds; smoothing techniques could then be applied directly to the maxima.

Smoothing has been introduced as a way to visualise structure in a time series. If this structure is to be interpreted as a trend, ideally there should be no other factors affecting the mean of the series. For example, if a time series contains a strong seasonal cycle then this may have a substantial effect on the results unless the weights $\{w_j\}$ are chosen carefully (see Section 2.3). In general, if other factors are present then a full analysis requires that they are all considered simultaneously. However, for exploratory purposes some simple adjustment may suffice, at least initially.

Example 2.6

Figure 2.7(a) shows the time series of deseasonalised monthly mean wind speeds at De Bilt, from 1961 to 2002. Denoting by y_{ij} the observed mean wind speed in month i of year j ($i = 1, \ldots, 12$; $j = 1961, \ldots, 2002$), these deseasonalised values (or *anomalies*) have been computed as

$$
y_{ij}^* = 100 y_{ij}/\bar{y}_i,
$$

---

[3] The R command `polyfil`, supplied as part of the software accompanying this book, can be used to find the weights corresponding to any local polynomial smoother for equally spaced data.

where $\bar{y}_i$ is the mean of all observations in month i. The anomalies thus transform each value to a percentage of the long-term mean for the corresponding month. There are far too many points on this plot to see any clear structure.

Figures 2.7(b) to (d) show the same monthly anomalies, with smoothed versions superimposed. Panel (b) uses a simple moving average of the form (2.6), with $w_j = 1/61$ for $j = -30$ to 30. The filter length of 61 months has been chosen to be as close as possible to five years (a length of exactly five years cannot be obtained since, as noted previously, the length of any symmetric filter must be an odd number). Panels (c) and (d) both use 15-point filters; that in panel (c) has $w_j = 1/15$ for $j = -7$ to 7 and panel (d) uses the weights given in (2.8).

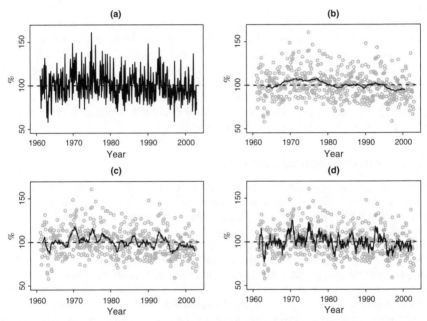

Figure 2.7   (a) Monthly mean wind speed anomalies (%) at De Bilt, 1961 to 2002, (b) with 61-point moving average, (c) with 15-point moving average, (d) with moving average based on the 15-point polynomial weights (2.8).

Although the three smooths all give an overall impression of a decreasing trend, the details are rather different in each case. As expected, the 61-point filter yields the smoothest result. However, panels (c) and (d) show that the choice of weights can affect the outcome even with filters of the same length – the simple moving average appears to smooth more than the local polynomial.

These plots also illustrate that, as noted previously, the filters cannot be applied over the entire range of the data. In panel (b), for example, smoothed values cannot be computed for the first and last 30 time points. This 'end effect' is less serious in panels (c) and (d), which use shorter filters.   ■

This example shows that the results of any smoothing exercise are highly dependent on the filter used. In the next section, we develop a framework within which to think about the implications of such a choice. This will also yield insights into some other applications of filters.

2.3   Linear filters

In everyday life, 'filtering' describes a process by which mixtures are separated into their component parts, often to remove impurities. It is no accident that the term is also used in the context of smoothing a time series. One might, for example, regard a particular series as consisting of an underlying trend with some irregular variation (the 'impurity') superimposed. Ideally, the effect of smoothing will be to remove the irregular variation, leaving just the trend (see Figure 2.8).

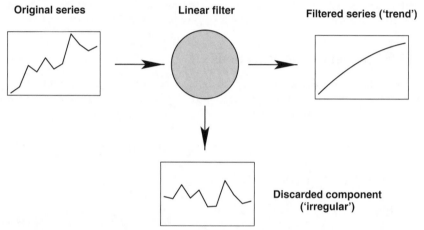

Figure 2.8   The use of a linear filter to separate a time series into 'trend' and 'irregular' components.

Continuing this analogy, the choice of filter for a particular application depends on what we want to extract. If, for example, we know that any trend will be linear, then we should use a filter through which linear trends will pass. Intuitively, it is fairly obvious that this can be achieved using a filter based on local linear smoothing, as illustrated in Figure 2.6. Similarly, if we think that a trend will be a polynomial then we could use a local polynomial filter of the same degree. In fact, however, there are many other filters that allow polynomials to pass through. It can be shown (Brockwell and Davis, 1991, Exercise 1.2) that if the original series is a polynomial of degree p and the weights satisfy the conditions

$$\sum_{j=-k}^{k} w_j = 1 \quad \text{and} \quad \sum_{j=-k}^{k} j^r w_j = 0 (r = 1, \ldots, p),$$

then the original and filtered series are identical. Consider, for example, *Spencer's 15-point filter*, which has weights $w_j = w_j^*/320$ and w_j^* as follows:

$$
\begin{array}{c|cccccccc}
j & 0 & \pm1 & \pm2 & \pm3 & \pm4 & \pm5 & \pm6 & \pm7 \\
w_j^* & 74 & 67 & 46 & 21 & 3 & -5 & -6 & -3
\end{array}
\tag{2.9}
$$

It is easy to check that these weights satisfy $\sum_{j=-7}^{7} w_j = 1$, $\sum_{j=-7}^{7} j w_j = 0$, $\sum_{j=-7}^{7} j^2 w_j = 0$ and $\sum_{j=-7}^{7} j^3 w_j = 0$. This filter will therefore allow linear, quadratic and cubic polynomials through; it could therefore be seen as an alternative to (2.8), if the aim is to extract such a polynomial.

This demonstrates that to choose an appropriate filter, more is required than simply the ability to extract a polynomial trend of specified degree: an infinite number of filters will achieve this. They differ in the other kinds of structure that will pass through: it would be useful to know what such structures might look like. Moreover, in an exploratory analysis it seems premature to choose methods based on assumed forms for the underlying trends – recall that one aim of such an exercise is to allow the data themselves to guide the subsequent analysis (a further point is that if the underlying trend really is a polynomial, it is best estimated using the methods described in Chapter 3 rather than using a filter). It is therefore necessary to develop a more general framework to understand the mechanisms of filtering. One way forward is to think about oscillations at different frequencies.

2.3.1   Frequency considerations

It is well known that any sequence $y_1, \ldots, y_T$ can be expressed as a sum of sine and cosine waves. Let $\lfloor T/2 \rfloor$ denote the integer part of $T/2$ (equal to $T/2$ if T is even and to $(T-1)/2$ if T is odd), and define

$$f_p = \frac{p}{T} \qquad (p = 0, 1, \ldots, \lfloor T/2 \rfloor). \tag{2.10}$$

Then, for $t = 1, \ldots, T$, it can be shown that

$$y_t = a_0 + \sum_{p=1}^{\lfloor T/2 \rfloor} \left(a_p \cos 2\pi f_p t + b_p \sin 2\pi f_p t \right), \tag{2.11}$$

where the *Fourier coefficients* $\{a_p, b_p\}$ are given by

$$a_p = \frac{1}{T} \sum_{t=1}^{T} y_t \cos 2\pi f_p \quad (p = 0, T/2),$$

$$a_p = \frac{2}{T} \sum_{t=1}^{T} y_t \cos 2\pi f_p t \quad (p \neq 0, T/2) \tag{2.12}$$

and

$$b_p = \frac{2}{T} \sum_{t=1}^{T} y_t \sin 2\pi f_p t \qquad \text{for all } p.$$

Equation (2.11) is called the *Fourier representation* of the sequence (y_t). Notice the following (see also Figure 2.9):

- When $p = 0$, $\cos 2\pi f_p t = \cos 0 = 1$ for all t; hence $a_0 = T^{-1} \sum_{t=1}^{T} y_t = \bar{y}$ is just the mean of the sequence.

- A graph of $a_p \cos 2\pi f_p t + b_p \sin 2\pi f_p t$ against t is a sine wave that repeats itself every T/p time units. This represents an oscillation around the overall mean of the sequence. The amplitude of the oscillation (i.e. the maximum departure from the mean) is

$$\sqrt{a_p^2 + b_p^2} = c_p, \text{ say.} \tag{2.13}$$

The quantity $f_p = p/T$ is called the *frequency* of the oscillation: it is the number of cycles per unit time.

- p can only take the value $T/2$ when T is even. In this case, $f_{T/2} = 1/2$ and $b_{T/2} = 2T^{-1} \sum_{t=1}^{T} y_t \sin \pi t = 0$, since $\sin \pi t = 0$ whenever t is an integer. Therefore, when T is even, only the cosine term contributes to the Fourier representation at the highest frequency.

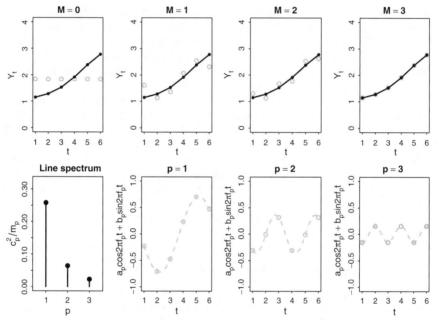

Figure 2.9   Fourier approximation of a sequence. Top row: specimen sequence (black), together with successive approximations. Bottom row, first panel: line spectrum for the specimen sequence. Remaining panels: individual components in the Fourier approximations, on a common scale.

Recall now that a trend, according to the definition in Chapter 1, represents long-term variation. In the Fourier representation of a time series, the sine waves associated with high frequencies (i.e. large values of p) oscillate rapidly; these frequencies are therefore largely irrelevant as far as trend is concerned. This suggests that the trend in a series could be approximated by retaining just the first few terms in (2.11):

$$\tilde{y}_t = a_0 + \sum_{p=1}^{M} \left(a_p \cos f_p t + b_p \sin f_p t \right)$$

for some value of $M \leq \lfloor T/2 \rfloor$. This is illustrated in the top row of Figure 2.9, which shows the Fourier representation of a sequence displaying a clear trend. Setting $M = 0$ yields $\tilde{y}_t = a_0$, which is constant; adding successive terms improves the accuracy of the approximation until, when all terms have been added, a perfect reconstruction is obtained. The final three plots in the bottom row show the individual sine waves added at each stage. As expected, the amplitude decreases for higher frequencies.

To visualise the relative contribution of each frequency in a Fourier representation, we could plot the amplitudes $\{c_p\}$ defined in (2.13). However, the *squared* amplitudes are usually of more interest. In statistical terms, this can be justified via the relationship (Chatfield, 2003, Section 7.3)

$$\sum_{t=1}^{T} (y - \overline{y})^2 = \sum_{t=1}^{T} (y - a_0)^2 = \sum_{p=1}^{\lfloor T/2 \rfloor} c_p^2 / m_p, \qquad (2.14)$$

where $m_{T/2} = 1$ and $m_p = 2$ for all other values of p. The left-hand side here is essentially the variance of the sequence (y_t); the right-hand side is a sum of contributions from each frequency. Therefore the quantity c_p^2 / m_p can be thought of as the variance explained by sinusoidal oscillations at frequency f_p. A plot of c_p^2 / m_p against p is called a *line spectrum*: see the first plot on the bottom row of Figure 2.9. In this example, the trend is evident from the large value at the lowest frequency.

Having established that trends contribute primarily to the low frequency components in a Fourier representation, it is natural to ask: How do other types of sequence manifest themselves? Obviously, different sequences have different frequency properties. Recall, however, that in some applications it may be appropriate to regard a series as consisting of 'trend' plus 'irregular variation'. In the absence of autocorrelation, the latter component could be regarded as a white noise sequence (see Section 2.1.3). It can be shown that, on average, the line spectrum of a white noise sequence is flat–the amplitudes of the components are expected to be the same at all frequencies. Therefore, we would expect a trend component to contribute primarily to the low frequencies whereas irregular variation will contribute to all frequencies.

2.3.2   The convolution theorem and filter design

We are now in a position to return to the problem of choosing a filter. It should now be clear what is required to extract a trend: we need a filter through which only low frequency components will pass. The key to finding such a filter is the *convolution theorem*, so called because an expression of the form (2.6) is known as a *convolution* of the sequences (y_t) and (w_j). A rigorous mathematical treatment is well beyond the scope of this book: we therefore give a heuristic presentation that sacrifices the finer points of detail, but hopefully conveys the main idea. A full development can be found in Percival and Walden (1993, Chapter 3).

If T is very large, the frequencies f_p in (2.10) will be close together and will effectively fill the range $(0, 1/2)$. In this case, it seems reasonable to approximate the line spectrum $\{c_p^2 / m_p\}$ by a continuous function $G(f)$ (imagine joining the points in the bottom left panel of Figure 2.9). We will refer to $G(f)$ simply as the *spectrum* of the sequence (y_t). In a similar vein, we could consider the sequence $\{w_j\}$ of filter weights in (2.6) and define a spectrum $H(f)$ proportional to $[a^*(f)]^2 + [b^*(f)]^2$, where $a^*(f) = \sum_{j=-k}^{k} w_j \cos(2\pi f)$ and $b^*(f) = \sum_{j=-k}^{k} w_j \sin(2\pi f)$. The essence of the convolution theorem is that the spectrum, $\tilde{G}(f)$ say, of the filtered sequence $(\tilde{y}_t)$ is proportional to $G(f) H(f)$. The 'weight spectrum' $H(f)$ is called the *squared gain* of the filter.

The first implication of the convolution theorem is that we can learn a lot about the properties of a filter by studying its squared gain. It also provides an opportunity to specify a desired form for $H(f)$ and to search for a filter of given length that will reproduce this as closely as possible. For example, if features on timescales of less than a couple of years are of little interest, one could attempt to construct a *low-pass* filter (i.e. a filter that only allows low frequencies through) for which $H(f)$ is effectively zero at frequencies corresponding to cycle lengths under two years, but close to 1 at lower frequencies. Methods for constructing such a filter are described by Percival and Walden (1993, Chapter 5). Perhaps the simplest is *least squares filter design*: for a low-pass filter allowing frequencies below f_0 through, the corresponding weights in (2.6) are proportional to

$$w_0 = 2f_0 \quad \text{and} \quad w_j = \frac{\sin 2\pi f_0 j}{\pi j} \ (j = \pm 1, \ldots, \pm k). \tag{2.15}$$

As might be expected, larger values of k give more flexibility and hence generally more accurate filters; however, recall from Example 2.6 that longer filters have more pronounced end effects. In practice therefore, the choice of k is a compromise.

Example 2.7
Figure 2.10 shows four different squared gains. Those in Figure 2.10(a) correspond to the filters used in the bottom panels of Figure 2.7. Apart from some wiggles, the squared gain of the 15-point moving average is effectively zero at frequencies greater than about 0.06 cycles per unit time. The 15-point cubic polynomial filter, however, allows frequencies up to about 0.1 to pass through. This explains why Figure 2.7(c) appears smoother than Figure 2.7(d) – the moving average removes more high frequency oscillations.

Figure 2.10(b) shows the squared gains for two 61-point filters; the horizontal axis is truncated at 0.25 to show the structure more clearly. The first filter is the five year moving average used in Figure 2.7(b); the second is designed to remove frequencies with cycle lengths below 24 time units (i.e. two years, if the data are monthly). The frequency corresponding to such a cycle length is $1/24 = 0.0417$ cycles per unit time; the filter weights are obtained by putting $f_0 = 1/24$ and $k = 61$ in (2.15). The squared gain for the corresponding 'ideal' filter is also shown. Although the designed filter approximates this reasonably well, notice that its squared gain exceeds 1 for frequencies between about 0.02 and 0.04 (cycle lengths between $1/0.04 = 25$ and $1/0.02 = 50$ time units). It will therefore produce a filtered series showing enhanced oscillations in this frequency range, which are associated more with the characteristics of the filter than with any properties of the underlying data. ∎

This example demonstrates that filtering can have unexpected consequences, such as the introduction of spurious oscillations in particular frequency ranges. Care is therefore required in the interpretation of a filtered series. For exploratory purposes, when a filter is used merely as an aid to visualisation, it probably suffices to be aware of the kinds of artefacts that can be introduced – a plot of the squared gain can be helpful in this respect – and to bear these in mind when interpreting the results.

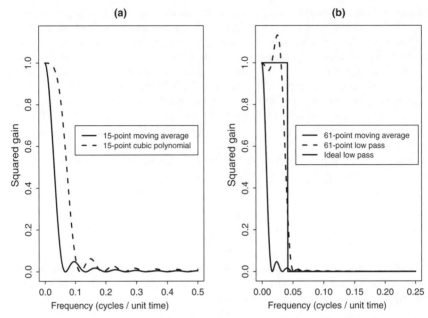

Figure 2.10   Squared gains for selected filters: (a) 15-point filters used in Figures 2.7(c) and (d), and (b) 61-point filter used in Figure 2.7(a), along with a 61-point low-pass filter designed to remove frequencies with cycle lengths below 24 time units.

2.3.3   Dealing with end effects

We have already noted that (2.6) yields smoothed values only between times $k + 1$ and $T - k$. The lack of output at the ends of the series is slightly unsatisfactory, particularly for longer filters. It is therefore customary either to modify the filter at each end of the series or to extend the data sequence artificially, so as to enable the calculation of smoothed values at all times between 1 and T. There is no single 'right' way to do this, and a variety of procedures are in use. Some of the more common are as follows:[4]

- At each time point, calculate a weighted average of just the available observations. For example, at time $t = 1$ the values $y_1, \ldots, y_{k+1}$ are available, so one could calculate the smoothed value

$$\tilde{y}_1 = \frac{\sum_{j=0}^{k} w_j y_{1+j}}{\sum_{j=0}^{k} w_j}. \tag{2.16}$$

- For local polynomial smoothing, at each time point use a polynomial fitted just to the available observations. Note that, in general, this is not the same as deriving the weights from a full local polynomial fit and then applying a formula such as (2.16). To see this, consider smoothing a hypothetical series in which the first three observations are $y_1 = 15$, $y_2 = 20$ and $y_3 = 25$, using local linear fitting to

---

[4] All of these are implemented in the R command `linear.fil`, supplied as part of the software accompanying this book.

successive groups of five observations. The observations available for smoothing at time $t = 1$ are y_1, y_2 and y_3, which fall on a straight line here; hence fitting a line to them will yield the result $\tilde{y}_1 = 15$. On the other hand, we saw earlier that when all required observations are present, local linear fitting places equal weight on all of them. In this case, equal weighting of just the available observations yields $\tilde{y}_1 = (15 + 20 + 25)/3 = 20$. In this simple example, the local linear fit seems preferable since it tracks the trend more closely. It can be shown that this is a feature of local polynomial fitting in general; this is discussed further in Chapter 4.

- Add k artificial data values to the beginning and end of the series. A simple strategy is to make the artifical values at the beginning all equal to the first observation y_1 and those at the end to the last observation y_T. In series with a strong seasonal component (and a seasonal period of S time units, say), this may be regarded as too simplistic and the artificial values could be obtained by replicating the first and last S observations as necessary to make up the required length.

- Use $y_{T-k+1}, \ldots, y_T$ in place of $y_{1-k}, \ldots, y_0$ and use $y_1, \ldots, y_k$ in place of $y_{T+1}, \ldots, y_{T+k}$. This is sometimes referred to as a 'circular' filter or as a *periodic extension* of the series–imagine wrapping the observations $y_1, \ldots, y_T$ round a circle, so that y_T and y_1 are next to each other. Clearly, this approach is not very suitable for application to series showing clear trends, since in this case the values at the beginning and end of the series will be very different. However, of the techniques described here this is the only one that is currently implemented in the built-in `filter` command in R.[5]

Of course, the properties of a filtered series obtained using any of these methods will be different at the ends than in the central portion. This is unavoidable, but needs to be accounted for in any analysis based on smoothing. This is discussed further in Chapter 4.

2.3.4   Other applications

Until now, we have discussed filters as a means of smoothing a series so as to see more clearly the nature of any apparent trends. However, their use extends considerably beyond this. We now consider two further applications: removal of trend and seasonality.

Recall from Section 1.2 that sometimes it is necessary to identify and remove any trends so as to focus on aspects of real interest. In such situations, with reference to Figure 2.8, interest lies in the 'irregular' component rather than the 'trend'. Of course, since the original sequence (y_t) is just the sum of the two components, it is easy to recover the irregular component given the smoothed version $\tilde{y}_t$: just calculate the difference $y_t - \tilde{y}_t$. However, from the definition of $\tilde{y}_t$ at (2.6), we have

$$y_t - \tilde{y}_t = y_t - \sum_{j=-k}^{k} w_j y_{t+j} = \sum_{j=-k}^{k} v_j y_{t+j}, \text{ say,} \qquad (2.17)$$

$$\text{where} \quad v_j = \begin{cases} 1 - w_j & j = 0, \\ -w_j & j \neq 0. \end{cases}$$

Notice that the expression for the 'detrended' series (2.17) looks identical to (2.6), except that the weights $\{w_j\}$ have been replaced by another collection $\{v_j\}$. Using the definition

---

[5] Version 2.2.0.

of v_j given above, and remembering that $\sum_{j=-k}^{k} w_j = 1$, we find that

$$\sum_{j=-k}^{k} v_j = 1 - \sum_{j=-k}^{k} w_j = 0.$$

This shows that trends can be removed from a series using a filter with weights that sum to zero. The squared gain for such a filter is close to zero at low frequencies, but not at high frequencies: it is a *high-pass filter*. It may be worth pointing out that since the weights do not sum to one, these filters are not weighted averages. It is therefore not appropriate to deal with end effects using a formula such as (2.16), since it is not clear what the denominator should be. In general, to avoid confusion it is probably easiest to calculate detrended series as $y_t - \sum_{j=-k}^{k} w_j y_{t+j}$ rather than $\sum_{j=-k}^{k} v_j y_{t+j}$, since end effects can be dealt with easily, using any of the methods described above, in the former case but not the latter. However, the interpretation of trend removal as high-pass filtering can be a useful aid to understanding.

Another use of filters is for the removal of seasonality. So far we have dealt with this in a simple way, subtracting or dividing each observation by the mean value for the time of year. For series showing little trend and a regular seasonal pattern, this is often perfectly adequate. However, in the presence of trends or if the seasonal cycle is not exactly regular, something more sophisticated may be needed. Consider first the case when a more or less regular seasonal cycle is present along with a clear underlying trend, as in Figure 2.11(a). In this situation, if the trend can be removed without disturbing the seasonal cycle then the latter can be estimated from the detrended series. For monthly data with an annual cycle, a rough estimate of the trend can be obtained using a filter of length 13, with weights $w_0 = w_{\pm 1} = \cdots = w_{\pm 5} = 1/12$, $w_{\pm 6} = 1/24$. This is effectively a moving average over a period of one year; the requirement for the filter length to be an odd number is accommodated by extending the length of the filter to 13 rather than 12 and halving the weight at each end. According to (2.17), the corresponding high-pass filter, yielding the detrended series, has weights

$$v_0 = 1 - 1/12 = 11/12, \quad v_{\pm 1} = \cdots = v_{\pm 5} = -1/12, \quad v_{\pm 6} = -1/24. \qquad (2.18)$$

The squared gain of this filter is shown in Figure 2.11(b). Notice that it passes higher frequencies, removes most lower frequencies and does not affect the frequency of the annual cycle, since the squared gain is 1 here.

The result of applying a filter such as (2.18) will be a series showing seasonality but no trend, as in Figure 2.11(c) (here, end effects have been dealt with by replicating the first and last 12 observations at the start and end of the series respectively, as described in Section 2.3.3). Denote the 'seasonal averages' of this detrended series by $\overline{m}_1^*, \ldots, \overline{m}_S^*$, where S is the number of observations per year (so that $S = 12$ for monthly data). These seasonal averages can themselves be regarded as estimates of the seasonal cycle; however, it is conventional to subtract their mean so that the result can be interpreted as an average oscillation about an 'underlying' level:

$$\overline{m}_i = \overline{m}_i^* - \frac{1}{S} \sum_{j=1}^{S} \overline{m}_j^* \quad (i = 1, \ldots, S).$$

The original series can now be 'deseasonalised' by subtracting the estimated seasonal cycle $\{\overline{m}_i : i = 1, \ldots, S\}$ from the original observations, as in Figure 2.11(d). If necessary, the

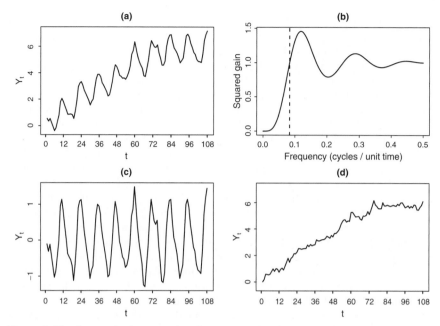

Figure 2.11   Removal of seasonality from an artificial monthly series containing both trend and seasonal components: (a) original series, (b) squared gain for the filter $\{v_j\}$ defined by (2.18) (dashed line indicates the frequency of the annual cycle), (c) detrended series, (d) original series after subtraction of the estimated seasonal cycle.

result could be used as the starting point for a more refined estimate of the trend and the procedure could be repeated.

In the above procedure, the use of seasonal averages to estimate the seasonal cycle is sensible if this cycle itself is fairly constant. However, in environmental contexts this is often unrealistic – for example, in many areas climate change is associated with changes in seasonal patterns of temperature and precipitation (for example Hulme *et al.*, 2002; Räisänen *et al.*, 2004), and environmental changes may lead to shifts in the seasonal cycle within ecosystems (Rosenzweig *et al.*, 2007, Section 1.3). In this case, an alternative is to estimate the changing cycle at time t using a filter that places all its weight on observations at times $t \pm S, t \pm 2S, \ldots$. For example, given monthly data the seasonal cycle for January of a particular year could be estimated as a weighted average of the detrended January observations in neighbouring years. Denoting by $\breve{y}_t$ the value of the detrended series at time t as in Figure 2.11(c), such an estimate of the seasonal cycle at time t is

$$\sum_{j=-\ell}^{\ell} w_j^{(s)} \breve{y}_{t-Sj}$$

for filter weights $\left\{ w_j^{(s)} : j = 0, \pm 1, \ldots, \pm\ell \right\}$ with $\sum_{j=-\ell}^{\ell} w_j^{(s)} = 1$. The superscript (s) is used to emphasise that these are weights in a 'seasonal' filter. The corresponding deseasonalised value is

$$\breve{y}_t - \sum_{j=-\ell}^{\ell} w_j^{(s)} \breve{y}_{t-jS}.$$

The squared gain for the filter $\left\{w_j^{(s)}\right\}$ will be close to zero for all frequencies except those associated with a cycle length of S time units. Such a filter, which only passes frequencies in a limited range, is called a *band-pass filter*.

The ideas discussed here form the basis for classical techniques designed to separate a series into 'trend', 'seasonal' and 'irregular' components (Kendall and Ord, 1990, Chapters 3 and 4; Brockwell and Davis, 1991, Section 1.4; Chatfield, 2003, Section 2.6), as well as more modern methods such as the 'STL' procedure to be discussed in Chapter 4. We have focused on the case when trend and seasonal components are removed by subtraction: for series such as the Dutch wind speed data, in which the variability increases with the mean, it may be more appropriate to use division instead (as in the earlier examples in this chapter). In this case, the detrended or deseasonalised series cannot be interpreted strictly as the output of a high-pass filter; nonetheless, the basic principle of separating components at different frequencies remains a helpful aid to understanding.

We conclude this discussion of filters with a caveat. As we have seen, the aim of filtering is to extract structures of interest: this is a useful aid to visualisation in the exploratory phase of an analysis. However, since filtering changes the properties of a series, care needs to be taken with any subsequent analyses. It may be tempting, for example, to smooth a series so as to visualise any apparent trends more easily and then to carry out one of the tests described in the next section to see if the trends in the smoothed series are 'genuine'. However, this would not be appropriate: even in the absence of underlying trends, smoothed series will show 'trend-like' features such as a concentration of variability at low frequencies, and this may lead to incorrect test results. Put another way: if a test shows evidence for a trend in a smoothed series, this could be due either to a genuine underlying trend or to an artefact of the smoothing procedure. In general, it is unwise to analyse filtered data without a clear understanding of the implications: at the very least, it is necessary to account somehow for the filtering when interpreting the results. As a general rule, it is probably best to avoid such problems altogether by using only the original data in any subsequent analyses. In this case, the aim of filtering is to provide insight into the types of structure that may be present: the results can be used to guide the choice of subsequent analysis.

2.4   Classical test procedures

Finally in this chapter, we discuss some procedures that are commonly used to test for the existence of trends in environmental time series. For further details, see El-Shaarawi (1993). The tests described here have been used extremely widely, with recent environmental applications including Wasmund and Uhlig (2003), Carslaw (2005), Hamed (2008) and Chen and Grasby (2009). The Mann–Kendall and seasonal Kendall tests (see, for example, Chen and Grasby, 2009; Esterby, 1993; Hirsch, Slack and Smith, 1982) are classic nonparametric tests used in the literature to assess monotonic trends. They are perhaps the most resilient survivors of a range of nonparametric tests that were once popular but, with the advent of modern computing and model based statistics, have fallen into disuse – see Kendall and Ord (1990, Chapter 2) for a review. They can handle missing data and do not make any distributional assumptions, since they use the relative magnitudes of the data rather than their absolute values, but they do in general assume that the observations are independent. In R, implementations can be found in the `Kendall` library (McLeod, 2005).

The original Mann–Kendall test for trend is a nonparametric test that applies to monotonic trends only. The null hypothesis is that the data are realised values of independent, identically distributed random variables. Clearly, in this case no trend is present. Given a data sequence $y_1, \ldots, y_T$, the differences $\{d\,(t_1, t_2) = y_{t_2} - y_{t_1} : t_2 > t_1\}$ between all pairs of observations are calculated. Positive differences are then assigned the value 1, negative differences are assigned the value -1 and zero differences are assigned the value zero. The test statistic is then the sum of these assigned values, i.e.

$$S = \sum_{t_1=1}^{T-1} \sum_{t_2=t_1+1}^{T} \text{sgn}\,[d\,(t_1, t_2)],$$

where

$$\text{sgn}\,[d\,(t_1, t_2)] = \begin{cases} 1 & \text{if}\,d\,(t_1, t_2) > 0, \\ 0 & \text{if}\,d\,(t_1, t_2) = 0, \\ -1 & \text{if}\,d\,(t_1, t_2) < 0. \end{cases}$$

From this definition it is clear that S can take any integer value between $-T(T-1)/2$ and $T(T-1)/2$. It can be shown that S is proportional to a nonparametric index of correlation, Kendall's τ (Wackerly, Mendenhall and Scheaffer, 2007, Section 15.10), between the observations and the time index t. In the absence of ties (i.e. if $d\,(t_1, t_2)$ is never zero), we have $S = T(T-1)\tau/2$. When $S = \pm T(T-1)/2$, we have $\tau = \pm 1$ and hence a perfect monotonic relationship between the observations and the time index. Under the null hypothesis of no trend, S is expected to be close to zero. Large negative values indicate downward trends and large positive values indicate upward trends. For small values of T, the null distribution of the test statistic can be tabulated exactly (Kendall, 1975); for larger values ($T \geq 40$ is sometimes suggested as a rule of thumb), this can be approximated by a normal distribution with mean zero and variance $T(T-1)(2T+5)/18$.

If the Mann–Kendall test indicates that a trend is present, it is natural to try and quantify its magnitude. The *Theil–Sen slope estimator* (Sen, 1968) is commonly used for this purpose: this is defined as the median of the pairwise slopes $\{[d\,(t_1, t_2)]\,/\,[t_2 - t_1] : t_2 > t_1\}$, and can be regarded as a 'typical' index of change over a unit time period.

Hirsch, Slack and Smith (1982) developed the seasonal Mann–Kendall test for trend analysis of monthly water quality data and Hirsch and Slack (1984) present a version of the test that accommodates serial dependence. Seasonality is handled by calculating a test statistic that is a sum of contributions from individual seasons and no assumption about the form of the distribution function of the data is made. Libiseller and Grimvall (2002) extend these techniques by developing the partial Mann–Kendall test, which allows the inclusion of covariates in the Hirsch and Slack test for trends with serially correlated data collected over several seasons.

In any trend test the probability of rejecting the null hypothesis in the presence of a genuine trend will depend on the sample size (i.e. record length), the magnitude of the trend and the variability of the data about the trend. For some test procedures, statistical *power calculations* can be used to evaluate the probability of rejecting the null hypothesis explicitly as a function of such factors (see Davison, 2003, Section 7.3.2 or Rice, 2006, Section 11.2.2 for some simple examples in a slightly different context). However, power calculations typically require precise specification of the form of the trend as well as the distribution of the observations: therefore they are not readily adapted to the nonparametric tests described above, except under some precisely specified parametric model that serves as a benchmark. Nonetheless, for any test procedure it will usually be difficult or impossible to detect trends with any confidence in very short time series. For ecological applications in particular, Wilson, Hammond and Thompson (1999) give some useful guidance on how to think about what data may be required for the reliable detection of trends. The ability to detect environmental trends has also been investigated by Gerrodette (1987) who considered linear trends in ecological data and by Frei and Schär (2001) who were concerned with the trends in extreme rainfall events.

In the context of environmental monitoring, it is sometimes required to examine the evidence for changes within a system on the basis of observations collected on a handful of sampling occasions but at several spatial locations. In such situations, rather than trying to draw any definitive conclusions about trend, perhaps the most that can be achieved is to determine whether there are any significant differences between the sampling occasions. If data from the different locations can be considered as statistically independent, a simple way to achieve this is using a two-way analysis of variance (ANOVA) as described by Manly (2001, Section 5.6). In the presence of spatial dependence, however, more sophisticated methods are required (see Section 6.1).

The tests described above are designed primarily for *retrospective* or *offline* analysis. However, it is often of interest to detect changes as they occur. For example, when monitoring river water quality near a sewage treatment works, it is desirable to detect any deterioration as quickly as possible so that appropriate corrective action can be taken. *Control charts* provide a simple tool for *online* monitoring of this type. The most common control charts are designed to monitor the mean level and variability of a series. For the mean, at each time point the running mean of the observations is computed: $\bar{y}_t = S_t/t$ say, where $S_t = \sum_{j=1}^{t} y_j$. In the absence of any change and under some assumptions about the nature of the underlying process (for example that the observations are drawn independently from normal distributions with a common variance), it is possible to define limits within which $\bar{y}_t$ should lie with specified probabilities. If these probabilities are chosen appropriately then the presence of a potential change is suggested whenever the values fall outside the limits. It is customary to display the results graphically, with the running means and control limits plotted against the time index. An alternative that typically allows changes to be detected more rapidly is to produce a *cusum chart* in which limits are set for the cumulative deviations from some target value. The variability of a process can be monitored in a similar way, but using the sample range or standard deviation instead of the mean. Manly (2001, Section 5.7) gives a brief introduction to control charts in an environmental monitoring context; for a more detailed discussion of the available methods, see Hawkins and Olwell (1997), Ledolter and Burrill (1999) or Montgomery (2009).

Test procedures such as those reviewed above are widely used. However, their effective use requires caution, good judgement and an awareness of the underlying logic. In particular, it is important to recognise that the null hypothesis for many of the tests is specified far more precisely than the innocuous-sounding 'absence of trend'. For example, although the Mann–Kendall test makes no assumptions about the form of the distribution from which the observations are generated, the null hypothesis is that the observations are independently and identically distributed. This in itself is an extremely strong assumption. According to the logic of significance tests (see Section 1.4.3), a statistically significant result suggests that the null hypothesis is false. It is not appropriate to conclude from this that the observations support the specific alternative of interest. If a significant Mann–Kendall test result is obtained, one may legitimately conclude that the observations are not independently and identically distributed. However, given the complexity of most environmental systems, such a conclusion is hardly groundbreaking. The test result may indeed indicate that a trend is present, but it could also indicate that the observations are autocorrelated (indeed, we have already seen that it can be difficult to distinguish between autocorrelation and trends when interpreting the sample ACF) or that they are influenced by other factors and hence not independently distributed. This is not to say that hypothesis testing is of no value in environmental applications: rather, it is important to understand precisely what the null hypothesis is and to ensure that it provides at least a plausible representation of the system. Rejection of a plausible null hypothesis gives much more credence to the alternative of interest. Often, therefore, tests for trend require rather more sophistication than these classical procedures.

A further caveat regarding trend tests is that often research questions arise in response to observations that are considered to be unusual. Suppose, for example, that an investigation into temperature trends is prompted by the observation that three of the warmest years in a particular century occurred in the final decade of the century. In this situation, it would be inappropriate to incorporate the observations from the final decade when testing for trends, since in this case the observations being used to test a hypothesis would include those that suggested it in the first place. Any random sequence will occasionally produce clusters of high values, and to test for trends only after observing such a cluster will naturally bias the results.[6]

2.5 Concluding comments

A wide range of techniques has been reviewed in this chapter. Many of them are well established in one or more areas of environmental science, and it is perhaps natural to ask why they have all been included under the heading 'exploratory analysis' – surely, after dealing with visualisation, serial dependence, smoothing, trend and seasonality removal and testing, there is little more to be said.

It is true that, to some extent, the ideas presented above form the basis for everything else in the book. However, there is a shift of emphasis in the next and subsequent chapters, insofar as all of the material is based on different types of statistical model, specified in terms of the probability distributions from which the data are assumed to be generated. As described in Section 1.1, such specifications enable scientific questions to be framed precisely and without ambiguity, and most modern statistical methods are based on them. One advantage of this approach is that it offers the potential to represent explicitly the complexity of the process under study, and hence to account properly for all relevant information. In contrast, the methods described above tend to focus on the various issues separately, which can lead to difficulties – for example, the interpretation of a sample autocorrelation function depends on whether or not a trend is present, but any of the standard tests for trend are themselves affected by the presence of autocorrelation. It is therefore unwise to draw conclusions solely on the basis of these methods: they are best used to indicate the issues that need to be considered in the quest for an appropriate model. On completion of a thorough exploratory analysis, however, the way forward should be reasonably clear.

---

[6] To avoid any ambiguity this is not an argument against global warming. The evidence for global warming is far more wide-ranging than a post hoc trend test.

References

Brockwell, P. J. and Davis, R. A. (1991) *Time Series: Theory and Methods*, 2nd edition. Springer-Verlag, New York.

Brockwell, P. J. and Davis, R. A. (2002) *Introduction to Time Series and Forecasting*, 2nd edition. Springer-Verlag, New York.

Carslaw, D. C. (2005) Evidence of an increasing NO_2/NO_X emissions ratio from road traffic emissions. *Atmospheric Environment*, **39**, 4793–4802.

Chatfield, C. (2003) *The Analysis of Time Series–An Introduction*, 6th edition. Chapman & Hall/CRC Press, Boca Raton, Florida.

Chen, Z. and Grasby, S. E. (2009) Impact of decadal and century-scale oscillations on hydroclimate trend analyses. *Journal of Hydrology*, **365**, 122–133.

Cleveland, W. S. (1994) *The Elements of Graphing Data* revised edition. Murray Hill, New Jersey.

Davison, A. C. (2003) *Statistical Models*. Cambridge University Press, Cambridge.

El-Shaarawi, A. (1993) Environmental monitoring, assessment and prediction of change. *Environmetrics*, **4**, 381–398.

Esterby, S. T. (1993) Trend analysis methods for environmental data. *Environmetrics*, **4**, 459–481.

Frei, C. and Schär, C. (2001) Detection probability of trends in rare events: theory and application to heavy precipitation in the Alpine region. *Journal of Climate*, **14**, 1568–1584.

Gerrodette, T. (1987) A power analysis for detecting trends. *Ecology*, **68**, 1364–1372.

Hamed, K. H. (2008) Trend detection in hydrologic data: the Mann-Kendall trend test under the scaling hypothesis. *Journal of Hydrology*, **349**, 350–363.

Hawkins, D. M. and Olwell, D. H. (1997) *Cumulative Sum Charts and Charting for Quality Improvement*. Springer-Verlag, New York. xvi + 247 pp.

Hirsch, R. and Slack, J. (1984) A nonparametric trend test for seasonal data with serial dependence. *Water Resources Research*, **20**, 727–732.

Hirsch, R. Slack, J. and Smith, R. (1982) Techniques of trend analysis for monthly water quality data. *Water Resources Research*, **18**, 107–121.

Hulme, M., Jenkins, G. J., Lu, X., Turnpenny, J. R., Mitchell, T. D., Jones, R. G., Lowe, L., Murphy, J. M., Hassell, D., Boorman, P., McDonald, R. and Hill, S. (2002) *Climate Change Scenarios for the United Kingdom: The UKCIP02 Scientific Report*. Tyndall Centre for Climate Change Research, School of Environmental Sciences, Norwich, UK. 120 pp. Also available from http://www.ukcip.org.uk/resources/publications/.

Kendall, M. G. (1975). *Rank Correlation Methods*, 4th edition. Charles Griffin, London.

Kendall, M. and Ord, J. (1990) *Time Series*, 3rd edition. Edward Arnold.

Ledolter, J. and Burrill, C. W. (1999) *Statistical Quality Control: Strategies and Tools for Continual Improvement*. John Wiley & Sons, Inc., New York. x + 526 pp.

Libiseller, C. and Grimvall, A. (2002) Performance of partial Mann–Kendall tests for trend detection in the presence of covariates. *Environmetrics*, **13**(1), 71–84.

McIlveen, R. (1992) *Fundamentals of Weather and Climate*. Chapman & Hall, London.

McLeod, A. (2005) *Kendall: Kendall Rank Correlation and Mann–Kendall Trend Test*. R package version 2.0.

Manly, B. F. J. (2001) *Statistics for Environmental Science and Management*. Chapman & Hall/CRC, Boca Raton, Florida.

Montgomery, D. C. (2009) *Statistical Quality Control: A Modern Introduction*, 6th edition. John Wiley & Sons, Inc., Hoboken, New Jersey. xiv + 734 pp.

Percival, D. B. and Walden, A. T. (1993) *Spectral Analysis for Physical Applications–Multitaper and Conventional Univariate Techniques*. Cambridge University Press, Cambridge.

Pinheiro, J., Bates, D., DebRoy, S., Sarkar, D. and the R Core Team (2008) *nlme: Linear and Nonlinear Mixed Effects Models*. R package version 3.1-89.

Priestley, M. B. (1981) *Spectral Analysis and Time Series*. Academic Press, New York.

Räisänen, J., Hansson, U., Ullerstig, A., Döscher, R., Graham, L. P., Jones, C., Meier, H. E. M., Samuelsson, P. and Willén, U. (2004) European climate in the late twenty-first century: regional simulations with two driving global models and two forcing scenarios. *Climate Dynamics*, **22**, 13–31. DOI: 10.1007/s00382-003-0365-x.

Ribeiro, P. J. and Diggle, P. J. (2001) geoR: a package for geostatistical analysis. *R-NEWS*, **1**(2), 14–18. ISSN 1609-3631.

Rice, J. (2006) *Mathematical Statistics and Data Analysis*, 3rd edition. Duxbury Press, Belmont, California.

Rosenzweig, C., Casassa, G., Karoly, D. J., Imeson, A., Liu, C., Menzel, A., Rawlins, S., Root, T. L., Seguin, B. and Tryjanowski, P. (2007) Assessment of observed changes and responses in natural and managed systems. In *Climate Change 2007: Impacts, Adaptation and Vulnerability. Contribution of Working Group II to the Fourth Assessment Report of the Intergovernmental Panel on Climate Change* (eds M. L. Parry, O. F. Canziani, J. P. Palutikof, P. J. van der Linden, and C. E. Hanson) Cambridge University Press, Cambridge. pp. 79–131.

Sen, P. K. (1968) Estimates of the regression coefficient based on Kendall's tau. *Journal of the American Statistical Association*, **63**(324), 1379–1389.

Wackerly, D., Mendenhall, W. and Scheaffer, R. (2007) *Mathematical Statistics with Applications*, 7th edition. Duxbury Press, Belmont, California.

Walther, G. (1997) Absence of correlation between the solar neutrino flux and the sunspot number. *Physical Review Letters*, **79**, 4522–4524.

Wasmund, N. and Uhlig, S. (2003) Phytoplankton trends in the Baltic Sea. *ICES Journal of Marine Science*, **60**(2), 177–186.

Webster, R. and Oliver, M. A. (2001) *Geostatistics for Environmental Scientists*. John Wiley & Sons, Ltd, Chichester.

Wilson, B., Hammond, P. S. and Thompson, P. M. (1999) Estimating size and assessing trends in a coastal bottlenose dolphin population. *Ecological Applications*, **9**, 288–300.

3

Parametric modelling – deterministic trends

As described in the previous chapter, preliminary exploration of the data is often a prelude to a more detailed analysis based on some form of statistical model. Models provide the opportunity to account simultaneously for all relevant structures, and hence to place trends in their proper context. Furthermore, writing down a model forces the analyst to formulate their questions precisely, usually by translating subject-matter concerns ('Is there evidence for a systematic decline in North Sea haddock stocks?') into questions about numerical quantities appearing in the model ('Is the trend slope negative?'). A related point is that a model based analysis is transparent: a fully specified model encapsulates all of the assumptions upon which the results are based so that these assumptions are immediately clear and, as we will see, there are usually well understood diagnostic techniques for checking them. Another advantage is that, for many models, particular statistical proce-dures are known to be optimal with respect to estimation and testing: this provides some reassurance that one is extracting the maximum possible information from hard-earned data. Finally, most (if not all) statistical models are probabilistic in nature: this provides the opportunity to quantify explicitly the uncertainty in conclusions. For reasons such as these, modern statistical thinking is fundamentally model-based. For some insight into this way of thinking, and an extremely comprehensive overview of modern statistics, the excellent text by Davison (2003) is strongly recommended.

In the specific context of a trend analysis, perhaps the simplest model one might consider is an adaptation of Equation (1.1) in Section 1.1. According to that equation, the observation at time t is considered as the realised value of a random variable Y_t with expected value μ_t. It follows that we can write

$$Y_t = \mu_t + \varepsilon_t, \tag{3.1}$$

where ε_t is a random variable with zero expectation. Equation (3.1) is a statistical model in which the observations are considered to be randomly scattered about the sequence

Statistical Methods for Trend Detection and Analysis in the Environmental Sciences, First Edition.
Richard E. Chandler and E. Marian Scott.
© 2011 John Wiley & Sons, Ltd. Published 2011 by John Wiley & Sons, Ltd.

(μ_t), which may be referred to as the 'trend function'. Taking expectations of both sides of (3.1), we see that μ_t is also the expected value of Y_t.

In this chapter, we consider estimation and testing for situations in which μ_t is assumed to have a particular mathematical representation: this is an example of a *para-metric model*. Some examples are given in Table 3.1. In each case, Greek letters denote the values of *parameters* that determine the exact behaviour of the trend, possibly given the values of other relevant variables such as the $\{x_{it}\}$ in models 3 and 4. We refer to such trends as *deterministic*, to distinguish them from an alternative class of parametric models to be considered in Chapter 5.

Table 3.1   Examples of parametric models for deterministic trends.

Model	Mathematical representation
1. Linear trend	$\mu_t = \beta_0 + \beta_1 t$
2. Growth curve (logistic model)	$\mu_t = \alpha \left[1 + \beta \exp\left(-\gamma t\right) \right]^{-1}$
3. Trend linearly associated with p other quantities – x_{it} is the value of the ith such quantity at time t	$\mu_t = \beta_0 + \sum_{i=1}^{p} \beta_i x_{it}$
4. Logarithm of trend linearly associated with changes in other quantities	$\log \mu_t = \beta_0 + \sum_{i=1}^{p} \beta_i x_{it}$

Coupled with the basic trend equation (3.1), the first three models in Table 3.1 are all regressions; it is therefore sometimes appropriate to refer to μ_t as the 'regression function' rather than the 'trend function'. The regression formulation is perhaps most obvious in the case of model 3, which is a standard multiple regression model. The linear trend (model 1) regresses the variable of interest upon the time index t: this model, with various modifications, is ubiquitous throughout the environmental science literature. Model 2 is again a regression upon t, but the trend is no longer linear: this is a simple model for phenomena such as population growth (e.g. Haddon, 2001, Section 2.4). Multiple regression (model 3) is useful for trying to identify specific factors that are linked to changes in the variable of interest over time. For example, to investigate an apparent change in the incidence of respiratory illness in a particular location, one might consider regressing monthly reported incidents upon indices of air quality. The nature of this model is slightly different from that of the previous two, in that the explanatory variables (i.e. the $\{x_{it}\}$) may themselves vary irregularly through time. However, we still refer to this as 'deterministic', because the trend is completely determined once all relevant quantities (i.e. the parameters and explanatory variables) are known.

A regression model is said to be *linear* if the parameters enter linearly into the regression function, as in models 1 and 3 in Table 3.1 (note the distinction between this and a linear *trend*, where the linearity refers to the time index rather than the parameters). Thus model 4 is also nonlinear, since here $\mu_t = \exp[\beta_0 + \sum_{i=1}^{p} \beta_i x_{it}]$ is a nonlinear function of the parameters $\beta_0, \ldots, \beta_p$. The reason for considering it separately is that the parameter nonlinearity can be removed by applying a simple transformation to μ_t (i.e. by taking logs). In such situations, models can often be fitted much more easily than for other types of nonlinear model. Notice that it is the trend μ_t that is transformed, not the data $\{y_t\}$. This is one of the key features of generalised linear models, which are considered further in Section 3.5.

As usual, when using regression methods in a trend analysis, most of the difficulties are caused by the potential presence of autocorrelation and other factors that affect the variable of interest. To illustrate the issues involved, in the next section we discuss the linear trend model in detail. This provides the necessary foundations for the more complicated models introduced subsequently. The discussion assumes operational familiarity with basic concepts of regression modelling, such as least squares fitting and testing hypotheses about regression coefficients. The necessary background material can be found, for example, in Draper and Smith (1998), Weisberg (2005), Rice (2006, Chapter 14) or Wackerly, Mendenhall and Scheaffer (2007, Chapter 11).

3.1   The linear trend

In this section, we focus mostly on the case when the observations are regularly spaced. This is purely for notational convenience, since the basic ideas can be applied with little or no modification to irregularly spaced series. The regular spacing allows us to write $y_1, \ldots, y_T$ for the observations and $Y_1, \ldots, Y_T$ for the corresponding random variables. The linear trend model is most commonly specified as

$$Y_t = \beta_0 + \beta_1 t + \varepsilon_t \qquad (t = 1, \ldots, T), \qquad (3.2)$$

where $\varepsilon_1, \ldots, \varepsilon_T$ is a sequence of random variables with zero mean. We follow standard practice in referring to these as 'errors', although this terminology is slightly pejorative: it suggests that in an ideal world they would not be present, whereas in the context of environmental time series they are a convenient and necessary device for representing variation that cannot be explained otherwise using the available data.

In the case when the errors all have the same variance, σ^2 say, and are uncorrelated (i.e. they form a white noise sequence – see the end of Section 2.1.3), (3.2) is just a linear regression of Y on t, and can be fitted easily using least squares in any standard software package. In the discussion below, it will be helpful to use a notation that extends directly to more general regression models. This can be achieved by writing the model as

$$Y_t = \beta_0 + \beta_1 x_t + \varepsilon_t \quad (t = 1, \ldots, T), \qquad (3.3)$$

where $x_t = t$.

If the errors are normally distributed, or if the sample size is large enough, a test for trend can be carried out using a t test of the null hypothesis $H_0 : \beta_1 = 0$ against the alternative $H_1 : \beta_1 \neq 0$. This procedure is exactly equivalent to a test for correlation between the observations $\{Y_t\}$ and the time indices $\{t\}$. In addition, standard linear regression theory enables the construction of confidence intervals for the parameters β_0 and β_1 (Draper and Smith, 1998, Section 1.4; Weisberg, 2005, Section 2.8). This theory is appropriate whenever the errors form a white noise sequence, in which case the least squares procedure also has appealing optimality properties (Draper and Smith, 1998, Section 5.2; Weisberg, 2005, Section 2.4).

Example 3.1
Consider the North Sea haddock data from Section 1.3.2. Figure 1.2 suggested that, on a logarithmic scale, there appears to be a linear trend in biomass. Accordingly, it seems reasonable to fit the model

$$\log Y_t = \beta_0 + \beta_1 t + \varepsilon_t,$$

where Y_t is the recorded biomass in the tth year of the record. If we define $Y_t^* = \log Y_t$ and $x_t = t$, this can be written as

$$Y_t^* = \beta_0 + \beta_1 x_t + \varepsilon_t, \tag{3.4}$$

which has exactly the same form as (3.3). Therefore, to fit this model all that is required is to use as data the logarithms, rather than the actual values, of the biomass series.

Table 3.2 shows the R output, slightly edited, from such a fit. The coefficient estimates are $\hat{\beta}_0 = 6.0583$ and $\hat{\beta}_1 = -0.0370$ to four decimal places; the associated standard errors are estimated as 0.1464 and 0.0065 respectively. The error variance, σ^2, is estimated as $0.4423^2 = 0.1957$. Notice that $\hat{\beta}_1$ has a much smaller standard error than $\hat{\beta}_0$: the slope of the trend line is estimated much more precisely than the intercept. We return to this point later.

Table 3.2   Edited R output for model (3.4) fitted to haddock biomass data.

```
Call:
lm(formula = logBiomass ~ Time, data = haddock.data)

Coefficients:
             Estimate Std. Error t value Pr(>|t|)
(Intercept)  6.058348   0.146390  41.385  < 2e-16
Time        -0.036992   0.006543  -5.653 2.02e-06

Residual standard error: 0.4423 on 36 degrees of freedom
Multiple R-Squared: 0.4703,   Adjusted R-squared: 0.4556
F-statistic: 31.96 on 1 and 36 DF,   p-value: 2.021e-06
```

The conventional t test of the hypothesis $H_0 : \beta_1 = 0$ is carried out by computing the ratio of the estimate $\hat{\beta}_1$ to its estimated standard error and comparing to a t distribution with the appropriate degrees of freedom. In Table 3.2, the value of the t statistic is -5.653, with an associated p-value of 2.02×10^{-6}. If the underlying assumptions are satisfied, this leads to an overwhelming rejection of the null hypothesis.

As discussed in Section 2.4, some care is required when interpreting this test result: strictly speaking, the only thing we have learned here is that the data are not a white noise sequence. However, the caveats that were discussed previously can now be framed rather more precisely, since in the current context they all correspond to failures of modelling assumptions. If these assumptions are satisfied (we will shortly explore methods for checking them) so that the data really are generated according to a process of the form (3.3), then we are justified in claiming the existence of a linear trend.

The output in Table 3.2 also contains the multiple and adjusted R^2 statistics, both of which measure the proportion of variation in the data $\{Y_t\}$ that is accounted for by the trend. The adjusted version is always the smaller of the two, and is arguably preferable since it takes into account the number of parameters in the model (Draper and Smith, 1998, Section 5.2). In general, however, it is wise to avoid overreliance on R^2 as a measure of model adequacy: in most applications, it is more important to know about the uncertainty associated with the use of the model, and this is measured much more directly using the residual variance or standard deviation. To illustrate the use of this in the present example, denote by $\hat{Y}_t = \hat{\beta}_0 + \hat{\beta}_1 t$ the *fitted value* at time t: then the residual standard deviation of 0.4423 suggests that under the assumption of normality, there is roughly a 95 % chance that the actual value Y_t will fall within the interval $\hat{Y}_t \pm 2 \times 0.4423 = \hat{Y} \pm 0.8846$. This indicates the uncertainty to be expected if the model is used for prediction, although in this particular case it is difficult to interpret directly since the model is on a logarithmic scale. In such situations, interpretation can be simplified by converting back to the original scale of measurement: here, the fitted value $\hat{Y}_t$ maps back to $\exp[\hat{Y}_t]$ and the range of uncertainty to

$$\left(\exp\left[\hat{Y}_t - 0.8846\right], \exp\left[\hat{Y}_t + 0.8846\right]\right) = \left(0.413\exp\left[\hat{Y}_t\right], 2.422\exp\left[\hat{Y}_t\right]\right).$$

If the model assumptions are valid, the model therefore enables us to predict total biomass in any year to within a factor of about two and a half. This might not seem particularly impressive; it is, however, more informative than knowing merely that the model explains 46 % of the variation in the data.

The final line of Table 3.2 summarises an F test for determining whether the model has significant explanatory power. Formally this tests the null hypothesis that, with the exception of β_0, the underlying values of all the regression coefficients are zero. Since this particular model contains only one regression term, in this case the test is identical to the earlier t test on $\hat{\beta}_1$ – in fact, the F statistic is just the square of the t statistic for $\hat{\beta}_1$ (Davison, 2003, Section 8.5) and yields the same p-value. This is easily verified from the output given. The F statistic is also closely related to the multiple R^2 (Davidson and Mackinnon, 2004, Section 4.4). Specifically, with a single regression term we have $F = (T - 2)R^2/(1 - R^2)$, which, again, is easily verified from the output ($T - 2$ is the residual degrees of freedom, given in the output as 36). For readers who do not share our reservations about the use of R^2, the F test can be regarded as testing whether the observed R^2 differs significantly from zero. ∎

3.1.1   Checking the model assumptions

As indicated already, it is necessary to check the assumptions of any model before interpreting the results. For many statistical models, checks are usually based on *residuals* that represent some measure of discrepancy between data and model. When model (3.3) is fitted to yield parameter estimates $\hat{\beta}_0$ and $\hat{\beta}_1$, the residual at time t is defined as

$$e_t = Y_t - \hat{\beta}_0 - \hat{\beta}_1 x_t. \tag{3.5}$$

If the estimates $\hat{\beta}_0$ and $\hat{\beta}_1$ were equal to the 'true' parameters β_0 and β_1, the residual e_t would be the same as ε_t in (3.3). In general, of course, this is not the case, and the properties of the residuals differ slightly from those of the 'underlying' errors $\{\varepsilon_t\}$. Nonetheless, it seems reasonable to expect the residuals to behave more or less like the errors if the model is adequate. Later, we will investigate this in more detail; for the moment, however, to fix ideas we proceed informally.

As noted above, least squares fitting is appropriate when the errors form a white noise sequence. In this case the residuals should appear to be uncorrelated, with zero mean and constant variance. A variety of graphical techniques, similar to those used for exploratory analysis in Chapter 2, can be used to check these assumptions. The following diagnostics can be useful:

- A time series plot of the residuals. If the model is adequate, this should show a scatter about zero with no obvious structure. Systematic patterns in such a plot may suggest that the underlying trend is not linear. If there are many data points on the plot, it can be helpful to add a smooth curve to aid interpretation – any of the techniques described in Chapters 2 or 4 could be used for this purpose.

- A time series plot of the squared residuals $\{e_t^2\}$ or of their absolute values $\{|e_t|\}$. Under the assumption of constant variance or *homoscedasticity*, if T is large enough the expected value of e_t^2 is approximately the same for all t; therefore, the structure in such a plot suggests a violation of the assumption. Note, however, that squaring things accentuates their differences: therefore, a few of the squared residuals are often much larger than all the others. Since these tend to dominate the visual impression when plotted, it is usually preferable to work with the absolute values instead, although we will see below that this does not entirely eliminate the problem. Again, it can be useful to add a smooth curve to aid interpretation: for this purpose, the use of a robust smoother such as lowess (see Chapter 4) helps to protect against artefacts caused by the large values.

- A correlogram of the residuals. If the model assumptions are adequate, this should be more or less consistent with the correlogram of a white noise sequence.

Strictly speaking, because the residuals are not equal to the underlying errors, they have slightly different properties and the usual calculations of confidence bands around a non-parametric smooth, or 95 % intervals for a correlogram, are incorrect. We will see below that the effect of this can be nonnegligible; however, for all but the smallest sample sizes such calculations can be used as an informal aid to interpretation. If required, a formal test for residual autocorrelation can be based on the *Durbin–Watson statistic*:

$$\text{DW} = \frac{\sum_{t=2}^{T} (e_t - e_{t-1})^2}{\sum_{t=1}^{T} e_t^2}, \tag{3.6}$$

providing the observations are regularly spaced. Under the null hypothesis of uncorrelated errors, DW is expected to take values around 2, although its precise distribution depends on the number of regression coefficients estimated. Positive autocorrelation tends to produce smaller values, because successive residuals tend to be more similar to each other than in the uncorrelated case. The Durbin–Watson test has conventionally been implemented via rather cumbersome tables of critical values (Greene, 2003, Section 12.7). However, the exact distribution under the null hypothesis is that of a linear combination of chi-squared distributions, for which accurate approximations are available in modern statistical software. In R, for example, the test can be carried out using the dwtest command in the lmtest package or using the durbin.watson command in the car package.

The diagnostics outlined above are designed to check the three fundamental assumptions of the linear trend model fitted by least squares. In addition to these, since the usual tests for the significance of regression coefficients assume either that the errors are normally distributed or that the sample size is 'large enough', it is useful to assess the normality of the residuals. The easiest way to do this is using a normal quantile–quantile plot (Fox, 2002, Section 1.2; Faraway, 2005, Section 4.1) in which the ordered residuals are plotted against their expected values under the normality assumption. If this assumption holds, the points on such a plot should fall roughly on a straight line.

Example 3.2
Figure 3.1 shows diagnostic plots for the model fitted previously to the haddock biomass data. The residual time series plot shows no obvious trends, although there is a suggestion

of a cyclical pattern with alternating peaks and troughs roughly every 15 years. This pattern is reflected in the plot of absolute values of the residuals, which possibly show a decreasing trend in addition. If this is correct, it would suggest that the variance is decreasing through time. However, on closer inspection the visual impression seems due largely to the configuration of the few large values on the plot (if the first observation was removed, the impression would not be as strong); perhaps, therefore, it should not cause undue concern at this stage.

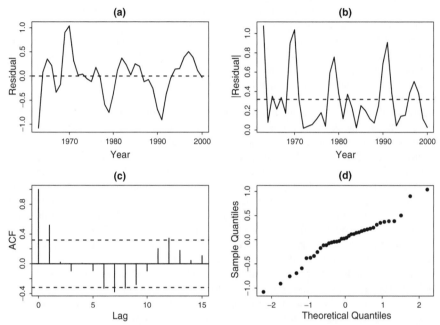

Figure 3.1   Residual plots for model (3.4) fitted to haddock biomass data: (a) residuals $\{e_t\}$, (b) absolute values $\{|e_t|\}$, (c) correlogram, (d) normal quantile–quantile plot. The dashed line in panel (a) is at zero; that in panel (b) is at the mean of the absolute values.

The residual ACF does, however, reveal a problem: there is strong evidence that the residuals are correlated. The oscillating pattern is consistent with the cycle suggested previously, but now the peak at lag 12 allows the cycle length, if genuine, to be pinned down more accurately to around 12 years. The reason for the caveat here is that it is possible for stationary processes (see the end of Section 2.1.3) to exhibit apparent cyclical behaviour; this is discussed further in Chapter 5. It is hardly necessary in this case to carry out a formal test for the significance of the residual autocorrelation structure; for illustrative purposes, however, we have computed the Durbin–Watson statistic (3.6). The value of the test statistic is 0.793, which is significantly less than the value of 2 expected under the null hypothesis of zero autocorrelation. The `dwtest` command in R gives the p-value for this test as 5.90×10^{-16}.

The final plot in Figure 3.1 is a normal quantile–quantile plot. It is probably premature to devote too much attention to this at present, due to the problems revealed by the ACF. Nonetheless, most of the residuals cluster more or less about a straight line, which suggests that the normality assumption may not be too unreasonable.   ∎

Having discussed how to check the assumptions underlying the linear trend model, it is natural to ask: What are the consequences if they are not satisfied? Failure of the first assumption – that the errors have constant zero mean – means that any trend in the data is not represented adequately by a linear function of time. Given the objectives of a trend analysis, the consequence of this is clear: a better representation of the trend is needed! It may appear that this is not so important if the aim of the analysis is purely descriptive, since in this case the slope of a linear trend merely provides a convenient index of overall change. However, if the underlying trend is not linear then successive observations will tend to fall on the same side of the fitted line. The effects of this are essentially the same as those that arise when the errors are autocorrelated.

In the presence of autocorrelation, least squares usually delivers reasonable parameter estimates. Indeed, for some models (including the linear trend considered here, as well as more general polynomial trends), least squares estimators remain optimal for a wide range of autocorrelation structures, at least in large samples (Priestley, 1981, Section 7.7). In general, however, least squares estimates of parameters in regression models are less precise than if the autocorrelation is correctly accounted for (Greene, 2003, Chapter 10). More seriously, the estimated standard errors of the parameters are incorrect, so that the usual hypothesis tests and confidence intervals are invalid. We will see in Section 3.3 that positive autocorrelation usually leads to underestimation of standard errors. This leads, in turn, to inflation of t statistics for testing hypotheses such as $H_0 : \beta_1 = 0$ (and hence to inflation of the corresponding F statistics for testing the explanatory power of a model). As a result, such hypotheses tend to be rejected too often when they are true. Another way to view this is that the null hypothesis must be interpreted within the context of the model structure: under model (3.3), the data must be white noise if $\beta_1 = 0$. In the presence of autocorrelation, by definition the data are *not* white noise and it is quite correct to reject the null hypothesis (see also the discussion in Section 2.4).

As far as standard errors and hypothesis tests are concerned, the effects of nonconstant variance (or *heteroscedasticity*) are similar to those of autocorrelation but may be less serious so long as the variance is not itself correlated with x_t (Davidson and Mackinnon, 2004, Section 5.5; Greene, 2003, Section 11.2). Heteroscedasticity is more problematic, however, if the fitted model is to be used for prediction. As in the haddock biomass example previously, an approximate 95 % prediction interval for an observation at time t can be computed, under the assumptions of normality and constant variance, as

$$\hat{Y}_t \pm 2\hat{\sigma}, \qquad (3.7)$$

where $\hat{\sigma}$ is the residual standard deviation. We will develop a more accurate version of this interval later; for the moment, however, note that if heteroscedasticity is present, (3.7) should read

$$\hat{Y}_t \pm 2\hat{\sigma}_t,$$

where now $\hat{\sigma}_t$ estimates the standard deviation at time t. If the heteroscedasticity is substantial, this interval could differ markedly from (3.7). If decisions are to be made on the basis of such predictions, it can be important to give realistic assessments of prediction uncertainty; the use of (3.7) in the presence of substantial heteroscedasticity could therefore have serious implications.

The relevance of normality is also mainly confined to prediction. As noted previously, least squares estimation is valid whether or not the errors are normally distributed, and test procedures and confidence intervals are not particularly sensitive to moderate departures from normality providing the sample size is large enough. However, expression (3.7) is completely unjustified unless the errors are normal.

This discussion of assumptions, and the consequences of their failure, has been necessarily brief. Greene (2003) and Davidson and Mackinnon (2004) give more details. We will shortly consider the options available when the assumptions break down; for the moment, however, we continue to study the basic linear trend model (3.3).

3.1.2   Choosing the time index

So far, we have assumed that the time index t starts at 1. In this case, β_1 can be interpreted as the average rate of change per unit time over the period of interest and β_0 as the 'underlying' level of the series at time $t = 0$, i.e. immediately before the start of observation. However, this choice of time origin is essentially arbitrary: in the haddock stocks example, we could equally have chosen to regress on the year instead of the observation number, in which case the time index would have run from 1963 to 2000 instead of from 1 to 38. It should be clear that such a relabelling does not affect the estimated slope $\hat{\beta}_1$ or its standard error; nor does it affect the fitted values $\{\hat{Y}_t\}$ or the residuals. However, the intercept $\hat{\beta}_0$ will change to an estimate of the underlying level at the new time origin. This can be useful when an aim of the analysis is to produce a prediction for a single specified time point (and when it is sensible to use the linear trend model for such a prediction): if this point is defined as the time origin then the predicted level, along with its standard error, can be read directly from the output corresponding to the intercept.

Example 3.3
Suppose that one of the objectives of the haddock stocks analysis was to predict biomass in the year 2005, assuming that the previous trend continued up to this point. This could be achieved by changing the definition of x_t in (3.4) to

$$x_t = \text{Year} - 2005.$$

When log biomass is regressed on this new time index, the estimated coefficients (to three significant figures) are $\hat{\beta}_0 = 4.468$ and $\hat{\beta}_1 = -0.0370$. The slope estimate is identical to that in Table 3.2, but the intercept has changed: at the new time origin (i.e. in 2005), the expected log biomass is 4.468 according to this model. The estimated standard error of $\hat{\beta}_0$ is 0.170, which would give some idea of prediction uncertainty if the model assumptions were satisfied. Although we have seen that this is not the case here, the example nonetheless illustrates that a judicious choice of origin can be helpful.   ■

We will see later that tricks such as this are not really necessary for prediction and extrapolation. However, in general it can save some effort if a model can be written in such a way that its parameters correspond directly to quantities of interest. In this context, another useful time transformation in the linear trend model is given by the centred index

$$x_t = t - \bar{t}, \tag{3.8}$$

where $\bar{t}$ is the sample mean of the original time indices. For regularly spaced time series, $\bar{t} = (T + 1)/2$. As before, the slope estimate $\hat{\beta}_1$ is unaffected by this transformation: the estimated intercept, however, becomes the sample mean of the Y values. This is shown in the next section.

3.1.3   An overview of least squares

So far in this chapter, we have avoided technicalities. Later, however, it will be helpful to have some insight into the underlying theory. Here we summarise a few key ideas.

In model (3.3), the least squares estimators are defined so as to minimise the quantity

$$S(\beta_0, \beta_1) = \sum_{t=1}^{T} (Y_t - \beta_0 - \beta_1 x_t)^2$$

with respect to β_0 and β_1. This is done by differentiating with respect to β_0 and β_1 and equating the derivatives to zero, to yield the *normal equations*:

$$\sum_{t=1}^{T} \left(Y_t - \hat{\beta}_0 - \hat{\beta}_1 x_t\right) = 0$$

and

$$\sum_{t=1}^{T} x_t \left(Y_t - \hat{\beta}_0 - \hat{\beta}_1 x_t\right) = 0. \tag{3.9}$$

A comparison with (3.5) shows that the first of these equations can be written in terms of the residuals as $\sum_{t=1}^{T} e_t = 0$: this shows that the residuals from a least squares fit of (3.3) will always sum to zero.

In most elementary treatments, the normal equations are solved to yield explicit formulae for the estimators. For example, the first equation can be rearranged to yield

$$\hat{\beta}_0 = \overline{Y} - \hat{\beta}_1 \overline{x},$$

where $\overline{Y}$ and $\overline{x}$ are the sample means of the Y and x values respectively; this can then be substituted into the second equation to find $\hat{\beta}_1$. If $\overline{x} = 0$, we have $\hat{\beta}_0 = \overline{Y}$: this is why, when the observations are regressed against the centred time index defined in (3.8), the estimated intercept is just the sample mean of the observations.

At this point, to lay the foundations for subsequent material we will use matrices to develop the theory further. Readers needing to refresh their knowledge of matrices and their properties may care to consult one of the references given after Section 1.6.

The key to the matrix approach is to notice that (3.3) actually defines T separate equations, one for each observation. We can stack these equations one on top of the other and write them as

$$\begin{pmatrix} Y_1 \\ \vdots \\ Y_T \end{pmatrix} = \begin{pmatrix} 1 & x_1 \\ \vdots & \vdots \\ 1 & x_T \end{pmatrix} \begin{pmatrix} \beta_0 \\ \beta_1 \end{pmatrix} + \begin{pmatrix} \varepsilon_1 \\ \vdots \\ \varepsilon_T \end{pmatrix}$$

or, more compactly, as

$$\mathbf{Y} = X\boldsymbol{\beta} + \boldsymbol{\varepsilon}, \tag{3.10}$$

where $\mathbf{Y}$ and $\boldsymbol{\varepsilon}$ are $T \times 1$ vectors, $\boldsymbol{\beta}$ is a 2×1 vector and X is a $T \times 2$ matrix called the *design matrix* of the regression. It will be convenient to write $\mathbf{x}_t$ for row t of this matrix:

$$\mathbf{x}_t = \begin{pmatrix} 1 & x_t \end{pmatrix}. \tag{3.11}$$

The fundamental assumption of the linear trend model is that the expected value of the error vector $\boldsymbol{\varepsilon}$ is $\boldsymbol{0}$. Let $\boldsymbol{\Sigma}$ denote the covariance matrix of this error vector. Then, from (3.10), we can write down the mean vector ($\boldsymbol{\mu}$, say) and covariance matrix of the observation vector $\mathbf{Y}$: denoting the covariance matrix by $\mathrm{Var}(\mathbf{Y})$, we have

$$\boldsymbol{\mu} = X\boldsymbol{\beta} \qquad \text{and} \qquad \mathrm{Var}\,(\mathbf{Y}) = \boldsymbol{\Sigma}. \tag{3.12}$$

Note that if the errors form a white noise sequence, with variance σ^2, say, then we can write $\boldsymbol{\Sigma} = \sigma^2 I_T$, where I_T is the $T \times T$ *identity matrix* with ones on the diagonal and zeros everywhere else.

We now return to the normal equations (3.9), which can be rearranged as

$$\sum_{t=1}^{T} \left(\hat{\beta}_0 + \hat{\beta}_1 x_t \right) = \sum_{t=1}^{T} Y_t$$

and

$$\sum_{t=1}^{T} x_t \left(\hat{\beta}_0 + \hat{\beta}_1 x_t \right) = \sum_{t=1}^{T} x_t Y_t. \tag{3.13}$$

It is tedious but straightforward to verify, using the rules of matrix multiplication (e.g. Poole, 2006, Section 3.1 or Healy, 2002, Section 1.4), that these equations can be rewritten as

$$X'X\hat{\boldsymbol{\beta}} = X'\mathbf{Y}. \tag{3.14}$$

This provides an explicit expression for the least squares estimators:

$$\hat{\boldsymbol{\beta}} = \left(X'X \right)^{-1} X'\mathbf{Y} = A\mathbf{Y}, \text{ say.} \tag{3.15}$$

The properties of the estimators now follow immediately from the following standard result (see, for example, Flury, 1997, Section 2.11 or Krzanowski, 1988, Section 7.1):

Result
Suppose $\mathbf{Y}$ is a $T \times 1$ vector of random variables, with mean vector $\boldsymbol{\mu}$ and covariance matrix $\boldsymbol{\Sigma}$. Let M be a $p \times T$ matrix, so that $\mathbf{Z} = M\mathbf{Y}$ is a $p \times 1$ vector of random variables. Then the expectation and covariance matrix of $\mathbf{Z}$ are $M\boldsymbol{\mu}$ and $M\boldsymbol{\Sigma}M'$ respectively. ∎

Applying this result to (3.15) and noting that the transpose of A is $A' = X(X'X)^{-1}$, the mean and covariance matrix of $\hat{\boldsymbol{\beta}}$ are

$$\mathrm{E}\left(\hat{\boldsymbol{\beta}} \right) = A\boldsymbol{\mu} = \left(X'X \right)^{-1} X'X\boldsymbol{\beta} = \boldsymbol{\beta} \tag{3.16}$$

and

$$\mathrm{Var}\left(\hat{\boldsymbol{\beta}} \right) = A\boldsymbol{\Sigma}A' = \left(X'X \right)^{-1} X'\boldsymbol{\Sigma}X \left(X'X \right)^{-1}. \tag{3.17}$$

Equation (3.16) shows that the least squares estimators are unbiased. In the white noise case when $\boldsymbol{\Sigma} = \sigma^2 I_T$, $X'\boldsymbol{\Sigma}X$ reduces to $\sigma^2 X'X$ and (3.17) becomes

$$\mathrm{Var}\left(\hat{\boldsymbol{\beta}} \right) = \sigma^2 \left(X'X \right)^{-1}. \tag{3.18}$$

The diagonal elements of this matrix give the variances of the individual components of $\hat{\boldsymbol{\beta}}$; taking square roots yields the corresponding standard errors.

Example 3.4

Consider the linear trend model (3.3) with $x_t = t - \bar{t}$ as in (3.8). If the observations are regularly spaced so that $\bar{t} = (T + 1)/2$, some elementary algebra shows that

$$X'X = T \begin{pmatrix} 1 & 0 \\ 0 & \left(T^2 - 1\right)/12 \end{pmatrix}.$$

The inverse of this matrix is

$$\left(X'X\right)^{-1} = \frac{1}{T} \begin{pmatrix} 1 & 0 \\ 0 & 12/\left(T^2 - 1\right) \end{pmatrix}.$$

Therefore, in the white noise case the variances of $\hat{\beta}_0$ and $\hat{\beta}_1$ are

$$\frac{\sigma^2}{T} \qquad \text{and} \qquad \frac{12\sigma^2}{T\left(T^2 - 1\right)}$$

respectively, and the correlation between them is zero. This lack of correlation is a result of the matrix $X'X$ being diagonal, and is convenient for interpretation since it enables us unambiguously to separate out the estimation of the overall mean level β_0 from the trend β_1. If the parameter estimates were correlated, this would not be possible. We revisit this issue in Section 3.2.

Notice also that for all but the smallest values of T, $12/(T^2 - 1)$ is much less than 1 so that β_1 is estimated much more precisely than β_0. This can be traced back to the large bottom right-hand element of $X'X$, reflecting the wide separation of the observations along the time axis. It seems intuitively reasonable that in these circumstances the slope can be estimated more precisely than the overall mean level (imagine plotting the true trend line by hand on graph paper, by marking two points and connecting them with a ruler – if the points are widely separated, any small errors will have a negligible effect on the slope of the plotted line). Moreover, as discussed earlier, the estimate $\hat{\beta}_1$ and its standard error are unaffected by shifting the labels on the time axis: this explains why, in Table 3.2, the standard error of $\hat{\beta}_1$ was so much smaller than that of $\hat{\beta}_0$. ■

We conclude this section by examining the properties of the residuals $\{e_t\}$. Notice first that, given the least squares estimates (3.15), the fitted values can be assembled into a vector

$$\hat{\mathbf{Y}} = X\hat{\boldsymbol{\beta}} = X\left(X'X\right)^{-1}X'\mathbf{Y} = H\mathbf{Y}, \text{ say.} \qquad (3.19)$$

The symmetric $T \times T$ matrix $H = X(X'X)^{-1}X'$ is usually called the *hat matrix*, because its effect is to put a 'hat' on the vector $\mathbf{Y}$. It is also sometimes called the *influence matrix*, for reasons that will become clear in Section 3.1.5. Perhaps unexpectedly, H is unchanged when multiplied by itself: $HH = H$. This is easily verified. Such a matrix is called *idempotent*.

Having defined the vector of fitted values, we can define the vector of residuals in a similar way:

$$\mathbf{e} = \mathbf{Y} - \hat{\mathbf{Y}} = \mathbf{Y} - H\mathbf{Y} = \left(I_T - H\right)\mathbf{Y} \qquad (3.20)$$

where, as before, I_T denotes the $T \times T$ identity matrix. One application of this is to derive an estimator of the error variance σ^2. It is natural to base such an estimator on the *residual sum of squares*: RSS $= \sum_{t=1}^{T} e_t^2$, which can be written in matrix notation as $\mathbf{e'e}$. The representation (3.20) can be used to show (Greene, 2003, Section 4.6; Davidson and Mackinnon, 2004, Section 3.6) that the expected value of $\mathbf{e'e}$ is $(T - k)\sigma^2$, where k is (usually) the number of regression parameters in the model, i.e. the length of the vector $\boldsymbol{\beta}$. The usual variance estimator

$$\hat{\sigma}^2 = \frac{1}{T - k} \sum_{t=1}^{T} e_t^2 \tag{3.21}$$

is therefore unbiased for σ^2.

Representation (3.20) also allows us to write down the mean vector and covariance matrix of the residuals using the Result given earlier in this section:

$$\mathrm{E}\,(\mathbf{e}) = (\mathbf{I}_T - \mathbf{H})\,\boldsymbol{\mu} = \left[\mathbf{I}_T - \mathbf{X}\left(\mathbf{X'X}\right)^{-1}\mathbf{X'}\right]\mathbf{X}\boldsymbol{\beta} = \mathbf{0} \tag{3.22}$$

and

$$\mathrm{Var}\,(\mathbf{e}) = (\mathbf{I}_T - \mathbf{H})\,\boldsymbol{\Sigma}\,(\mathbf{I}_T - \mathbf{H})'. \tag{3.23}$$

We are now in a position to examine in detail the behaviour of the residuals from the linear trend model in the white noise case. Here, (3.23) reduces to

$$\mathrm{Var}\,(\mathbf{e}) = \sigma^2\,(\mathbf{I}_T - \mathbf{H})\,(\mathbf{I}_T - \mathbf{H})' = \sigma^2\left(\mathbf{I}_T - \mathbf{H} - \mathbf{H'} + \mathbf{HH'}\right) = \sigma^2\,(\mathbf{I}_T - \mathbf{H})\,, \tag{3.24}$$

the last step following because $\mathbf{H}$ is symmetric (i.e. $\mathbf{H'} = \mathbf{H}$) and idempotent. Recall now that the residuals are the same regardless of the choice of time origin. For convenience, therefore, we will take the time origin to be $\bar{t}$. In this case, the matrix $\mathbf{X'X}$ was derived in Example 3.4. From this, it can be shown that the (t, s)th element of the hat matrix is

$$h_{ts} = \frac{1}{T}\left[1 + \frac{12\left(t - \bar{t}\right)\left(s - \bar{t}\right)}{T^2 - 1}\right].$$

From (3.24), the variance of e_t is simply $\sigma^2\,(1 - h_{tt})$ and the covariance between e_t and e_s is $-\sigma^2 h_{ts}$ for $t \neq s$. We therefore have

$$\mathrm{Var}\,(e_t) = \sigma^2\left\{1 - \frac{1}{T}\left[1 + \frac{12\left(t - \bar{t}\right)^2}{T^2 - 1}\right]\right\} \tag{3.25}$$

and

$$\mathrm{Cov}\,(e_t, e_s) = -\frac{\sigma^2}{T}\left[1 + \frac{12\left(t - \bar{t}\right)\left(s - \bar{t}\right)}{T^2 - 1}\right] \qquad (s \neq t). \tag{3.26}$$

Notice from (3.25) that the variances of the residuals are not constant. In fact, they range from a minimum value of

$$\sigma^2\left[1 - \frac{4T - 2}{T(T + 1)}\right]$$

at $t = 1$ and $t = T$ to a maximum of $\sigma^2(1 - T^{-1})$ at $t = \bar{t}$. If T is large enough, the second term in both expressions is effectively negligible so that the variances can be regarded as constant for all practical purposes – for example, when $T = 100$ the range is from $0.96\sigma^2$ to $0.99\sigma^2$. For small T, however, the differences are larger. One way to deal with this when checking models is to work with *standardised residuals*, defined as

$$r_t = \frac{e_t}{\hat{\sigma}\sqrt{1 - h_{tt}}}. \tag{3.27}$$

In the white noise case, the standardised residuals all have the same variance, which is approximately 1. For the haddock biomass example, diagnostic plots of the standardised residuals (not shown here) look almost identical to those shown later in Figure 3.4. This indicates that for informal model checking purposes, standardisation is not really necessary for this particular example.

Given that the residuals always sum to zero, it is perhaps to be expected that they are negatively correlated, as indicated by the sign of the covariance at (3.26). The covariances range from $-\sigma^2/T$, when either $s = \bar{t}$ or $t = \bar{t}$, to

$$-\sigma^2 \left[\frac{4T - 8}{T(T + 1)} \right]$$

for the first and last pair of time points. The corresponding correlations are smallest for pairs of residuals with low covariance and high variance. These pairs occur close to $\bar{t}$, where the variances are roughly $\sigma^2(1 - T^{-1})$ and the covariance is $-\sigma^2/T$ so that the correlation is approximately $-1/(T - 1)$. At the other extreme, for pairs at the ends of the series with low variance and high covariance, a similar argument shows that the correlation is approximately $-4/(T - 1)$. Therefore, for the haddock biomass example, even if the error process was white noise some of the inter-residual correlations could be as large as $-4/37 = -0.11$. Unfortunately this correlation cannot be eliminated via any simple device such as the use of standardised residuals.

The conclusion from this exercise is that for small sample sizes, the properties of the residuals may differ noticeably from those of the underlying errors. Care should therefore be taken with the interpretation of residual plots: in particular, devices such as the addition of the usual 95 % limits to a residual correlogram should be treated as at best informal aids to interpretation. This is directly analagous to the problems noted in Chapter 2, where the process of detrending via filtering was seen to change the properties of a series. When using parametric models, however, many of the problems recede given enough data.

3.1.4   Extrapolation

In Section 1.2, we saw that one of the reasons for carrying out a trend analysis may be to extrapolate past patterns into the future. Such extrapolation is particularly simple if the trend can be represented plausibly over the period of interest using a parametric model. For example, if a 50-year sequence of observations exhibits a clear linear trend then, in the absence of information to the contrary, it may be reasonable to suppose that this will continue at least in the short term. Indefinite extrapolation would, of course, be unwise: trends are likely to change over time, reflecting changes in the environment controlling the process of interest, as well as the fact that the unlimited increases or decreases suggested by a linear trend are often not physically realisable. Methods for allowing trends to change over time are discussed in Chapter 5. In the meantime, however, we will proceed on the assumption that a linear trend model has been fitted to data $y_1, \ldots, y_T$ and is judged to be a suitable basis for extrapolation up to time $T + \ell$.

The actual value at time $t + \ell$ will be the realised value of a random variable $Y_{t+\ell}$. Under model (3.3), we have $Y_{t+\ell} = \beta_0 + \beta_1 x_{t+\ell} + \varepsilon_{t+\ell}$, where $x_{t+\ell} = t + \ell$. It is natural to use the estimates of β_0 and β_1 for extrapolation: $\hat{Y}_{t+\ell} = \hat{\beta}_0 + \hat{\beta}_1 x_{t+\ell}$, say. Of course, the actual value of $Y_{t+\ell}$ will not be equal to $\hat{Y}_{t+\ell}$, even if it is generated according to model (3.3). This is firstly because the value of $\varepsilon_{t+\ell}$ cannot be predicted in advance and secondly because the extrapolation uses the parameter estimates $\hat{\beta}_0$ and $\hat{\beta}_1$ rather than the true underlying values. It is therefore useful to be able to say something about the likely magnitude of any error associated with the prediction.

The prediction error is $Y_{t+\ell} - \hat{Y}_{t+\ell} = (\beta_0 + \beta_1 x_{t+\ell} + \varepsilon_{t+\ell}) - (\hat{\beta}_0 + \hat{\beta}_1 x_{t+\ell})$. This can be rewritten as $(\beta_0 - \hat{\beta}_0) + (\beta_1 - \hat{\beta}_1)x_{t+\ell} + \varepsilon_{t+\ell}$ or, in matrix notation,

$$\begin{pmatrix} 1 & x_{t+\ell} \end{pmatrix} \begin{pmatrix} \beta_0 - \hat{\beta}_0 \\ \beta_1 - \hat{\beta}_1 \end{pmatrix} + \varepsilon_{t+\ell} = \mathbf{x}_{t+\ell} \left(\boldsymbol{\beta} - \hat{\boldsymbol{\beta}} \right) + \varepsilon_{t+\ell}.$$

From equations (3.16) and (3.18), we find that the vector $\boldsymbol{\beta} - \hat{\boldsymbol{\beta}}$ has mean $\mathbf{0}$ and covariance matrix $\sigma^2 (X'X)^{-1}$. Next, the Result in Section 3.1.3 tells us that the quantity $\mathbf{x}_{t+\ell} (\boldsymbol{\beta} - \hat{\boldsymbol{\beta}})$ has expectation zero and variance $\sigma^2 \mathbf{x}_{t+\ell} (X'X)^{-1} \mathbf{x}'_{t+\ell}$. Now, since $\varepsilon_{t+\ell}$ has expectation zero and variance σ^2, and is independent of $\hat{\boldsymbol{\beta}}$, we find that the expected prediction error is zero and that its variance is

$$\sigma^2 \left[1 + \mathbf{x}_{t+\ell} \left(X'X \right)^{-1} \mathbf{x}'_{t+\ell} \right]. \tag{3.28}$$

This expression is a sum of two terms, which may be thought of as the respective contributions from 'random' and 'estimation' error. The former is due to $\varepsilon_{t+\ell}$ and the latter to $\boldsymbol{\beta} - \hat{\boldsymbol{\beta}}$.

If the variance σ^2 is known and the $\{\varepsilon_t\}$ are normally distributed, (3.28) enables us to construct a *prediction interval* for $Y_{t+\ell}$. For example, a 95 % interval is given by

$$\hat{Y}_{t+\ell} \pm 1.96\sigma \sqrt{1 + \mathbf{x}_{t+\ell} \left(X'X \right)^{-1} \mathbf{x}'_{t+\ell}}.$$

As usual, in the more realistic situation when the variance is unknown and has been estimated, the multiplier 1.96 here should be replaced by the upper 2.5 % point of the appropriate t distribution.

If the model is correct and prediction intervals are calculated for several future time points, we expect 95 % of them to contain the actual values when these become available. The matrix based derivation above makes it particularly simple to calculate several intervals simultaneously: simply replace the vector $\mathbf{x}_{t+\ell}$ in (3.28) with a matrix, the rows of which correspond to the time points of interest. The result will be the covariance matrix of all the prediction errors; the diagonal elements of this matrix give the variances needed to form prediction intervals. This procedure is much simpler (operationally, at least) than the device suggested in Section 3.1.2, of choosing the time origin to coincide with a particular time point of interest. Note also that the emphasis here is on predicting future observations, whereas previously we considered only estimation of the underlying trend at the origin (i.e. the parameter β_0). Uncertainty in prediction is greater than that in estimation, because of the error terms $\{\varepsilon_t\}$. The variance of the estimation error can be obtained by dropping the first term in (3.28). The result can be used to calculate confidence intervals for the underlying trend at time points of interest.

Example 3.5

Consider using the linear trend model in Table 3.2 to predict log haddock biomass for each of the years 2001–2005. The `predict` function in R implements the theory outlined above. The results are summarised in Table 3.3. The predictions $\{\hat{Y}_{\text{Year}}\}$ decrease linearly as expected. The row entitled 'standard error of fit' gives the standard deviation corresponding to the estimation (rather than prediction) error, i.e. the square root of $\hat{\sigma}^2 \mathbf{x}_{t+\ell} \left(\mathbf{X}'\mathbf{X} \right)^{-1} \mathbf{x}'_{t+\ell}$. In general, for a regression model of the form (3.3), these standard deviations increase with the distance $|x_t - \bar{x}|$; here, therefore, the standard errors increase as the predictions move further into the future. Notice also that the standard error for the fit in 2005 is 0.170; this is the same as the value obtained in Example 4.3 by reparameterising the model.

Table 3.3 Predictions of North Sea haddock biomass for 2001–2005, according to the linear trend model summarised in Table 3.2.

Year	2001	2002	2003	2004	2005
Fitted trend, $\hat{Y}_{\text{Year}}$	4.616	4.579	4.542	4.505	4.468
Standard error of fit	0.146	0.152	0.158	0.164	0.170
95 % limits for trend line	(4.319, 4.913)	(4.270, 4.887)	(4.221, 4.862)	(4.173, 4.837)	(4.124, 4.812)
95 % limits for log biomass	(3.671, 5.561)	(3.630, 5.527)	(3.589, 5.494)	(3.548, 5.461)	(3.507, 5.429)
95 % limits for biomass	(39.3, 260.0)	(37.7, 251.5)	(36.2, 243.3)	(34.7, 235.4)	(33.3, 227.8)

Table 3.3 also shows the difference between confidence intervals for the trend line and prediction limits for log biomass; the former are much narrower, because they do not account for the error term in the model. Finally, notice that the prediction limits for log biomass can easily be converted back to limits for the actual biomass (in thousands of tonnes); these limits are shown in the last two rows of the table.   ■

Of course, any exercise in extrapolation is only warranted insofar as the underlying model holds. For the haddock stocks data we have already seen that this is not the case, whence the predictions in Table 3.3 should not be taken particularly seriously (although the qualitative features of the example are quite general). However, even if the fitted model appears to pass all conceivable diagnostic tests, any extrapolation into the future is ultimately predicated on the strong, but unavoidable, assumption that the model structure will continue to hold over the time period of interest. This assumption will rarely hold exactly, although it may be a fair approximation in the short term, particularly if the guidelines in Section 1.4 are followed during the model-building process. In general, a prudent interpretation of any extrapolation should recognise that uncertainty about the future is often greater than the prediction intervals from a single model would suggest, simply because such intervals take no account of the possibility that the model is wrong.

3.1.5   Influential observations

Consider the following hypothetical scenario: a three year research project aims to monitor radioactivity levels in marine sediments, offshore of a nuclear waste reprocessing plant. The plan is to make quarterly trips in a research vessel, to gather sediment samples for subsequent analysis. Unfortunately, at the end of the first year the research vessel is damaged and the repairs take 21 months, so that it is only possible to make one further trip before the end of the project. The measured caesium-137 (^{137}Cs) concentrations in the samples taken during quarters 1, 2, 3, 4 and 12 are 13, 12, 15, 13 and 18 Bq kg^{-1} respectively. If a linear trend is fitted to these data, the estimated slope is 0.49 Bq kg^{-1} per quarter and is significant at the 5 % level (the associated p-value is 0.04). The residuals show no obvious problems (which is hardly surprising given so few observations). However, a time series plot shows clearly that the appearance of a trend is almost entirely dominated by the final observation. If this observation is removed and the model is refitted, the trend estimate changes to 0.3 Bq kg^{-1} per quarter and is no longer significant (the p-value becomes 0.69).

Clearly, it is undesirable that the individual observations should have such a strong influence on the results of any analysis. In general, therefore, it is worth checking that such problems are not present. A widely used *influence measure* is derived by considering, for each observation in turn, what would be the effect of its deletion upon the fitted values for the remaining observations. Let $\hat{y}_s$ denote the fitted value at time s and let $\hat{y}_s^{(-t)}$ denote the fitted value obtained if the model is refitted after discarding the observation at time t (we use lower case letters here, to indicate that we are considering the particular values that are observed). Then the overall change in the fit when observation t is omitted can be summarised (Cook and Weisberg, 1999, Section 15.2) as

$$D_t = \frac{1}{k\hat{\sigma}^2} \sum_{s=1}^{T} \left(\hat{y}_s^{(-t)} - \hat{y}_s \right)^2, \tag{3.29}$$

where, as previously, k is the number of regression parameters in the model and $\hat{\sigma}^2$ is the estimated error variance defined at (3.21).

The quantity D_t is called *Cook's distance*. At first glance, it appears infeasible to calculate this routinely for every observation in a data set, because of the need to refit the model to obtain $\hat{y}_s^{(-t)}$. However, it can be shown that such refitting is not necessary: an alternative formula for D_s is

$$D_t = \frac{r_t^2}{k} \frac{h_{tt}}{1 - h_{tt}}, \tag{3.30}$$

where r_t is the standardised residual defined at (3.27) and h_{tt} is the tth diagonal element of the hat matrix as before. This shows that an observation may have a large influence either if it has a large residual (r_t) or if the quantity $h_{tt}/(1 - h_{tt})$ is large. From the definition of the hat matrix at (3.19), the latter quantity depends (in the context of the linear trend model) only on the observation times, and not on the observations themselves. The quantities $\{h_{tt}\}$ are called *leverages*, and can be regarded as measuring the effect of each observation on the estimated regression coefficients (as opposed to the Cook's distances, which measure the effect on the fitted values). It is necessary to examine both leverages and Cook's distances, along with residuals, to obtain a full understanding of a fitted model. For example, it is possible for an observation with a Cook's distance of zero to have a substantial effect on the estimated regression coefficients (Davison, 2003, Section 8.6).

Leverages and Cook's distances are usually interpreted informally. It is common, for example, to regard any leverage greater than $2k/T$ and any Cook's distance greater than $8/(T - 2k)$ as giving potential cause for concern (in the case of irregularly spaced data, T here denotes the total number of observations). In such cases, unless T is very large it may be worth refitting the model without the corresponding observations to see if there are any substantive changes in the conclusions. Of course, if T is large then it is almost inevitable that some observations will have leverages or Cook's distances exceeding $2k/T$ and $8/(T - 2k)$ respectively; however, in absolute terms the effect of any individual observation on the fit is usually negligible.

Example 3.6
Figure 3.2(a) shows Cook's distance for each observation in the hypothetical ^{137}Cs data set, along with the threshold $8/(T - 2k)$. The final observation has the largest Cook's distance, but according to this plot it does not have much overall influence on the fitted values. In contrast, Figure 3.2(b) shows that it does have a high leverage, which is why the regression coefficient changes so much when the observation is deleted. This illustrates the need to consider Cook's distances and leverages separately.

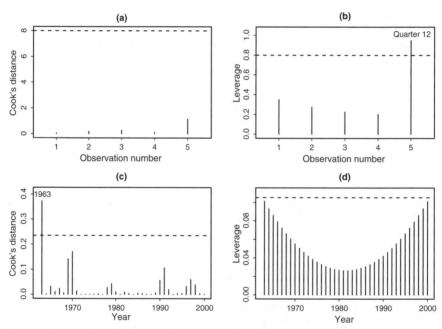

Figure 3.2   Cook's distances and leverages for linear trend models fitted to the hypothetical ^{137}Cs data (a, b) and to the log haddock biomass series (c,d).

Figures 3.2(c) and 3.2(d) show the same diagnostics for the haddock biomass model of Table 3.2. Here, the regular spacing of the observations ensures that there are no excessively high leverage points; however, the observation for 1963 seems to be influential according to the Cook's distance plot. This is because it has a large residual (as already noted when discussing Figure 3.1) coupled with a moderately high leverage. Since the series is relatively short, it may be worth investigating the effect of refitting the model without this first observation. If this is done, the slope of the trend changes from -0.037 to -0.042 and the residual variance decreases from 0.1957 to 0.1642. However, the residual plots for the refitted model look almost identical to those in Figure 3.1. Clearly, the overall conclusion of a decreasing linear trend with substantial residual autocorrelation is unaffected by removing the first observation, but the combination of a change in slope and a reduction in error variance could have a noticeable impact upon, for example, prediction intervals for future observations. In this particular example, at this stage the main problem with prediction intervals is that the assumption of uncorrelated errors is clearly inadequate, and therefore it is probably premature to worry about the effect of the 1963 observation; however, the exercise reinforces the comments made previously about the need for caution when making predictions.   ∎

As an alternative to removing the influential observations, we could consider the use of *robust fitting methods*, which aim to use all of the data but to limit the effect of individual observations on the results. In general, the price to pay for robustness is some loss in precision (and hence in the ability to detect weak trends when they exist) when the least squares assumptions are satisfied.

For parametric models, *least absolute deviations* (LAD) regression is a popular robust method: here, the regression coefficients are estimated by minimising the quantity

$$\sum_{t=1}^{T} |y_t - \beta_0 - \beta_1 x_t| \, .$$

By comparison with least squares, the use of absolute rather than squared differences reduces the weight attached to large deviations from the trend line. In fact, the underlying statistical model differs slightly from (3.2), since it may be shown that the LAD regression line is an estimate of trend in the median of the observations, rather than in the mean. This is discussed in more detail in Section 4.3.5.

Santer *et al*. (2000) carried out a comparison between least squares and LAD regressions for the linear trend model applied to atmospheric temperature series and found that the two methods sometimes yielded rather different results. In such situations, it is important to understand the implications of the different estimation methods: if the influential observations are considered to be an intrinsic part of the system under study, then one might reasonably consider the least squares estimates to be more relevant, and vice versa.

For further details of robust regression methods and their implementation in R, see Faraway (2005, Section 6.4).

3.1.6   Other methods of model fitting

The principle behind least squares estimation is intuitive and easily understood: to choose values of the model parameters that yield the best predictions (in some sense) of the available data. However, least squares is not the only method available for fitting statistical models. In this section we briefly introduce two alternative methods, both of which are widely applicable. These are the *maximum likelihood* and *Bayesian* methods respectively.

In maximum likelihood estimation, the parameter estimates are the values for which the probability (or, in many cases, the density) of the observations is maximised. The intuition is that these values are, in some sense, more consistent than any others with the observations. In many situations the method has desirable optimality properties, at least in large samples (Garthwaite, Jolliffe and Jones, 2002, Chapter 3; Azzalini, 1996, Section 3.3; Krzanowski, 1998, page 40): it is therefore the method of choice in much modern statistical work.

To illustrate the basic ideas, consider the linear trend model (3.3). Under this model, the observation y_t is considered to be drawn from a normal distribution with mean $\mu_t = \beta_0 + \beta_1 x_t$ (where $x_t = t$) and variance σ^2. Thus, the density of this observation is given by the probability density function of the corresponding normal distribution:

$$f_t\left(y_t; \boldsymbol{\beta}, \sigma^2\right) = \frac{1}{\sigma\sqrt{2\pi}} \exp\left[-\frac{(y_t - \mu_t)^2}{2\sigma^2}\right],$$

where $\boldsymbol{\beta} = (\beta_0 \ \beta_1)'$ is the vector of regression coefficients as before. Furthermore, since the observations are all considered to be independent, their joint density is the product of their individual marginal densities:

$$f\left(\mathbf{y}; \boldsymbol{\beta}, \sigma^2\right) = \prod_{t=1}^{T} f_t\left(y_t; \boldsymbol{\beta}, \sigma^2\right) = \left(2\pi\sigma^2\right)^{-T/2} \exp\left[-\frac{1}{2\sigma^2}\sum_{t=1}^{T}(y_t - \mu_t)^2\right]. \quad (3.31)$$

Treating this joint density as a function of $\boldsymbol{\beta}$ and σ^2 yields the *likelihood function* for the parameters. The *maximum likelihood estimates* (MLEs) are the parameter values that maximise this expression. Alternatively, because the values that maximise (3.31) also maximise its logarithm, we can work with the *log-likelihood function*

$$\ell\left(\boldsymbol{\beta}, \sigma^2; \mathbf{y}\right) = \log f\left(\mathbf{y}; \boldsymbol{\beta}, \sigma^2\right) = \sum_{t=1}^{T}\log f_t\left(y_t; \boldsymbol{\beta}, \sigma^2\right)$$

$$= -\frac{T}{2}\log 2\pi\sigma^2 - \frac{1}{2\sigma^2}\sum_{t=1}^{T}(y_t - \mu_t)^2, \quad (3.32)$$

which is easier to maximise than the likelihood function itself. Noting that the coefficient vector $\boldsymbol{\beta}$ enters only through the means $\{\mu_t\}$, maximisation with respect to $\boldsymbol{\beta}$ is equivalent to minimising $\sum_{t=1}^{T}(y_t - \mu_t)^2$: thus the MLEs of the regression coefficients are the same as the least squares estimates.

For the variance σ^2, we can maximise the log-likelihood by differentiating and equating to zero. We have

$$\frac{\partial\ell}{\partial\sigma^2} = -\frac{T}{2\sigma^2} + \frac{1}{2\sigma^4}\sum_{t=1}^{T}(y_t - \mu_t)^2,$$

which is equal to zero when $\sigma^2 = T^{-1}\sum_{t=1}^{T}(y_t - \mu_t)^2$. It can be verified that this does indeed correspond to a maximum of (3.32). Notice that this estimate of σ^2 is not the same as the usual unbiased least squares estimate (3.21), where the divisor $T - k$ was used in place of T ($k = 2$ being the number of elements of $\boldsymbol{\beta}$). Clearly, if T is very large then there will be very little difference between the two estimates, but in small samples the MLE of the error variance may be biased. Further discussion of this and related issues is given in Section 3.5.

Since the data $\mathbf{y}$ are considered as generated from a probability model for the associated random variables $\mathbf{Y}$, the log-likelihood function and associated quantities can themselves be regarded as the realised values of random variables: for example, (3.32) is the realised value of $-\left[T\log 2\pi\sigma^2 - \sigma^{-2}\sum_{t=1}^{T}(Y_t - \mu_t)^2\right]/2$. Thus it is meaningful to talk about the expected value of such quantities. A key role in the theory of maximum likelihood is played by the *expected information matrix*, defined as $J = -\Delta$, where Δ is the matrix of expected second derivatives of the log-likelihood with respect to the parameters. It can be shown (Davison, 2003, page 118) that, in large samples, for many models the distribution of maximum likelihood estimates is approximately multivariate normal with the covariance matrix equal to J^{-1}. This can be used, for example, to calculate approximate (and, in some cases, exact) standard errors and confidence intervals for individual parameters. For linear regression models, including the linear trend model, the likelihood based standard errors for the regression coefficients are the same as those obtained by least squares.

Maximum likelihood estimation aims to use the data to learn about the values of model parameters (β_0, β_1 and σ^2 in the case of the linear trend model) that are considered as fixed but unknown. Bayesian estimation takes a very different perspective: specifically, that the role of the data is to update the analyst's existing knowledge of the parameters. Before observing $\mathbf{y}$ the analyst may have some idea, based on past experience with similar data or from some understanding of the system being studied, of roughly what the parameter values should be. A Bayesian approach requires that this knowledge is represented via a joint probability distribution for the parameters, called the *prior distribution*: if the parameters are collected into a single vector $\boldsymbol{\theta}$, say, then the density of the prior distribution is usually written as $\pi(\boldsymbol{\theta})$. In this context, the 'classical' interpretation of probabilities (as long-run frequencies in a notional sequence of repetitions of some experiment) is unhelpful: rather, probability statements about $\boldsymbol{\theta}$ are seen as a convenient device for representing the analyst's uncertainty. Note also that the assignment of a probability distribution to the parameters represents a significant conceptual shift from the idea that they are fixed quantities.

In a Bayesian setting, knowledge of $\boldsymbol{\theta}$ after observing $\mathbf{y}$ is expressed by the conditional probability distribution of $\boldsymbol{\theta}$ given that $\mathbf{Y} = \mathbf{y}$. This conditional distribution is called the *posterior distribution* for $\boldsymbol{\theta}$, and is denoted by $\pi(\boldsymbol{\theta}|\mathbf{y})$. Bayes' theorem (for example Davison, 2003, Section 11.1) gives

$$\pi(\boldsymbol{\theta}|\mathbf{y}) = \frac{f(\mathbf{y}|\boldsymbol{\theta})\pi(\boldsymbol{\theta})}{f(\mathbf{y})}.$$

Here, $f(\mathbf{y}|\boldsymbol{\theta})$ is the density of $\mathbf{y}$ given $\boldsymbol{\theta}$. Regarded as a function of $\boldsymbol{\theta}$ rather than $\mathbf{y}$, this is just the likelihood for $\boldsymbol{\theta}$ as defined above: $f(\mathbf{y}|\boldsymbol{\theta}) = L(\boldsymbol{\theta}|\mathbf{y})$, say. Moreover, $f(\mathbf{y})$ is the unconditional density of $\mathbf{y}$, which does not vary with $\boldsymbol{\theta}$. We therefore have $\pi(\boldsymbol{\theta}|\mathbf{y}) \propto L(\boldsymbol{\theta}|\mathbf{y})\pi(\boldsymbol{\theta})$ or, in words,

$$\text{Posterior} \propto \text{Likelihood} \times \text{Prior.} \tag{3.33}$$

This relation is at the centre of all Bayesian inference. The constant of proportionality is defined by the requirement that $\int \pi(\boldsymbol{\theta}|\mathbf{y}) \, d\boldsymbol{\theta} = 1$, which must hold since $\pi(\boldsymbol{\theta}|\mathbf{y})$ is a probability distribution. In the Bayesian framework, the posterior distribution completely summarises the available information about $\boldsymbol{\theta}$. The *Bayes estimator* of $\boldsymbol{\theta}$ is defined as the mean of this posterior distribution: $E(\boldsymbol{\theta}|\mathbf{y}) = \int \boldsymbol{\theta}\pi(\boldsymbol{\theta}|\mathbf{y}) \, d\boldsymbol{\theta}$, and percentiles of the marginal posterior distributions for each component of $\boldsymbol{\theta}$ can be used to construct *Bayesian credible intervals*, which are the Bayesian analogue of confidence intervals.

For some simple models it is possible to calculate the posterior distribution explicitly for specific choices of prior; see, for example, Davison (2003, Section 11.1) or Gelman *et al.* (1995, Chapters 3 and 8). However, there is a limited range of models and prior distributions for which such calculations are possible. Thus, Bayesian estimation and inference is often carried out using *Markov chain Monte Carlo* (MCMC) computer algorithms that are designed to provide a simulated sample from the posterior distribution, from which any quantities of interest can be estimated (see Gelman *et al.*, 1995, Part III for an overview). MCMC methods are also extremely useful for the fitting of complex models in situations where direct calculation of the likelihood is infeasible. In R, libraries R2WinBUGS (Sturtz, Ligges and Gelman, 2005) and BRugs (Thomas *et al.*, 2006) provide interfaces to packages such as WinBUGS (Lunn *et al.*, 2000) and OpenBUGS (Thomas *et al.*, 2006), which can be used to carry out MCMC computations for a wide range of Bayesian analyses. BayesX (Belitz *et al.*, 2009) is another freely available package for Bayesian computation; this is used in Chapter 10 of the present volume.

In many environmental situations, a Bayesian approach provides an appealing means of incorporating genuine subject-matter knowledge into an analysis via the prior distribution. For example, Leith and Chandler (2010) carry out a Bayesian analysis of simulated climate data for the end of the twenty-first century, in which physical limits on the climate system are incorporated by using the ranges of historical climate observations to set prior distributions for regression coefficients representing future means and linear trends. If, on the other hand, little subject-matter knowledge is available then this can be represented by using prior distributions with very large variances, although such 'noninformative' or 'diffuse' priors can have unexpected implications in some circumstances (Davison, 2003, Section 11.1.3).

The need to specify a prior distribution is sometimes seen as a disadvantage of Bayesian methods because it introduces an element of subjectivity: two analysts, fitting the same model to the same data but with different priors, could reach different conclusions. However, it can be shown that in large samples irrespective (within reason) of the choice of prior, the results from Bayesian and likelihood based analyses are almost the same: for example, the Bayes estimator and credible intervals are very similar to the MLE and corresponding confidence intervals (Davison, 2003, Section 11.2). Therefore, given enough data, our two analysts should be able to resolve their differences. If they cannot do this and their priors are both justifiable on subject-matter grounds, the implication is that the available data do not contain enough information to discriminate between alternative plausible scenarios.

3.2   Multiple regression techniques

An obvious extension to the linear trend model (3.2) is to supplement, or replace, the time index t with the values of other quantities that may be responsible for changes in the variable of interest. Such quantities are referred to as *covariates*, and the variable of interest is often called the *response variable*. The simplest way to model the effect of several covariates is via a multiple regression model of the form

$$Y_t = \beta_0 + \sum_{i=1}^{p} \beta_i x_{it} + \varepsilon_t = \mu_t + \varepsilon_t, \text{ say} \qquad (t = 1, \ldots, T). \qquad (3.34)$$

Here, x_{it} denotes the value of the ith covariate at time t.

The multiple regression model can once again be written in the matrix form (3.10), the only differences being that $\boldsymbol{\beta} = \begin{pmatrix} \beta_0 & \beta_1 & \cdots & \beta_p \end{pmatrix}'$ now has length $k = p + 1$ and that X has dimensions $T \times k$. As before, the intercept β_0 is accommodated by filling the first column of X with ones. A consequence of the matrix representation is that all of the results from the previous section are applicable here as well. Techniques for identifying influential observations can be used without modification; the diagnostics described in Section 3.1.1 can also be used to check the model. Additionally, it can be useful to plot the residuals against each of the covariates individually to check for unmodelled structure. Some effort can be saved by plotting the residuals against the fitted values, rather than against each covariate individually. The plot.lm function in R produces a variety of diagnostics for multiple regression models; straightforward summaries of the main ideas can be found in Davison (2003, Section 8.6) and Faraway (2005, Chapter 4). For more extensive discussion and details of more sophisticated diagnostics, see Cook and Weisberg (1999, Chapter 14) and Fox (2002, Chapter 6).

When the underlying assumptions are satisfied, multiple regression provides the ability to represent all of the processes affecting the quantity of interest within a single model. This contrasts with the common approach of standardising time series data prior to analysis so as to remove structure that is not of direct interest. A disadvantage of the latter approach is that any form of adjustment, such as the removal of seasonality, is a form of preprocessing and therefore needs to be accounted for subsequently. By way of illustration, consider a hypothetical example involving the association between air pollution and human mortality. Mortality time series typically show seasonal fluctuations, some but not all of which may be attributable to seasonal variation in pollution (Schwartz, 1994). If seasonality is removed from a mortality time series prior to analysis, for example by subtracting monthly means, it is likely that some of the pollution effect will be removed inadvertently at the same time. Moreover, if the resulting *anomalies* are regressed on raw or deseasonalised pollution levels, the standard errors of regression coefficients will tend to be underestimated because the analysis does not allow for the possibility that pollution is responsible for some of the discarded seasonal structure. The problem can be avoided entirely by fitting a multiple regression model to the raw data, containing covariates that represent seasonality explicitly as described below.

In principle, multiple regression models can also be used for extrapolation, using the methodology described in Section 3.1.4. However, in practice this is only possible if future values of the covariates are available. One way to achieve this is by using lagged values of the covariates in the model, if such lagged values have any explanatory power at the time horizon of interest. An alternative is to base extrapolations on 'scenarios', whereby the effect of a prespecified sequence of covariates is investigated. Scenario based extrapolation is useful in situations where the future values of the covariates are, at least nominally, under the control of policymakers – examples of such covariates might include levels of industrial sulfur emissions and fisheries quotas.

3.2.1   Representing seasonality in regression models

Seasonality is often one of the most important factors controlling environmental processes at sub-annual timescales. In some cases, it arises mainly due to dependence on one or more seasonally varying covariates, and can be accounted for by including these covariates in a multiple regression model. However, if there are no plausible covariates to which seasonality can be attributed, or if data on such covariates are not available, a different approach is required.

A crude way to handle seasonality is to fit separate models for different times of year. However, the multiple regression framework offers the possibility of representing seasonal controls explicitly via the use of 'dummy' covariates. The simplest option is perhaps to define binary *indicator variables* for each time period. For example, for quarterly data one could define variables x_{1t}, x_{2t}, x_{3t} and x_{4t}, such that x_{jt} takes the value 1 for observations in quarter j and zero otherwise. A multiple regression model involving just these covariates takes the form

$$Y_t = \beta_0 + \sum_{j=1}^{4} \beta_j x_{jt} + \varepsilon_t. \tag{3.35}$$

In this model, the fitted value for quarter j will be $\beta_0 + \beta_j$, since $x_{jt} = 1$ during this quarter and the other covariates are all zero. A least squares fit will, in principle, equate $\beta_0 + \beta_j$ with the mean of the observations for quarter j ($j = 1, \ldots, 4$). However, this reveals a problem: since there are only four quarterly means, it is not possible to estimate the five coefficients β_0 to β_4. The model is said to be *overparameterised*. The difficulty is usually resolved by imposing a constraint on the coefficients, for example by setting one of them to zero. The precise choice of constraint does not affect the fitted values from the model, but it does affect the interpretation of the coefficients. Consider, for example, setting $\beta_0 = 0$ in (3.35). In this case, the fitted value for quarter j is just β_j, which can therefore be interpreted as the mean level for that quarter. If instead we set $\beta_1 = 0$, then the fitted value for quarter 1 is β_0 and the fitted value for any other quarter j is $\beta_0 + \beta_j$. In this case therefore, β_0 is the mean for quarter 1 and, for quarter $j > 1$, β_j is the difference between the means for quarters 1 and j.

The use of indicator variables can also be regarded as a means of adjusting for seasonality. The residuals from model (3.35) are precisely the anomalies that would be obtained by subtracting the quarterly means prior to analysis. If the purpose of the analysis is to assess the effect of some other covariate on the response, this can be quantified by fitting an extended model incorporating the extra covariate in addition to the seasonal indicators. The fitted values from such a model will be the same as those from a separate analysis of the anomalies, but the regression coefficient corresponding to the covariate of interest, and its standard error, may be rather different since they take into account all of the available information.

In the discussion above, the dummy covariates x_{1t} to x_{4t} effectively code for a single variable 'quarter', which defines four separate groups or categories. Such grouping variables are called *factors*; the separate groups are referred to as *levels*. Regression software will usually handle factors automatically, providing they are defined as such (correct behaviour can be guaranteed in R by using characters, rather than numbers, to represent the different groups). The issue of overparameterisation is, however, always present and, to interpret software output, it is necessary to know what constraints have been imposed. In R, the default behaviour for unordered factors is to use 'corner-point' constraints, in which the coefficient associated with the first level (β_1 in the discussion above) is set to zero. Another option is to constrain all of the coefficients associated with a factor to sum to zero. In model (3.35), if there were equal numbers of observations in each quarter then, under a sum-to-zero constraint, the estimate of β_0 would be the overall mean of the series and β_j would be the average deviation from this overall mean in quarter j. The interpretation is less straightforward with differing numbers of observations per quarter. For further details of factor coding in general, see Dobson (2001, Section 2.4). Fox (2002, Chapter 4) and Venables and Ripley (1999, Section 6.2) give a comprehensive account of the facilities available in R.

The factor based approach to modelling seasonality is similar in spirit to the practice of fitting separate models to different subgroups of observations. In both cases the need to make an essentially arbitrary choice of groups, coupled with sampling variability within each group, means that the estimated seasonal structure can be rather a rough approximation to reality, particularly when (as is often the case) the underlying seasonal cycle is expected to be fairly smooth. As an alternative, therefore, one might consider defining dummy seasonal covariates that explicitly represent seasonality in a smooth manner. An obvious candidate is a cosine function:

$$\cos\left(2\pi\frac{t-\tau}{S}\right),\tag{3.36}$$

where S is the number of time units in a complete seasonal cycle (e.g. 12 if t is measured in months) and τ defines a time point at which the seasonal cycle achieves its maximum value; this is sometimes referred to as the *phase* of the cycle. Notice that τ is not uniquely defined since, by definition, the seasonal cycle has at least one maximum every year. However, the resulting cosine function will be the same regardless of the year in which τ falls. Of more concern, perhaps, is that in general the exact value of τ is unknown. At first glance this appears problematic, since τ is not a regression coefficient and hence cannot be estimated using the methods discussed so far. Usually in such situations, nonlinear regression techniques are required (see Section 3.4). Notice, however, that (3.36) can be rewritten using the standard trigonometric identity $\cos(\omega_1 - \omega_2) = \cos(\omega_1)\cos(\omega_2) + \sin(\omega_1)\sin(\omega_2)$:

$$\cos\left(\frac{2\pi t}{S} - \frac{2\pi\tau}{S}\right) = \cos\left(\frac{2\pi t}{S}\right)\cos\left(\frac{2\pi\tau}{S}\right) + \sin\left(\frac{2\pi t}{S}\right)\sin\left(\frac{2\pi\tau}{S}\right)$$

$$= A\cos\left(\frac{2\pi t}{S}\right) + B\sin\left(\frac{2\pi t}{S}\right),\text{ say,}$$

where A and B are constants depending on the unknown value of τ. If, therefore, we fit a multiple regression model including two covariates $\cos(2\pi t/S)$ and $\sin(2\pi t/S)$, the unknown constants A and B will simply be subsumed into the regression coefficients.

Of course, in many cases the seasonal cycle is not exactly of the form (3.36). However, a consequence of the Fourier representation discussed in Section 2.3 is that any periodic function can be represented as a sum of such cosine waves, at frequencies that are integer multiples of the frequency of the annual cycle. Such frequencies are called *harmonics*. It follows that if (3.36) does not produce an adequate representation of seasonality, improvements may be obtained by adding covariates $\cos(4\pi t/S)$ and $\sin(4\pi t/S)$, then $\cos(6\pi t/S)$ and $\sin(6\pi t/S)$ and so forth. Our experience is that one or two additional harmonics are enough to capture the seasonal structure of most environmental series. Notice that two dummy covariates are associated with each harmonic: hence, with regularly spaced data at least, at most $S/2$ harmonics are available. The Fourier representation of seasonality is therefore of limited value for coarse-resolution (e.g. quarterly) regularly spaced data. It is much more useful, however, for data at finer resolution. For example, to fit an annual sinusoid plus one harmonic to a monthly time series involves the estimation of four parameters, rather than the 11 that would be required by treating 'month' as a factor. Furthermore, the Fourier representation can be applied directly to irregularly spaced data since the dummy covariates are defined exactly at all time points. The alternative – to group the data into, say, months and treat these as factors – results in some loss of information since it effectively coarsens the temporal resolution of the data.

3.2.2 Interactions

Suppose now that a quantity of interest depends on two covariates, but the effect of one of these depends on the value of the other as in the following model:

$$Y_t = \beta_0 + \beta_1 x_{1t} + \beta_{2t} x_{2t} + \varepsilon_t \tag{3.37}$$

$$\text{with}\quad \beta_{2t} = \gamma_0 + \gamma_1 x_{1t}, \text{ say.} \tag{3.38}$$

In this case the covariates are said to *interact*. For example, x_{1t} and x_{2t} may represent respectively some aspect of seasonality and a linear trend; then (3.38) would imply that the slope of the trend varies with the time of year.

Substituting (3.38) into (3.37), we obtain

$$\begin{aligned} Y_t &= \beta_0 + \beta_1 x_{1t} + (\gamma_0 + \gamma_1 x_{1t}) x_{2t} + \varepsilon_t \\ &= \beta_0 + \beta_1 x_{1t} + \gamma_0 x_{2t} + \gamma_1 x_{1t} x_{2t} + \varepsilon_t, \end{aligned} \tag{3.39}$$

which has the same form as (3.37) but with an extra term added, containing the product $x_{1t} x_{2t}$. This illustrates the general procedure for accommodating an interaction in a multiple regression model: simply add an extra *interaction term*, defined using the product of the interacting covariates. Notice that interaction terms are symmetric in their covariates: (3.38) implies that x_{1t} modulates the effect of x_{2t}, but (3.39) would still be obtained if the roles were reversed. In the trend – seasonality example this role reversal would be expressed by saying that the seasonal cycle changes through time. The preferred interpretation of an interaction will often depend on the context.

In (3.39), the terms involving single covariates are called *main effects*. It is worth considering the interpretation of the associated regression coefficients. From (3.38) it is clear that γ_0 is the coefficient of x_{2t} when $x_{1t} = 0$. Now suppose that we change the units of measurement of x_{1t}, replacing it with $x_{1t}^* = x_{1t} - c$, say, for some arbitrary constant c. Clearly, this will not affect the model's ability to represent the relationships between the covariates and response (it is effectively the same as changing the time index in the linear trend model, as in Section 3.1.2), but some of the coefficients in the model must change to accommodate it. In particular, the new coefficient of x_{2t} will represent the effect when $x_{1t}^* = 0$ or equivalently when $x_{1t} = c$. Depending on the value of c, this may be very different from γ_0 – it may even have the opposite sign. This shows that in the presence of interactions, the coefficient corresponding to the main effect of one interacting covariate depends on the units of measurement of the other! The lesson from this is that main effects should be interpreted with care in the presence of interactions.

Interaction terms offer the possibility of representing relatively complex relationships within a single model. They are not restricted just to pairs of covariates, although those of order higher than three can be difficult to interpret. In a trend analysis, interactions involving a trend function are an economical, and aesthetically appealing, alternative to the common practice of estimating trends separately for different subsets of data (e.g. different months of the year); they also offer increased flexibility, and hence the opportunity to represent more realistically the structure of any trends that may be present. If, for example, seasonality is represented via a smooth Fourier representation, as discussed above, a seasonal – trend interaction implies that the estimated trends vary smoothly throughout the year rather than changing suddenly between months. The interaction based approach also reduces the temptation to test for the significance of trends separately in different subsets of data: apart from the inherent dangers of carrying out multiple significance tests (Manly, 2001, Section 4.9; see also Cox, 2006, Section 5.15 for a clear and succinct summary of the scientific issues at stake), such procedures are easily misinterpreted and can lead to confusing conclusions, such as that 'genuine' trends exist in (for example) June and August but not in July.

A potential disadvantage of using a single model to represent the structure of an entire data set is that assumptions such as constant error variance may hold approximately within subsets of the data but not overall. For example, there is clear seasonal structure in the variances of the daily wind speed data illustrated in Figure 2.2, but within any individual month the variances are likely to be much more nearly constant. We discuss how to address such issues in Section 3.3.

3.2.3   Model building and selection

Multiple regression provides a flexible and powerful way to represent structure in a data set. The cost of this flexibility is the need to think carefully about which covariates and interactions to include in a model. The most common way to decide whether a term should be included is to examine the t statistic for the associated coefficient estimate. However, if two or more covariates are highly correlated (or *collinear*) these t statistics can be difficult to interpret unless the sample size is extremely large: it can appear that neither covariate is necessary, when exploratory analysis clearly indicates that at least one of them is (see Chandler, 1998, for an example of this). The reason for such apparent anomalies is that if either covariate is dropped from the model, the estimated coefficient of the other can be adjusted to compensate. Another way to view the problem is that if two covariates are highly collinear, it is difficult on the basis of the data alone to determine which (if either) is genuinely responsible for changes in the response; this uncertainty will be reflected in large standard errors for the coefficient estimates. The increase in uncertainty due to correlation between covariates can be quantified using *variance inflation factors*, as described by Fox (2002, Section 6.5) and implemented in the car library in R (Fox, 2009). See also the contribution by Zuur *et al.* in Chapter 9 of the present volume.

If the covariates in a regression model are mutually uncorrelated, t statistics are usually much easier to interpret. Usually this can only be achieved by controlling the covariate values carefully in a designed experiment; the experimental design is then said to be *orthogonal*. Unfortunately, as noted in Chapter 1, such experimental control tends to be the exception rather than the rule in environmental studies. Nonetheless, it is sometimes possible to write a model in such a way that at least some of the covariates are uncorrelated. In a trend analysis, perhaps the most useful example of this is the use of *orthogonal polynomials* to represent moderately nonlinear trend functions. Suppose, for example, that after fitting a linear trend model of the form (3.2), a time series plot of residuals showed some curvature. In this case one might want to try and improve the fit by including a quadratic term:

$$Y_t = \beta_0 + \beta_1 t + \beta_2 t^2 + \varepsilon_t.$$

In this model, the covariates $x_{1t} = t$ and $x_{2t} = t^2$ both increase as t runs from 1 to T; hence collinearity is likely to be a problem. If, however, $x_{1t} = t$ and $x_{2t} = t^2$ are replaced by new covariates $x_{1t}^* = x_{1t} - \overline{x}_1$ and $x_{2t}^* = x_{2t} + ax_{1t} + b$, with

$$a = -\frac{\sum_{t=1}^{T} x_{1t}(x_{2t} - \overline{x}_2)}{\sum_{t=1}^{T} x_{1t}(x_{1t} - \overline{x}_1)} \quad \text{and} \quad b = -(\overline{x}_2 + a\overline{x}_1), \tag{3.40}$$

then it can be verified that the new covariates x_{1t}^* and x_{2t}^* both have mean zero and are uncorrelated (i.e. $\sum_{t=1}^{T} x_{1t}^* x_{2t}^* = 0$). Therefore the t-statistics from a model fitted using x_{1t}^* and x_{2t}^* can be used to distinguish unambiguously between them (notice that x_{1t}^* is the centred time from Example 3.4, where its use resulted in a convenient diagonal covariance matrix for the regression coefficient estimates). Furthermore, they are interpretable: x_{1t}^* is a 'linear' component of trend, whereas x_{2t}^* is a quadratic component. Cubic and higher-order polynomials can be constructed in a similar way; in R, this is implemented in the command `poly`.

The development leading to (3.40) applies to both regularly and irregularly spaced time series. Indeed, it can be used to transform any pair of correlated covariates into an uncorrelated pair. Notice, however, that it treats the covariates asymmetrically – x_{1t}^* depends only on x_{1t}, whereas x_{2t}^* is a combination of x_{1t} and x_{2t} – and this is often difficult to justify. In more general settings, *principal components regression* is an alternative approach to dealing with collinearity, which treats all covariates on an equal footing. A good discussion of this can be found in Faraway (2005, Chapter 9).

An alternative way to assess the importance of an individual covariate is to drop it from the model and investigate the effect upon model performance as measured by the residual sum of squares. Let RSS_1 denote the sum of squares for the original model and RSS_0 that for the *reduced model* (i.e. after dropping the covariate); then the comparison is usually based on the F statistic

$$F = \frac{(RSS_0 - RSS_1)/m}{RSS_1/(T-k)}, \qquad (3.41)$$

where $m = 1$ is the number of terms dropped from the original model and, as previously, k is the number of regression coefficients in this model. Under the null hypothesis that the data were in fact generated from the reduced model, (3.41) follows an F distribution with m and $T - k$ degrees of freedom; large values should be taken as evidence against the null hypothesis.

To some extent, when dealing with individual covariates the use of F tests does not achieve anything new since, as noted in Section 3.1, the resulting p-value is identical to that from the t test for the regression coefficient. More generally, however, F tests can be used to compare any sequence of *nested models*, by which we mean a sequence in which each successive model is a generalisation of the one before. For example, we might start by fitting a model containing just seasonal covariates, then add a linear trend term, followed by terms representing seasonal – trend interactions, at each stage carrying out an F test to see if the increase in model complexity is justified. The results of such a sequence of tests are usually displayed in an analysis of variance (ANOVA) table, showing the reduction in sums of squares achieved at each stage (in R, this is achieved using the `anova` command). Here, a useful generalisation over the usual t tests is that it is possible to test for the significance of several terms simultaneously (the quantity m in the numerator of (3.41) represents the number of extra coefficients estimated in the more complex model). Simultaneous testing may be required, for example, when using a Fourier representation of seasonality (see the end of Section 3.2.1) or when considering interaction terms. More details are given below.

The concept of a nested sequence of models is helpful insofar as it forces the analyst to think in a structured way about the model-building process. A possible disadvantage is that the final model selected can depend on the order in which terms are considered. However, in environmental problems an understanding of the underlying process often dictates a natural ordering. The following guidelines may be helpful when building regression models:

- Start with a simple model including just covariates that are obviously relevant (on the basis either of prior understanding of the process or of exploratory analysis), as well as those that are not of direct scientific interest but for which adjustment may be required. Such a model serves as a baseline against which to judge successive, more complex, models.

- The reason for including covariates that are not of direct scientific interest is to adjust for their effects. To ensure that such effects have been properly accounted for, it may be worth keeping such covariates in a model even though they do not meet the usual criteria for statistical significance. Here, there is a link with the common practice of adjusting the data to remove possible unwanted structure, without formally testing for its significance, prior to analysis (see Section 3.2). Clearly there is a balance to be struck, since the inclusion of too many such covariates will generally reduce the precision with which the effects of scientific interest can be estimated (Davidson and Mackinnon, 2004, Section 3.7). However, given enough data this is unlikely to be a major problem.

- Build up a model gradually, using an understanding of the underlying processes to suggest what may be the most important new terms to add at each stage. It is also worth examining diagnostics such as plots of residuals against potential covariates, to indicate the most promising directions in which to expand the model.
  If this guideline is followed, functions of the time index will often be among the last terms to be added to a model since, almost by definition, these represent trends that cannot be explained in any other way using the available data.

- It is possible for individual covariates to operate primarily by controlling the effect of others. In such situations, one would expect an insignificant main effect but significant interactions. In general, therefore, an insignificant main effect does not necessarily mean that the associated covariate is irrelevant; it may be necessary to explore the interactions more deeply. One may then want to assess simultaneously the significance of a main effect together with all its interactions; this is an example of a situation where an F test can be used but an ordinary t test cannot.
  If a covariate has significant interactions but an insignificant main effect, the latter should usually be retained in the model. This is sometimes called the *marginality principle*, and can be justified by considering that when interactions are present, the main effects depend on the choice of origin on the covariate measurement scale (see Section 3.2.2). Therefore, unless the origin has a clear interpretation in the context of the problem at hand, it is scientifically meaningless to test whether a main effect is 'truly' zero. Moreover, unless the main effects are all included, fitted values and predictions are origin-dependent. Nelder (2000) gives an example of a model in which temperature is used to predict ozone concentrations but, because it does not respect the marginality principle, yields different predictions depending on whether temperature is measured in degrees Celsius or Fahrenheit. This is clearly unsatisfactory.
  A similar comment applies when using a Fourier representation of seasonality: the use of sine and cosine terms is merely a convenient device for representing the function (3.36), and the relative magnitudes of the associated coefficients depend

solely upon the choice of time origin. Usually, therefore, sine and cosine terms should be considered in pairs.

- As a modelling exercise develops, covariates that appeared significant in the early stages may become less so. At appropriate points it may therefore be appropriate to delete some of these before continuing to expand the model. In general, however, it is probably best to err on the side of caution, at least until all of the covariates of scientific interest have been considered. Once this has been done, the effect of deleting terms from the model can be explored; the choice of terms to delete might be based on the t statistics for the regression coefficients, for example. However, when deleting more than one term simultaneously it will once again be necessary to use an F test to examine the effect.

It is sometimes desirable to compare models that are not nested, because each contains covariates that do not appear in the other. In this case an F test cannot be used for the comparison. A widely used alternative, which can be used for both nested and non-nested comparisons, is the *information criterion* introduced by Akaike (1974) and now generally referred to as Akaike's Information Criteria (AIC). This attempts to select models automatically by minimising some measure of (lack of) performance while taking into account the complexity of the model. The measure is defined as minus twice the log-likelihood for the model (see Section 3.1.6), plus twice the number of parameters estimated. 'Good' models yield high log-likelihoods with relatively few parameters; hence, under this measure, if two models yield values AIC_1 and AIC_2 respectively and $AIC_1 < AIC_2$, the first model is preferable. Various alternatives to AIC have been proposed (see Davison, 2003, Section 4.7), but they all operate along similar lines. The R command drop1 can be used to see what would be the effect upon the AIC of removing individual covariates from a regression model.

The model-building process described above requires quite a high level of informed decision-making on the part of the analyst. It can be tempting to try and avoid this by using an 'automatic' method of model selection, of which *stepwise regression* (implemented via the step command in R) is one of the most popular. *Backward* stepwise regression works as follows: start by fitting a model containing all covariates, then see which of these appears least relevant (using AIC or some other criterion), drop it from the model, refit and then repeat the process. If a second covariate is deleted, it becomes necessary to check that the first should not be re-inserted. Stepwise regression thus consists of a sequence of deletions and additions, until a model is obtained that cannot be improved upon according to the chosen criterion.

As an alternative to backward stepwise regression, one could start with no covariates and gradually increase the size of the model; this is called *forward* stepwise regression. Note that neither is guaranteed to find an optimal model (if such a thing exists), and automatic methods will rarely compete with an informed analysis in which the user brings their understanding to bear on the problem. Uncertainty also tends to be under-represented in models chosen by automatic methods (see Chatfield, 2000, Section 8.4.2 for a sobering account of this, albeit in a slightly different context). The main value of such methods is in situations where many routine analyses are required on a regular basis and it is infeasible to carry out each one manually.

Finally, on the subject of model selection, two models can only be compared formally if they have been fitted to the same set of observations $\{Y_t\}$. In many environmental problems, some covariate values are missing occasionally. In such situations, software packages often fit models using just the cases for which all covariates, and the response variable, are available (note that this carries an implicit assumption that the missingness itself is noninformative – for example that there is no tendency for unusually high or low values of the response variable to be missing). This raises the possibility that two models, involving different covariates, may be fitted to different subsets of data because fewer 'complete cases' are available for the larger model. In this case the comparisons described above are not valid. A correct analysis should use only the subset of cases for which both models can be fitted.

Example 3.7

The alkalinity data from the Round Loch of Glenhead (Section 1.3.3) provide a good opportunity to illustrate the use of multiple regression techniques in a trend analysis. Recall that the questions of interest here concern the impact of nonmarine sulfate deposition on the alkalinity of the loch, but that this can also be affected by marine deposition events for which chloride is an indicator and by weather conditions represented by daily temperature and rainfall. We saw in Figure 2.5 that autocorrelation in the alkalinity series could conceivably be explained by trends, and hence by dependence on the other chemicals; hence it is appropriate to start by building regression models under the assumption that the errors are uncorrelated. If this is incorrect, there will be a tendency to overstate the significance of relationships, as discussed below in Section 3.3; hence we may be led to a model that is unnecessarily complex. However, in the first instance this is perfectly acceptable: the model can always be simplified later if necessary.

Consideration of the processes involved suggests that at each sampling time the recorded alkalinity will depend primarily on weather conditions over some previous time interval. The duration of this interval depends upon a variety of factors, such as the size of the loch and the hydrological properties of the surrounding catchment. Accurate assessment of the most appropriate interval is difficult without extensive data on catchment characteristics and geology. We have therefore adopted the simple device of aggregating the meteorological variables separately over weekly, monthly and quarterly time windows (actually 7, 30 and 91 days) prior to each sampling time. Temperatures were aggregated by taking averages, whereas rainfalls were aggregated by calculating cumulative totals. Each of the aggregates was then standardised, using the scale command in R,[1] to have zero mean and unit variance. This does not affect the fit of any regression model; it does, however, ensure that the magnitudes of the regression coefficients are directly comparable in terms of their effect on the alkalinity, which is useful since one of the aims of the analysis is to determine the relative importance of the different covariates. The standardised weekly, monthly and quarterly temperatures are referred to below as Wtemp, Mtemp and Qtemp respectively, and the rainfalls as Wrain, Mrain and Qrain.

For the sulfate and chloride covariates (denoted respectively, after standardisation, by xSO4 and Cl), the cross-correlation analysis in Section 2.1.5 suggested weak association between alkalinity and chloride but was less clear about the association with sulfate. Further preliminary analysis, not shown here, suggested that of all the potential covariates in the data set, quarterly temperature (Qtemp) has the strongest relationship with alkalinity. This is consistent with weather being the primary driver of variations in water chemistry.

On the basis of these considerations, we start by developing a baseline model in which alkalinity is regressed on the meteorological variables only. The effect of rainfall may itself be temperature-dependent since, among other things, warmer temperatures lead to enhanced evaporation and hence a reduction in the amount of rainwater reaching the loch.

---

[1] This simply subtracts the mean of a variable and divides by its standard deviation.

Therefore, as well as considering the rainfall and temperature variables individually we examine interactions between them, and fit a model containing all six weather variables together with a rainfall – temperature interaction at each timescale. This is obviously far too complex a model: indeed, apart from the intercept only the Qtemp coefficient is significant at the 5 % level. We therefore use stepwise regression (the step function in R, with default settings) to simplify the model in the first instance. This removes the Qrain term, together with the Qtemp:Qrain interaction. Examination of the result suggests that all of the weekly terms are also redundant. If these terms are removed, we obtain the model given in Table 3.4. Notice here that Mtemp and Mrain do not appear significant at the 5 % level; however, the interaction between them is highly significant, so the main effects are retained, as discussed previously.

Table 3.4   Edited R output for the baseline alkalinity model.

```
             Estimate Std. Error t value Pr(>|t|)
(Intercept) -12.3722     0.5401 -22.906  < 2e-16 ***
Qtemp         4.1581     0.9413   4.417 4.96e-05 ***
Mtemp        -1.9794     1.0250  -1.931  0.05882 .
Mrain         0.2127     0.5483   0.388  0.69967
Mtemp:Mrain  -1.7977     0.5852  -3.072  0.00335 **
---
Signif. codes: 0 '***' 0.001 '**' 0.01 '*' 0.05 '.' 0.1

Residual standard error: 3.525 on 53 degrees of freedom
Multiple R-Squared: 0.4255, Adjusted R-squared: 0.3821
F-statistic: 9.812 on 4 and 53 DF, p-value: 5.142e-06
```

The F statistic reported in Table 3.4 is for testing the hypothesis that the model has no predictive power (see Section 3.1). While it is reassuring that this hypothesis is emphatically rejected, there are perhaps more interesting questions to be asked at this stage. An obvious one is whether, in dropping five terms from the initial model, we have inadvertently oversimplified it. This can be answered by comparing the two models using an F test. Using anova in R, the value of the test statistic (3.41) is obtained as 0.7487, on 5 and 48 degrees of freedom (the latter because there are $T = 58$ observations in total and 10 parameters were estimated in the more complex model). The p-value for the test is 0.591: this gives no evidence to reject the null hypothesis, so we conclude that simplification is justified (and, in passing, that the initial use of stepwise regression was not very effective in finding a reasonable model).

The time series of residuals for the baseline model in Table 3.4 (not shown) shows essentially the same trend as the original data (see Figure 1.3). This is unsurprising, given the absence of pronounced trends in either the temperature or rainfall series. Interestingly, however, the mean residuals in each quarter are close to zero and the autocorrelation function of the residuals looks almost identical to that for the deseasonalised alkalinity series in Figure 2.5. This indicates that seasonality in the alkalinity series can be explained entirely via weather conditions, which seems plausible.

Having established a reasonable baseline model, we investigate the effects of xSO4 and Cl. On the basis of the cross-correlation analysis in Section 2.1.5, and of our understanding of the processes involved, we consider only simultaneous, rather than lagged, relationships with these covariates. We do, however, consider the possibility that they may interact with the meteorological variables. We therefore proceed by adding xSO4 and Cl to the baseline model, along with each of their interactions with Qtemp, Mtemp and Mrain. A summary of the resulting model reveals that all the chemical – meteorological interaction terms appear insignificant; we therefore delete them all and check that this is justified using an F test, which yields a p-value of 0.310. The simplified model is summarised in Table 3.5. Both xSO4 and Cl appear significant, with evidence of quite a strong negative xSO4 effect (recall that, at least for the main effects, the regression coefficients are directly comparable since we have standardised all of the covariates) – this is consistent with the idea that reduced acidity will follow from a reduction in sulfur deposition. There is also a considerable improvement in the explanatory power of the model: the adjusted R^2 has risen from 38.2 % to 52.2 %. A formal comparison with the baseline model indicates that the improvement is significant: the F statistic is 8.987 on 2 and 53 degrees of freedom, with an associated p-value of 5.2×10^{-4}.

Table 3.5   Edited R output for the alkalinity model including sulfate and chloride effects.

	Estimate	Std. Error	t value	Pr(>|t|)	
(Intercept)	-12.1845	0.4801	-25.377	< 2e-16	***
Qtemp	3.8333	0.8724	4.394	5.64e-05	***
Mtemp	-1.7117	0.9369	-1.827	0.07357	.
Mrain	0.1941	0.4830	0.402	0.68939	
xSO4	-1.4347	0.4139	-3.467	0.00108	**
Cl	-1.0826	0.4418	-2.451	0.01773	*
Mtemp:Mrain	-1.4031	0.5350	-2.623	0.01148	*

Signif. codes: 0 '***' 0.001 '**' 0.01 '*' 0.05 '.' 0.1

Residual standard error: 3.101 on 51 degrees of freedom
Multiple R-Squared: 0.5723, Adjusted R-squared: 0.5219
F-statistic: 11.37 on 6 and 51 DF, p-value: 4.97e-08

It remains to check the model. A normal quantile–quantile plot (not shown) shows that the assumption of normality appears to hold reasonably well; further, the Cook's distances and leverages (see Section 3.1.5) reveal no particularly influential observations. However, Figure 3.3 shows that not all of the model assumptions appear to be satisfied. In the first instance, some trend remains in the residual time series in Figure 3.3(a) – this is apparent from the superimposed lowess smooth (see Section 4.3.1). It therefore appears that the model does not account for all of the trend in the alkalinity. Furthermore, Figure 3.3(b) shows that while there is no seasonal structure in the mean residuals, the assumption of constant variance may not hold: the residuals in quarter 4 appear more variable than those in the other three quarters. Finally, the residuals still appear weakly autocorrelated, according to the ACF in Figure 3.3(c). The correlation is weaker than for the original data and for the baseline model (see Figure 2.5), but the coefficients are still predominantly positive to a lag of around two years. A Durbin–Watson test (see Section 3.1.1) yields a p-value of 0.014, indicating that the residual autocorrelation is significant. In view of these results, it is probably unwise to overinterpret the output in Table 3.5. We will return to this example after discussing how to deal with failures of assumptions in regression models.

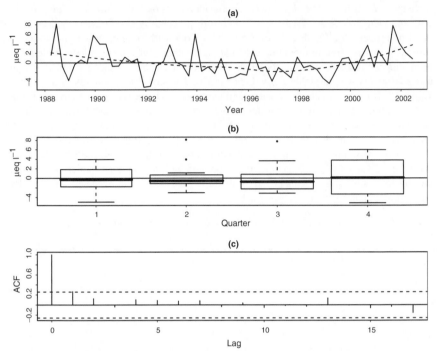

Figure 3.3   Residual plots for the alkalinity model in Table 3.5: (a) time series of resid-uals, with lowess smooth (dashed line), (b) residual distributions by quarter, (c) residual autocorrelation function.

3.3   Violations of assumptions

We have so far discussed how to fit and check linear models and how to use them for extrapolation if the model assumptions appear to be satisfied. We now turn to the question of how to proceed when this is not the case.

3.3.1   Dealing with heteroscedasticity

Heteroscedasticity (i.e. nonconstant variance) is easier to deal with than either non-normality or autocorrelation. The simplest approach is to work with transformed data, where the transformation is chosen so as to induce a constant variance. For example, the variance of the untransformed haddock biomass series decreases through time, but this feature is not present in the log-transformed series. Other commonly used transformations include the square root and cube root. However, if appropriate, the logarithmic transfor-mation is often particularly convenient for interpretation. For example, a linear trend with slope β_1 on the log scale corresponds to an average proportional increase of $\exp(\beta_1)$ per unit time on the original scale. For the haddock stocks model in Table 3.2, this estimated proportional increase is $\exp(-0.0370) = 0.96$; the implication of this model is therefore that North Sea haddock biomass decreased at an average rate of 4 % per year between

1963 and 2000. This is a very natural way of expressing the result, and is arguably more meaningful in biological terms than an analysis based on untransformed data (Buckland *et al.*, 2004, Section 5.3). Of course, the logarithmic transformation can only be used for positive-valued series, but this is applicable in many environmental applications (species numbers, pollutant concentrations and so forth) and, in such cases, a further advantage is that the fitted values from a regression model are guaranteed to be positive after back-transformation to the original measurement scale.

Unfortunately, it may not always be possible to find a transformation that stabilises the variance adequately throughout the series. Even when it is possible, it may not be desirable. One reason for this is that transforming the data will change the nature of any trend that is present – if the original data follow a linear trend, their logarithms will not. Another reason is that the interpretation of model parameters such as regression slopes can be difficult, except for a few special cases such as the logarithmic transformation. If transformation is not possible, or is undesirable, the most appropriate resolution to the problem depends on the nature of the heteroscedasticity. We distinguish between the following scenarios:

(a) The variance of Y_t is

$$\text{Var}\,(Y_t) = \frac{\sigma^2}{w_t}, \tag{3.42}$$

where σ^2 is unknown but the relative precisions $w_1, \ldots, w_T$ are all known. This might sound unrealistic but arises in practice, for example, when the data are estimates of some quantity of interest derived from repeated surveys (as in wildlife abundance monitoring). In the simplest case, the variance of the estimate at time t will often be proportional to $1/n_t$, where n_t is the sample size for the corresponding survey.

(b) The variance of Y_t appears to be related to the mean level so that, for example, large values of Y_t tend to be more variable than small values.

(c) The variance of Y_t appears to be related to other factors, for example the time of year. Of course, these factors may themselves influence the mean of the series, but perhaps not in such a way as to induce a straightforward relationship between the variance and the mean.

(d) The variance seems to fluctuate gradually through time, but not in a systematic way.

Scenarios (b) and (c) are most easily dealt with using generalised linear models (Section 3.5). Scenario (d) is encountered frequently in financial applications; for example, the variability (or *volatility*) of stock market returns shows short-term fluctuations depending on market conditions. In the financial and econometric literature, such series are often modelled using *generalised autoregressive conditionally heteroscedastic* (GARCH) models. In environmental applications, however, the processes controlling variability are usually rather more tangible than this. We therefore do not consider scenario (d) further in this book, beyond noting that Tol (1996, 1997) developed GARCH models for temperature and wind speed data and that Franses, Neele and van Dijk, (2001) extended these models to allow for asymmetric distributions in temperature data. The interested reader is referred to these references for further details.

Returning therefore to scenario (a): when the relative precisions are known, it is natural to attach more importance to the most precise observations when estimating parameters. In fact, if the assumptions of normality and independence are satisfied, the optimal parameter estimates in a linear model are derived by minimising the weighted sum of squares

$$S_w(\boldsymbol{\beta}) = \sum_{t=1}^{T} w_t \left(Y_t - \beta_0 - \sum_{i=1}^{p} \beta_i x_{it} \right)^2.$$

Most computer implementations work on the principle that $S_w(\boldsymbol{\beta})$ can be rewritten as

$$\sum_{t=1}^{T} \left[Y_t \sqrt{w_t} - \left(\beta_0 \sqrt{w_t} + \sum_{i=1}^{p} \beta_i x_{it} \sqrt{w_t} \right) \right]^2, \tag{3.43}$$

whence the weighted least squares estimates can be obtained from a least squares regression in which each observation Y_t, along with the corresponding row $\mathbf{x}_t$ of the design matrix X, is first multiplied by $\sqrt{w_t}$.

Some care is required with the interpretation of software output for weighted least squares. For example, the estimate of σ^2 in (3.42) depends on the absolute values of the weights: if all of the weights are doubled, their relative values (and hence the fitted model) will be unchanged but the estimate of σ^2 will be halved. A further point is that each residual should be multiplied by the square root of the corresponding weight before producing residual plots; under the model, these scaled residuals should have constant variance. In R currently, some but not all of the built-in linear model diagnostics use these scaled residuals – for example, the normal quantile–quantile plot uses them, but the default plot of residuals against fitted values does not.

To understand the properties of weighted least squares estimators, it is convenient to adopt a matrix based approach. The analogue of the normal equations (3.14) can be written as

$$\left[X'WX \right] \hat{\boldsymbol{\beta}} = X'W\mathbf{Y}, \tag{3.44}$$

where W is a diagonal matrix containing the weights $\{w_t\}$. The properties of the estimators can be derived from this matrix representation in exactly the same way as before – for example, in the absence of autocorrelation the covariance matrix of $\hat{\boldsymbol{\beta}}$ is now $\sigma^2 \left(X'WX \right)^{-1}$. Compare this with the expression given at (3.18).

3.3.2   Dealing with non-normality

As with heteroscedasticity, often the simplest way to deal with non-normality is to transform the data. In particular, the assumption of normality is at best an approximation for quantities that are necessarily non-negative, because the normal distribution can take negative as well as positive values. An obvious way around this particular difficulty is to work with the logarithms of the data – although, of course, the mere fact that the logarithms can take any real value is no guarantee of normality. Logarithmic, square root and cube root transformations are often used in the hope that the result will not only have constant variance but will also yield residuals with a distribution that is close to normal. In this context, these transformations are all special cases of the *Box–Cox transform*, defined as

$$y^*(\lambda) = \begin{cases} \log y, & \lambda = 0, \\ \left(y^\lambda - 1 \right) / \lambda, & \text{otherwise.} \end{cases} \tag{3.45}$$

In our haddock stocks analysis, by working with log transformed data we have effectively chosen to fix the value $\lambda = 0$. In principle, we could estimate the 'optimal' value of λ using the maximum likelihood (see Section 3.1.6) and test the hypothesis $H_0 : \lambda = 0$ as a means of checking that the log transform is appropriate. In R, this can be achieved using the boxcox command; see Faraway (2005, Chapter 7) and Venables and Ripley (1999, Section 6.8) for details.

Some arguments for and against data transformation were given in the previous subsection. For some types of data, however, no transformation will induce normality, even approximately. The most obvious example of this is when the observations are binary-valued, as when recording the presence or absence of some phenomenon of interest (for instance whether or not it rains, or whether a particular species is sighted during a survey). It is sometimes possible to circumvent such problems by working with aggregated data – for example, the distribution of the total number of rainy days in a year will be approximately normal in most locations – but this is unsatisfactory unless the aggregated data are of scientific interest in their own right. These and other limitations can be overcome to a large extent through the use of generalised linear models (GLMs), which are discussed in Section 3.5.

3.3.3   Dealing with autocorrelation

As noted in Section 3.1.1, ordinary least squares may deliver reasonable estimates of parameters in a regression model even if the errors are autocorrelated, but the associated standard errors will be incorrect. This suggests that a simple way to deal with autocorrelation is to adjust the standard errors after fitting a model using least squares. A very crude form of adjustment is based on the idea that the information content of T autocorrelated observations is in some sense equivalent to that of T^* independent observations, where $T^* < T$ is the *effective sample size*; therefore one could compute standard errors and tests by replacing T in the usual formulae with T^*. This approach is discussed in von Storch and Zwiers (1999, Section 17.1), Santer *et al.* (2000) and references therein. For example, suppose the errors are generated according to a *first-order autoregressive process* or *AR(1)* (see Chapter 5):

$$\varepsilon_t = \phi\varepsilon_{t-1} + \delta_t, \tag{3.46}$$

where $|\phi| < 1$ and (δ_t) is a white noise sequence. Then it can be shown that the effective sample size for estimating a mean is

$$T^* = T \times \frac{1 - \phi}{1 + \phi}.$$

In general, however, the 'effective sample size' approach is rather unsatisfactory since its justification is somewhat heuristic and its implications are not always clear. When adjusting standard errors in regression models, it is preferable to start by considering the exact expression for the covariance matrix of the parameter estimates, which is given by Equation (3.17). This expression shows that if $\hat{\Sigma}$ denotes an estimate of the error covariance matrix Σ, we can estimate the covariance matrix of $\hat{\beta}$ as

$$\left(X'X\right)^{-1} X' \, \hat{\Sigma} \, X \left(X'X\right)^{-1}. \tag{3.47}$$

Adjusted standard errors can then be obtained as the square roots of the diagonal elements of this matrix. Note that the (t, s) element of Σ is the covariance between ε_t and ε_s.

For any regression model that fits the data at all well in a time series context, it is usually reasonable to assume that (ε_t) is a stationary process and hence has a theoretical autocorrelation function $\{\rho(k)\}$ (see Section 2.1.3). In this case, the (t, s)th element of Σ is just $\sigma^2 \rho(t - s)$. For regularly spaced series, this gives rise to a banded diagonal (or *Toeplitz*) structure:

$$\Sigma = \sigma^2 \begin{pmatrix} 1 & \rho(1) & \rho(2) & \cdots & \rho(T-1) \\ \rho(1) & 1 & \rho(1) & \cdots & \rho(T-2) \\ \rho(2) & \rho(1) & 1 & \cdots & \rho(T-3) \\ \vdots & \vdots & \vdots & \ddots & \vdots \\ \rho(T-1) & \rho(T-2) & \rho(T-3) & \cdots & 1 \end{pmatrix}.$$

This suggests estimating Σ by replacing the theoretical autocorrelations with their sample counterparts (i.e. the residual autocorrelations) and using the residual variance $\hat{\sigma}^2$ in place of σ^2. Unfortunately, this tends not to work very well because it requires the estimation of $T - 1$ autocorrelations and one variance from T observations (after fitting a regression model with k parameters!). A better alternative may be to fit a time series model (see Chapter 5) to the error process (ε_t) and to base the estimate of Σ on the autocorrelation function for the fitted model. The exact choice of model may be based on an inspection of the residual autocorrelation function. It is common, however, to proceed as though the errors form an AR(1) process of the form (3.46). For such a process, the autocorrelation at lag k is just $\rho(k) = \phi^k$, and the parameter ϕ can be estimated using the residual autocorrelation at lag 1, $r(1) = \hat{\phi}$, say. The estimate of Σ is therefore

$$\hat{\Sigma} = \hat{\sigma}^2 \begin{pmatrix} 1 & \hat{\phi} & \hat{\phi}^2 & \cdots & \hat{\phi}^{T-1} \\ \hat{\phi} & 1 & \hat{\phi} & \cdots & \hat{\phi}^{T-2} \\ \hat{\phi}^2 & \hat{\phi} & 1 & \cdots & \hat{\phi}^{T-3} \\ \vdots & \vdots & \vdots & \ddots & \vdots \\ \hat{\phi}^{T-1} & \hat{\phi}^{T-2} & \hat{\phi}^{T-3} & \cdots & 1 \end{pmatrix}. \tag{3.48}$$

For irregularly spaced series the same ideas can be applied (see Section 5.4), but $\hat{\Sigma}$ no longer has a Toeplitz structure.

For obvious reasons, a covariance matrix estimator of the form (3.47) is called a *sandwich estimator*. As well as adjusting for autocorrelation, sandwich estimators can be used, via suitable choices of Σ, to allow for other aspects of model misspecification such as heteroscedasticity (Davidson and Mackinnon, 2004, Section 5.5) or if a linear trend model has been fitted when the underlying trend is not truly linear. Providing the sample size is not too small, such estimators generally protect well against the possibility of drawing erroneous conclusions from an incorrect model. However, for small samples with relatively strong dependence, they may still be inaccurate (Kauermann and Carroll, 2001).

Rather than merely adjusting for autocorrelation, one may consider accounting for it explicitly during the model-fitting process. For some time, the method of Cochrane and Orcutt (1949) provided a popular and easily implemented means of achieving this. Although this has largely been superseded by *generalised least squares* as discussed below, it is still helpful to consider the basic argument. Suppose that in the multiple regression model (3.34), the error terms follow the AR(1) process (3.46). Now consider the quantity

$$
\begin{aligned}
Y_t - \phi Y_{t-1} &= \beta_0 + \sum_{i=1}^{p} \beta_i x_{it} + \varepsilon_t - \phi \left[\beta_0 + \sum_{i=1}^{p} \beta_i x_{i(t-1)} + \varepsilon_{t-1} \right] \\
&= (1 - \phi)\beta_0 + \sum_{i=1}^{p} \beta_i \left(x_{it} - \phi x_{i(t-1)} \right) + \varepsilon_t - \phi \varepsilon_{t-1} \\
&= (1 - \phi)\beta_0 + \sum_{i=1}^{p} \beta_i \left(x_{it} - \phi x_{i(t-1)} \right) + \delta_t.
\end{aligned}
\tag{3.49}
$$

This has the form of a regression equation in which the quantities $Y_t - \phi Y_{t-1}$ are regressed on covariates $\{x_{it} - \phi x_{i(t-1)}\}$ and the error term is δ_t. However, (δ_t) is a white noise sequence. Therefore, if ϕ were known we could fit (3.49) by ordinary least squares to obtain estimates of $\beta_1, \ldots, \beta_p$, together with their standard errors (notice that the first observation would have to be omitted, since we cannot compute $Y_1 - \phi Y_0$ without observing Y_0.). To obtain an estimate of β_0, we could then divide the constant term in the resulting regression (along with its standard error) by $1 - \phi$. In practice of course, ϕ is rarely known, but as a preliminary estimate we could use the lag 1 autocorrelation from an ordinary least squares fit of the original regression model. If necessary, the estimate of β from (3.49) could then be used to construct a revised set of residuals, and hence an updated estimate of ϕ, and the whole procedure could be iterated to convergence (which is guaranteed – see Davidson and Mackinnon, 2004, Exercise 7.18). The effect of estimating the autoregressive coefficient ϕ is not accounted for in the standard errors obtained from the procedure, but the effect of this is negligible in large enough samples providing the covariates do not include previous values of the response variable (Davidson and Mackinnon, 1993, pages 338–339).

As noted above, the Cochrane–Orcutt procedure has now been largely replaced by generalised least squares (GLS), although we will shortly see that there is a close connection between the two. If the covariance matrix Σ is known, the GLS estimator of β minimises the quadratic form

$$
(\mathbf{Y} - \mathbf{X}\boldsymbol{\beta})' \, \boldsymbol{\Sigma}^{-1} \, (\mathbf{Y} - \mathbf{X}\boldsymbol{\beta})
\tag{3.50}
$$

with respect to β. This is equivalent to ordinary least squares when the observations are uncorrelated and homoscedastic; in this case, the diagonal elements of Σ are all equal to σ^2 and the remaining elements are zero. Moreover, in the presence of heteroscedasticity but no autocorrelation, minimising (3.50) is equivalent to weighted least squares. The strongest argument for using GLS is that the minimiser of (3.50) shares the usual optimality properties of the ordinary least squares estimator, but in a much wider variety of settings (Brockwell and Davis, 2002, Section 6.6). When the errors are normally distributed, GLS in fact delivers maximum likelihood estimators of the regression coefficients (see Section 5.1). The routine computation of GLS estimates has, however, only become feasible relatively recently.

On the face of it, it seems that even with modern computing power, minimising (3.50) will be a formidable task if T is large: not only must we store the $T \times T$ matrix Σ, but we must also invert it. However, things are not as bad as they appear. Since Σ is a covariance matrix, it is positive definite and hence has a Choleski decomposition (Horn and Johnson, 1985, page 114): $\Sigma = c\mathbf{L}\mathbf{L}'$, say, where $\mathbf{L}$ is a lower triangular matrix and the arbitrary nonzero constant c is included for ease of exposition below. Writing $\boldsymbol{\Gamma} = \boldsymbol{L}^{-1}$, we have $\boldsymbol{\Sigma}^{-1} = c^{-1}\boldsymbol{\Gamma}'\boldsymbol{\Gamma}$, and the GLS criterion (3.50) is proportional to

$$(\boldsymbol{Y} - \boldsymbol{X\beta})' \boldsymbol{\Gamma}'\boldsymbol{\Gamma} \, (\boldsymbol{Y} - \boldsymbol{X\beta}) = \left[\boldsymbol{\Gamma} \, (\boldsymbol{Y} - \boldsymbol{X\beta})\right]' \left[\boldsymbol{\Gamma} \, (\boldsymbol{Y} - \boldsymbol{X\beta})\right],$$

which is just the sum of squared elements of the vector $\boldsymbol{\Gamma}(\boldsymbol{Y} - \boldsymbol{X\beta})$. The GLS estimator can therefore be computed using ordinary least squares, by regressing the vector $\boldsymbol{\Gamma Y}$ on the columns of the matrix $\boldsymbol{\Gamma X}$. In Section 5.1 we will see that, when the error process can be modelled using one of the stationary time series models considered there, the matrix $\boldsymbol{\Gamma}$ can be calculated remarkably easily. In particular, in the AR(1) case with $|\phi| < 1$ it can be shown (Davidson and Mackinnon, 1993, Section 10.6) that for a suitable choice of c,

$$\boldsymbol{\Gamma} = \begin{pmatrix} \sqrt{1 - \phi^2} & 0 & 0 & \cdots & 0 & 0 \\ -\phi & 1 & 0 & \cdots & 0 & 0 \\ 0 & -\phi & 1 & \cdots & 0 & 0 \\ \vdots & \vdots & \vdots & \ddots & \vdots & \vdots \\ 0 & 0 & 0 & \cdots & 1 & 0 \\ 0 & 0 & 0 & \cdots & -\phi & 1 \end{pmatrix}. \tag{3.51}$$

For $t = 2, \ldots, T$, the tth element of the resulting vector $\boldsymbol{\Gamma Y}$ is given by $Y_t - \phi Y_{t-1}$, with a similar transformation applied to the rows of the X matrix. This transformation is identical with that used in the Cochrane–Orcutt procedure: however, GLS enables the first observation to be incorporated as well. This can make a surprisingly large difference if the sample size is small or if $|\phi|$ is close to 1 – see Davidson and Mackinnon (1993, Table 10.1). Note that (3.51) cannot be used if $|\phi| \geq 1$. This is because it is based on an expression for $\boldsymbol{\Sigma}$ derived from the AR(1) autocorrelation function, and the latter is only defined when $|\phi| < 1$ (see Section 5.1).

The transformation defined by (3.51) was first suggested by Prais and Winsten (1954), and is therefore sometimes called a *Prais–Winsten transformation*. The effect is to adjust each observation for autocorrelation by subtracting contributions from previous time points; this is true for more general error structures as well, thanks to the lower triangular form of the matrix $\boldsymbol{\Gamma}$ (which itself follows from the lower triangular form of $\boldsymbol{L}$).

In practice, as with the Cochrane–Orcutt method, the parameters in the correlation model must themselves be estimated when using GLS. There are various methods for doing this. In R, GLS can be performed using the `arima` command (to be discussed further in Chapter 5) as well as using the `gls` command in the `nlme` library (described in Pinheiro and Bates, 2000). These both offer the possibility of estimation based on maximum likelihood (as discussed in Section 5.1) or variants thereof. This is particularly convenient since it enables different models to be compared using criteria such as likelihood ratio tests (Section 3.5) or the AIC (Section 3.2.3). However, likelihood based GLS procedures make explicit use of the normality assumption, which therefore needs checking. Furthermore, it is worth noting that model comparison procedures, confidence and hypothesis tests for GLS are based on large-sample approximations and should therefore be interpreted with caution when small numbers of observations are available.

To check regression models fitted by GLS, perhaps the most straightforward approach is to examine the residuals from the regression of $\boldsymbol{\Gamma Y}$ on $\boldsymbol{\Gamma X}$; under the model, these should be approximately uncorrelated, homoscedastic and (if using likelihood based estimation) normally distributed. In case of concern over the influence of outliers, it may be desirable to combine this kind of approach with a robust fitting method (see Section 3.1.5).

Accessible methods for achieving this are currently relatively undeveloped; however, a useful review and comparison of some simple techniques can be found in Fried and Gather (2005). A discussion of prediction and extrapolation is deferred until Section 5.1.

The Cochrane–Orcutt and GLS techniques account for autocorrelation via an appropriate model for the structure of the error process (ε_t) in (3.34). This carries with it the implication that the autocorrelation is due to dependence among the errors. As an alternative, one might consider that autocorrelation is due to dependence between the observations themselves. In this case, it could be modelled by incorporating previous values of the response variable into the regression equation, for example by writing

$$Y_t = \beta_0 + \sum_{i=1}^{p} \beta_i x_{it} + \phi Y_{t-1} + \delta_t \tag{3.52}$$

in place of the original model, where (δ_t) is a white noise sequence. This will be justified using likelihood based arguments in Section 3.5.

Using techniques to be discussed in Chapter 5 it can be shown that, providing $|\phi| < 1$, (3.52) can be written as

$$Y_t = \frac{\beta_0}{1 - \phi} + \sum_{i=1}^{p} \beta_i \sum_{j=0}^{\infty} \phi^j x_{i(t-j)} + \sum_{j=0}^{\infty} \phi^j \delta_{t-j}.$$

This is in the same form as (3.34) but with a different constant term, covariates and error. The covariates are now

$$\sum_{j=0}^{\infty} \phi^j x_{i(t-j)} \qquad (i = 1, \ldots, p) \tag{3.53}$$

and the error term is $\varepsilon_t = \sum_{j=0}^{\infty} \phi^j \delta_{t-j}$. In Chapter 5 we will see that this is in fact an alternative representation of the AR(1) process (3.46). The 'lagged response' model (3.52) is therefore equivalent to a multiple regression with autocorrelated errors, but with a different set of covariates. This in turn means that the regression coefficients have a different interpretation, whence one should beware of fitting a model such as (3.52) and then interpreting the results as though they were obtained from (3.34). Nonetheless, if interpretation is not of paramount importance (for example, if the main aim of the analysis is extrapolation) or if the autocorrelation is relatively weak, then the inclusion of lagged response variables can be a convenient device. In particular, if the form of the autocorrelation is unknown, it can usually be accommodated simply by including enough previous Y values into the regression equation. This is because many stochastic processes can be approximated reasonably using autoregressions of sufficiently high order (Chatfield, 2003, Section 13.5). Notice, however, that when fitting lagged response models it will usually be necessary to discard the first few observations. Y_1 cannot be used to fit (3.52), for example, since Y_0 is not available. In such cases, for comparison purposes, care needs to be taken that all models are fitted to the same subset of data as discussed in Section 3.2. A further cautionary note is that, as with GLS, inference for lagged response models is based on large-sample approximations.

There is one special situation in which a lagged response model is directly equivalent to a multiple regression model, albeit with different parameters. This is when the covariates themselves are all polynomial functions of time: $x_{it} = t^i$, say. In this case, the trend represented by (3.52) is a polynomial of degree p in t, and the new covariates defined by

(3.53) are also polynomials. Therefore, although the models are parameterised differently, at a fundamental level the only difference between (say) a GLS fit of (3.2) and an ordinary least squares fit of (3.52) will be due to the omission of initial observations in the latter.

Example 3.8

We now return to the alkalinity series from the Round Loch of Glenhead. In Section 3.2 we fitted a multiple regression model and found evidence of trend, heteroscedasticity and autocorrelation in the residuals. It is possible that the autocorrelation is itself a reflection of the unexplained trend; however, here we will take the opportunity to illustrate the use of the methods described above to account for it.

Table 3.6 shows the results from refitting the previous model using the `gls` command in R, with an AR(1) correlation structure and with all parameters estimated using maximum likelihood. The autogressive parameter ϕ in (3.46) is estimated as 0.343. Before examining the remaining output in detail, it is worth carrying out a formal comparison with the previous version of the model in Table 3.5. Although the models are nested (the previous model is a special case of the current one in which ϕ was assumed to be zero), we cannot compare them using an F test since the new parameter relates to the error structure rather than to an additional covariate. We can, however, compare their AIC values: that for the original model is 304.4 and that for the GLS version is 301.1. The reduction suggests that the GLS fit is preferable, which might have been expected given the result of the earlier Durbin–Watson test. As an alternative to the AIC, we could use a likelihood ratio test (see Section 3.5.2) to compare the two models; this again supports the GLS version, with a p-value of 0.021 leading to rejection of the simpler model at the 5 % level.

Table 3.6   Edited R output for alkalinity model from Table 3.5, fitted using GLS.

```
Correlation Structure: AR(1)
 Parameter estimate(s):
      Phi
0.3432038


Coefficients:

              Value  Std.Error    t-value  p-value
(Intercept)-12.280443 0.6243128 -19.670339  0.0000
Qtemp        3.724922 0.8106471   4.594998  0.0000
Mtemp       -1.688423 0.8586464  -1.966378  0.0547
Mrain        0.063415 0.4161168   0.152397  0.8795
xSO4        -0.995508 0.5330224  -1.867666  0.0676
Cl          -1.242109 0.4957674  -2.505426  0.0155
Mtemp:Mrain -1.676306 0.4641772  -3.611349  0.0007

Residual standard error: 2.952969
Degrees of freedom: 58 total; 51 residual
```

Having established that the GLS fit improves upon the original, we examine it in more detail. First we examine the coefficient estimates. These are mostly similar to those in the original model, with the exception of those relating to xSO4 and to the

Mtemp:Mrain interaction. Indeed, the GLS fit suggests that xSO4 has a much smaller effect than previously: the coefficient has changed from −1.43 to −1.00. Examination of the standard errors is also instructive. A comparison between Tables 3.5 and 3.6 reveals that although some have increased in line with expectation, those relating to meteorological variables are smaller than previously. This illustrates that standard errors in regression models cannot generally be adjusted using the 'effective sample size' approach discussed at the beginning of this section.

The most important consequence of the changes is that the coefficient of xSO4 is now (just) insignificant at the 5 % level, whereas before it appeared highly significant. It is rather unfortunate that this should be the coefficient that is most sensitive to a change in the model assumptions, since this is precisely the coefficient of most scientific interest! At first sight, given that the residual autocorrelation in Figure 3.3 seemed very weak, it is surprising that the use of GLS should make such a difference to the results. An explanation can be found in the fact that the xSO4 series is itself more highly autocorrelated than any of the other covariates. Without explicitly allowing for autocorrelation in the alkalinity series, the least squares fit in Table 3.5 had essentially to ascribe all of the autocorrelation to a relationship with xSO4. However, the use of GLS removes this limitation, and our formal model comparisons suggest that the autocorrelation structure is better described through an AR(1) residual process that is unrelated to xSO4.

Figure 3.4 shows residual plots for the GLS fit. A comparison with the corresponding plots in Figure 3.3 shows a marked improvement: not only has the autocorrelation effectively vanished, but there is also less evidence of heteroscedasticity now. There is still some residual trend, with the dashed lowess smooth rising at the end of the series in Figure 3.4(a). For comparison, however, the corresponding smooth for the baseline model in Table 3.4 is also presented. It is clear that the inclusion of Cl and xSO4 has removed a large part of the trend, albeit without eliminating it entirely.

As an alternative to the use of GLS, we could account for autocorrelation by including the previous quarter's alkalinity in the regression equation. As discussed above, the first observation must be discarded to fit the resulting model, and the interpretation of the coefficients changes. When the model is fitted, the output in Table 3.7 is obtained. In this model, the coefficients of both xSO4 and Cl are almost equal and differ significantly from zero at the 5 % level, as does the coefficient of Y_{t-1} (estimated as 0.259 and labelled as Lag1 in Table 3.7). Some of the other coefficients have changed somewhat by comparison with their values in Tables 3.5 and 3.6, as expected, given that the model has a slightly different interpretation.

Since the lagged response and GLS models yield different conclusions regarding the effect of nonmarine sulfate (at least, if the p-values are interpreted literally at the 5 % level), it is quite important to try and distinguish between the two. This is not possible on the basis of residual plots, which look almost identical for the two models. Furthermore, formal comparison of the models is not possible at present since the first observation could not be used when fitting the lagged response model. The simplest solution is to refit the GLS model without the first observation and to compare the lagged response model to the refitted GLS. Since the models are not nested, we use AIC for the comparison: the lagged response and revised GLS fits yield AIC values of 297.06 and 297.01 respectively. The difference is really too small to be helpful; furthermore, the same number of parameters have been estimated in both models, so no other formal comparison measure will enable us to discriminate between them.

From a scientific point of view, the outcome here is perhaps rather unsatisfactory: the two models fit equally well and, if interpreted literally, lead to conflicting conclusions

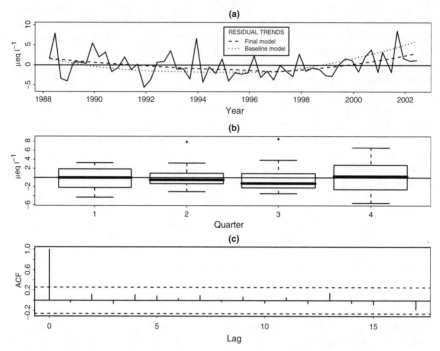

Figure 3.4   Residual plots for the alkalinity model in Table 3.6: (a) residual series, with lowess smooths for this series and for the baseline model residuals, (b) residual distributions by quarter, (c) residual autocorrelation function.

Table 3.7   Edited R output for the model including lagged alkalinity in addition to the covariates in Table 3.5.

	Estimate	Std. Error	t value	Pr(>|t|)	
(Intercept)	-9.34532	1.42554	-6.556	3.24e-08	***
Qtemp	3.31265	0.89050	3.720	0.000513	***
Mtemp	-0.86690	0.99508	-0.871	0.387900	
Mrain	-0.04638	0.48276	-0.096	0.923855	
xSO4 ·	-0.97832	0.45522	-2.149	0.036592	*
Cl	-0.98323	0.43482	-2.261	0.028219	*
Lag1	0.25919	0.12184	2.127	0.038448	*
Mtemp:Mrain	-1.63096	0.53317	-3.059	0.003595	**

```
---
Signif. codes:  0 '***' 0.001 '**' 0.01 '*' 0.05 '.' 0.1

Residual standard error: 3.018 on 49 degrees of freedom
Multiple R-Squared: 0.6099, Adjusted R-squared: 0.5542
F-statistic: 10.95 on 7 and 49 DF, p-value: 3.158e-08
```

regarding the relationship between alkalinity and nonmarine sulfate. We have, however, learned that seasonal variations in alkalinity may be attributed to temperature and rainfall, and that marine deposition plays an important role. The lagged response model suggests that the nonmarine sulfate effect is comparable with that of marine deposition (since the xSO4 and Cl coefficients are almost equal and all covariates were normalised prior to analysis); the GLS fit suggests that marine deposition events are rather more important, and indeed that the effect of nonmarine sulfate may be insignificant at this location. Ultimately, more data are required to discriminate between these two scenarios. As well as continuing monitoring to extend the length of the existing time series, it would be worth considering other covariates that may explain the remaining residual trend in both models. One possible candidate is nitrate, the concentration of which depends on complex catchment processes and which can contribute to the acidity of upland waters (Alliot *et al.*, 1995). ■

3.4   Nonlinear trends

In the linear trend and multiple regression models of the previous sections, we assumed that any trends in the data could either be explained by dependence on covariates or considered as linear or polynomial functions of the time index t. Obviously, although these models are all linear (see the definition in the beginning of this chapter), the trends are not. There are, moreover, other types of trend that cannot be represented using linear models. For example, when a species colonises a new habitat the population growth is typically slow at first, then increases rapidly and finally slows down to a more or less stable level as the capacity of the habitat is reached. The resulting S-shaped curve is often described as *sigmoid(al)*. Consideration of the simplified dynamics of such a situation can suggest particular parametric forms for the trend; for example, if the birth and death rates in a population are linearly dependent on the population size, then the logistic model in Table 3.1 is obtained (Haddon, 2001, Section 2.4). This is a nonlinear model since the parameters enter nonlinearly into the trend function: for all such models, the corresponding trend is also nonlinear. In this section we first consider a general approach to the fitting of nonlinear models and then discuss some specialised techniques for use with individual trend functions.

3.4.1   Nonlinear least squares

In principle, nonlinear models can be fitted using least squares in the same way as linear models providing the usual assumptions hold. The underlying theory is essentially the same as for linear models, except that in the nonlinear case the results are based on large-sample approximations. Standard errors, model comparisons and predictions can be calculated and interpreted in the usual way. The main difficulty is that there is usually no explicit solution for the parameter estimates in the nonlinear case. Minimising the sum of squares must therefore be done numerically. This can be a difficult computational task; however, a variety of reliable algorithms are now available. Some of these are implemented in the nls function in R, which also provides approximate standard errors for the parameter estimates, allows the comparison of nested models using anova and handles prediction. The facilities available for nonlinear modelling in R are extensive and powerful; for a full discussion with plenty of examples, see Venables and Ripley (1999,

Chapter 8). Autocorrelation and heteroscedasticity can be accommodated via the use of devices such as those discussed in Section 3.3, which enable a weighted or generalised least squares problem to be transformed to one involving the minimisation of a simple sum of squares.

In the environmental sciences, explicit parametric models for nonlinear trends tend to be applicable only in rather specialised settings such as the population growth example discussed above. Otherwise, it is usually difficult to justify the use of an explicit parameteric representation of a trend that is markedly nonlinear; such situations may be handled better using the nonparametric methods of Chapter 4. We therefore do not discuss nonlinear least squares further here. For more extensive treatment, the reader is referred to Bates and Watts (1998). An extensive discussion of growth curve models can be found in Piegorsch and Bailer (2005, Section 2.4).

3.4.2   Cycles

Occasionally, it is of interest to investigate cyclical phenomena in environmental time series. The most obvious example is seasonality, which has already been discussed in Section 3.2. Although this could hardly be described as a trend according to the definition in Section 1.1, the basic approach can be extended to any case where the trend is periodic with known cycle length: build regression models including sine and cosine terms at the corresponding frequency, together with harmonics as necessary.

Of course, except for the seasonal cycle, the frequency of any potential periodic structure is generally unknown and must be estimated. Consider the simple sinusoidal model

$$Y_t = \beta_0 + \beta_1 \cos\left[2\pi\,(t - \tau)\,f\right] + \varepsilon_t, \tag{3.54}$$

in which f is the frequency of the trend, corresponding to a cycle length of $1/f$ time units. In this model, f could be estimated using nonlinear least squares. However, for regularly spaced observations an easier way is to use the *periodogram*, which is essentially the line spectrum (see Section 2.3) of the data:

$$I(f) = T\left[A(f)^2 + B(f)^2\right] \tag{3.55}$$

where $A(f) = \dfrac{1}{T}\displaystyle\sum_{t=1}^{T} Y_t \cos 2\pi f t$ and $B(f) = \dfrac{1}{T}\displaystyle\sum_{t=1}^{T} Y_t \sin 2\pi f t.$

Some authors scale the periodogram differently: for example, Chatfield (2003, Section 7.3) multiplies by $T/2\pi$ in place of T in (3.55) and Priestley (1981, Section 6.1) multiplies by $2T$. We use (3.55) for compatibility with the function spec.pgram in R, although the default settings for this function must be changed in order to compute the periodogram as presented here (specifically, the arguments taper=0, detrend=FALSE and fast=FALSE are required).

The periodogram is usually evaluated at the Fourier frequencies $\{f_p = p/T : p = 1, \ldots, \lfloor T/2 \rfloor\}$. For these frequencies, it can alternatively be computed from the sample autocorrelation function (Chatfield, 2003, Section 7.3) and vice versa: the periodogram and the ACF contain equivalent information. It can be shown (Priestley, 1981, Section 3.1) that if $\beta_1 \neq 0$ in model (3.54), the expected value of the periodogram in the neighbourhood of frequency f is proportional to T, whereas for all other frequencies it is approximately σ^2, the variance of the error process. If data are generated according to (3.54), a plot of the periodogram should therefore show a large peak close to the true frequency f. The frequency corresponding to the largest periodogram ordinate can be taken as an estimate of f. If desired, the estimate can be refined using techniques described in Priestley (1981, pages 413–414); see also Hannan (1973) and Quinn (1996). If the underlying trend is periodic but not sinusoidal, the Fourier representation discussed in Section 2.3 guarantees that it can be represented as a sum of sinusoids at frequencies $f, 2f, 3f, \ldots$. In this case the periodogram would be expected to have additional peaks corresponding to the harmonics of f.

To test for the existence of genuine cycles in the model (3.54), if the errors satisfy the usual assumptions then a periodogram based test due to Fisher (1929) can be used. The test statistic is given by

$$g = \max_{p} \left[I\left(f_p\right) \right] / \sum_{p=1}^{\lfloor T/2 \rfloor} I\left(f_p\right),$$

and has a distribution that can be evaluated exactly under the null hypothesis $H_0 : \beta_1 = 0$. For large T, the p-value for this test is approximately $\lfloor T/2 \rfloor \exp(-g\lfloor T/2 \rfloor)$, where $\lfloor T/2 \rfloor$ denotes the integer part of $T/2$.

Example 3.9
In the haddock biomass example from Section 3.1, the residuals from the linear trend model (3.4) showed an apparent cyclical structure. On the basis of the residual ACF in Figure 3.1, a cycle length of around 12 years was suggested. A periodogram analysis offers the potential to investigate this more formally.

Figure 3.5 shows the periodogram of the $T = 38$ residuals from the linear trend model. There is a peak at the third frequency, corresponding to a cycle length of $38/3 = 12.7$ years. The horizontal lines on the plot show the approximate critical values for Fisher's test at 5 % and 1 % levels. Although this test is not strictly applicable here because the 'data' are in fact residuals from a model, we saw in Section 3.1 that the structure induced by fitting this model is relatively weak. The critical values therefore provide a reasonable basis for judging the magnitude of the observed peak. Since this comfortably exceeds the nominal 5 % critical value, we may be fairly confident in rejecting the null hypothesis.

This analysis appears to confirm the presence of a cycle in the residuals with a period of around 12 years. However, closer inspection of Figure 3.5 reveals a problem. Under model (3.54), the expected magnitude of the periodogram ordinates is the same at all frequencies apart from those associated with the periodic structure; however, in this example there is a clear tendency for the magnitudes to decrease with frequency. This cannot be an artefact of fitting a linear trend to the original data, since the effect of this is to remove structure at low, rather than high, frequencies (Section 2.3). It therefore appears that a simple periodic function is not adequate to describe the residual structure of this series. We will return to this example in Chapter 5. ∎

Although periodogram analysis can be useful for detecting genuine periodicities, its capabilities are limited (at least in the form presented here). This is mainly because the test for sinusoidal components is applicable only when the simple model (3.54) is potentially appropriate. Modifications are possible in the presence of autocorrelation; see Priestley (1981, Section 8.3) and references therein. For more complicated structures, for example when there is dependence upon other covariates in addition to the cyclical structure

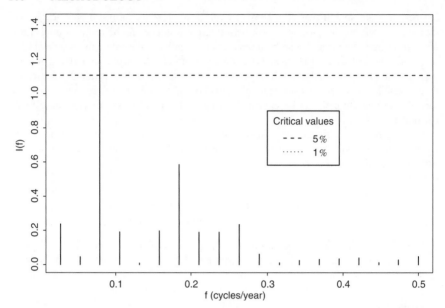

Figure 3.5 Periodogram for residuals from the linear trend model fitted to haddock biomass data. Horizontal lines indicate approximate critical values for Fisher's test of the maximum periodogram ordinate, at the levels specified.

or when these covariates modulate the amplitude or frequency of the oscillation, it will generally be necessary to resort to nonlinear least squares. Carter and Kohn (1997) propose an alternative procedure for identifying cycles in the presence of both autocorrelation and polynomial trends: at a fundamental level their method is based on state space models (see Section 5.5) for the sequence of periodogram ordinates.

3.4.3   Changepoints and interventions

In the statistical literature, a *changepoint* is defined as a point in time at which the properties of a process change abruptly. This may involve a sudden shift in the mean, the variance or in some other more subtle aspect such as the autocorrelation structure. Changepoints may result from known events such as the damming of a river. They may also reflect (possibly undocumented) changes in instrumentation or recording practice, or in the location of a monitoring station: all of these features can give rise to apparent trends that are associated with data artefacts rather than genuine changes in the system of interest (see Yang *et al.*, 2006 for an example of this) and must therefore be accounted for in any analysis. We note in passing that the word 'changepoint' has also been used to mean a point at which a trend is markedly nonlinear (Fewster *et al.*, 2000); this is not the sense discussed here.

We consider first the case when the time of a potential change is known. This may arise when studying the response of a system to some known *intervention*, for example the construction of a new sewage treatment works on a river: here, the potential change-point would be the time at which the intervention was implemented. When the time of a potential changepoint is known, it can be accommodated straightforwardly into a regression framework by defining a dummy covariate that takes the value zero before the changepoint and one afterwards. The coefficient of this covariate in a fitted model is then an estimate of the shift in the mean of the process. Changes in some other aspects can be investigated using interaction terms. For example, a change in the autocorrelation structure can be modelled by including lagged values of the response variable in the model, as discussed in Section 3.3, together with interactions between the lagged responses and the changepoint indicator.

The discussion in the previous paragraph assumes that the effect of an intervention is instantaneous and permanent. One could imagine situations in which this is not the case; for example, an offshore oil spill will have an immediate impact on the surrounding marine ecosystem but full or partial recovery will usually take place over time. It may be possible to represent the effect of such interventions using a plausible parametric model, which can be fitted using nonlinear least squares. In the oil spill example, if Y_t denotes the abundance of some affected species at time t, a natural model may be

$$Y_t = \begin{cases} \mu + \varepsilon_t, & t \le \tau, \\ \mu + \zeta + \xi \exp\left[-\lambda(t - \tau)\right] + \varepsilon_t, & t > \tau, \end{cases}$$

where (ε_t) is a white noise sequence as usual. The unknown parameters in this model are μ, the mean level before the intervention at time τ; ζ, the long-term change in the mean level; ξ, the additional instantaneous change at time τ; and $\lambda > 0$, the rate of recovery after the intervention.

Matters are much more complicated when the presence of a changepoint is suspected but its location is unknown. In this case several statistical approaches are available to test for the existence of one or more changepoints and to estimate their locations if present. A simple procedure is a variant of the Mann–Kendall test discussed in Section 2.4 (Kundzewicz and Robson, 2004). However, it is subject to all of the caveats associated with the use of such tests in general, and a model based approach is more consistent with the spirit of modern statistics. Consider, for example, the *mean-shift model* with unknown changepoint τ:

$$Y_t = \begin{cases} \mu + \varepsilon_t, & t \le \tau, \\ \mu + \zeta + \varepsilon_t, & t > \tau. \end{cases} \tag{3.56}$$

Here, μ and ζ represent the initial mean level and magnitude of the change, as before. If the (ε_t) form a white noise sequence and are, in addition, normally distributed then it can be shown (Hinkley, 1970) that the maximum likelihood estimate of τ is obtained by maximising the function $Z(\tau) = \tau(T - \tau)\left(\overline{Y}_{\le \tau} - \overline{Y}_{> \tau}\right)^2 / T$, where $\overline{Y}_{\le \tau}$ and $\overline{Y}_{> \tau}$ denote respectively the sample means of the observations before and after time τ. Moreover, the function $Z(\tau)$ achieves its minimum value either at one of the observation times or at time $T/2$. Therefore, to estimate the changepoint in model (3.56), all that is required is to calculate the quantities $Z(1), \dots, Z(T)$ and $Z(T/2)$, and to record the time at which the maximum value is achieved.

Unfortunately, the estimation of τ is the only straightforward aspect of model (3.56). The standard theory of statistical estimation breaks down for this and other changepoint models when the time of change is unknown. For example, it is tempting to consider testing for the existence of a changepoint in (3.56) by fitting models both with and without a change, and using a standard F test to compare them. It can be shown that this procedure is not valid: under the null hypothesis, the usual test statistic does not have an F distribution. The reasons for this are very technical; see Hinkley (1970), who also derives approximate procedures for use in this situation. However, these procedures are cumbersome to implement, and modern computer-intensive methods offer a more appealing alternative. Here we summarise one possible approach to testing for a changepoint using the *bootstrap* (see Section 3.6). The approach is essentially that of Julious (2001), who provides a clear introduction to the area. The procedure is as follows:

1. Fit a model without a changepoint; call this model 0.

2. Fit a model with a changepoint (model 1) and store its residuals.

3. Calculate the usual F statistic for comparing the two models.

4. Generate a new series by sampling T values, with replacement, from the residuals calculated in step 2 and add these to the fitted values from model 0.

5. Fit models both with and without a changepoint to the new data set created in step 4; compute the F statistic for comparing these two models.

6. Repeat steps 4 and 5 a large number of times, each time storing the resulting F statistic.

The idea here is that in step 4 we generate an artificial data set that satisfies the null hypothesis of no changepoint and also has residuals with distribution 'as close as possible' to that for the observed data. The distribution of the F statistic under the null hypothesis can therefore be approximated by the empirical distribution of the values computed in step 5, and hence the null hypothesis will be rejected at level α if the F statistic from the observations falls in the upper 100α % of this empirical distribution.

An alternative approach to the changepoint problem is to borrow ideas from the area of industrial quality control, where detection of changepoints is required to ensure, for example, that the quality of manufactured items does not fall below acceptable limits. In this context, changes in the mean level of a process are often signalled using a *cumulative sum* or *cusum* test. The idea is that in the absence of a changepoint, the cumulative sums

$$S_t = \sum_{j=1}^{t} \left(Y_t - \overline{Y} \right) \qquad (t = 1, \dots, T)$$

all have expectation zero. To test for the existence of a changepoint, one can therefore use the largest value of $|S_t|$; the estimate of the changepoint can be taken as the time at which this largest value is obtained, although refinements are possible (Hinkley, 1971). Once again, the distribution theory is rather complicated and bootstrap methods may be appropriate. The implementation of the bootstrap here is exactly the same as that described above, except that for each bootstrap sample we compute the cusum statistic $\max |S_t|$ instead of the F statistic.

The cusum test just described is suitable for retrospective detection of changepoints from a sequence of observations. In some applications, such as environmental monitoring, online detection is more relevant, in order that appropriate action can if possible be taken as soon as a change is detected. Indeed, online detection is much closer to the original purpose of cusum schemes in industrial applications. A brief introduction to the idea, with references, has been given in Section 2.4.

The literature on the changepoint problem is vast and often very technical, and we can only hope to give a brief introduction here. For a useful and cautionary review of changepoint detection methods in environmental applications, see Jarušková (1997). Another review, with an ecological focus, is given by Ficetola and Denoel (2009). Julious (2001) and Muggeo (2003) consider the problem of fitting simple regression models described by a set of piecewise linear sections (here the changepoints correspond to the joins, where the slope of the regression relationship changes abruptly). The latter author provides an R package `segmented` that is able to fit such models efficiently, although it is not clear that the standard errors and model comparison statistics reported by this package are correct since they rely on assumptions that do not hold in the changepoint estimation problem: bootstrap procedures may be preferable. The breakdown of classical theory in changepoint problems is widely underappreciated: see Lund and Reeves (2002), who also propose procedures for the detection of multiple changepoints in a single series. Sequential detection of multiple changepoints is also considered by Menne and Williams (2005) and by Tomé and Miranda (2004). The latter authors consider detection of changes in the slope of a linear trend, which amounts to fitting a piecewise linear trend function; they do not consider formal techniques for assessing the number of changes, however. Inclán and Tiao (1994) consider retrospective detection of changes in variance, using extensions of the cusum scheme. Range charts can be used for online detection of changes in variability (Manly, 2001, Chapter 5). Changepoint analysis in the presence of autocorrelation is considered by Garisch and Groenewald (1999), who apply their techniques to river flow data from the Nile. Finally, Fried and Imhoff (2004) develop an interesting method, based on ideas similar to the cusum scheme, for online detection of a changepoint where an autocorrelated process starts to drift away from a steady state.

3.5   Generalised linear models

So far in this chapter we have dealt exclusively with models based on the normal distribution. We now consider an extended class of regression models in which this assumption is relaxed. Generalised linear models (GLMs) have become a cornerstone of much statistical practice since their introduction by Nelder and Wedderburn (1972). The starting point is to note that the multiple regression model (3.34) can be regarded as a model for the distribution of Y_t given the covariates at time t:

$$Y_t \sim N\left(\mu_t, \sigma^2\right), \qquad \text{where} \quad \mu_t = \beta_0 + \sum_{i=1}^{p} \beta_i x_{it} = \eta_t, \text{ say.}$$

The generalisation is firstly to allow the conditional distribution to be non-normal and secondly to allow a nonlinear relationship between the mean μ_t and the *linear predictor*, η_t. Thus, in a GLM, the equivalent of the multiple regression model is to say that each observation is drawn from the same type of distribution (for example normal, Poisson or gamma) and that, given $x_t = t$, Y_t has mean μ_t with

$$g\left(\mu_t\right) = \beta_0 + \sum_{i=1}^{p} \beta_i x_{it} = \eta_t \qquad (3.57)$$

for some monotonic function $g(\cdot)$ called the *link function*. The usual regression model is thus a GLM with normal distributions and an identity link function: $g(\mu_t) = \mu_t$. Other commonly used GLMs include the loglinear Poisson model for count data and the logistic regression model for binary data. In the former, Y_t is taken to have a Poisson distribution with mean μ_t and a logarithmic link function is used:

$$\log \mu_t = \eta_t \qquad \Leftrightarrow \qquad \mu_t = \exp(\eta_t).$$

Logistic regression is used for binary-valued data, coded (without loss of generality) as 0 and 1. In this case, μ_t is just the probability that Y_t takes the value 1, and a logistic link function is used:

$$\log\left(\frac{\mu_t}{1 - \mu_t}\right) = \eta_t \qquad \Leftrightarrow \qquad \mu_t = \left[1 + \exp(-\eta_t)\right]^{-1}. \tag{3.58}$$

This should not be confused with the logistic growth model discussed in Section 3.4. Both models involve the use of a logistic transform, but are otherwise unrelated.

In these examples, the means $\{\mu_t\}$ can always be recovered from the linear predictors $\{\eta_t\}$, because the link function is monotonic. The use of a link function is superficially similar to data transformation; the important distinction is that in a GLM the transformation applies to the modelled means rather than to the data themselves.

In principle, any sufficiently 'well-behaved' distribution could be used within this model structure. However, the distributions in GLMs are restricted to belong to the *exponential family*, which includes many common distributions such as the normal, Poisson, binomial,[2] exponential and gamma. This restriction is made primarily for mathematical and computational convenience, since the exponential family has many appealing properties. Some of these are indicated below; for further details, see McCullagh and Nelder (1989) or Dobson (2001). An important feature is that for any distribution in the exponential family, the variance and mean are related: specifically, if the mean is μ then the variance is

$$\sigma^2 = \psi V(\mu), \tag{3.59}$$

where $V(\cdot)$ is called the *variance function* of the distribution and ψ is a *dispersion parameter*. The glm function in R has built-in facilities for modelling using several different distributions; these, along with their variance functions, are given in Table 3.8.

Table 3.8   Means, variance functions and dispersion parameters for some common distributions in the exponential family.

Distribution	μ	$V(\mu)$	ψ	Comment
Normal (μ, σ^2)	μ	1	σ^2	Constant variance
Binomial (n, p)	p	$p(1 - p)$	n^{-1}	–
Poisson (μ)	μ	μ	1	Variance equal to mean
Gamma (α, λ)	α/λ	μ^2	$1/\alpha$	Constant coefficient of variation
Inverse Gaussian (μ, λ)	μ	μ^3	λ^{-1}	–

[2] In basic statistics courses, the binomial distribution is usually introduced as a model for counts. In the context of exponential families, however, it is mathematically more convenient to work with the corresponding proportions, so the binomial is used as a distribution for data taking values between 0 and 1. Notice, for example, that in Table 3.8 the mean of the distribution is given as p rather than the more familiar np.

Most of these distributions will be familiar to readers, with the possible exception of the inverse Gaussian: this arises when considering the time taken for a one-dimensional Brownian motion (see Section 5.4) to reach a specified threshold. It has rarely been used in environmental applications, with the exception of Freidlin and Pavlopoulos (1997), who suggested on the basis of physical considerations that it could be appropriate for modelling durations of rain events and the intervening dry periods.

Applications of GLMs in environmental trend analysis have so far been relatively limited. A notable exception is the work of Frei and Schär (2001), who used logistic regression to examine trends in numbers of extreme rainfall events. Loglinear Poisson models have also been used to model trends in species abundance (see Fewster *et al.*, 2000, for a review), although in this type of application it is usually necessary to adopt a nonparametric representation of the trend function. This can be achieved within the framework of generalized additive models (GAMs), which are described in Chapter 4; examples of their application include Fewster *et al.* (2000) and Freeman, Baillie and Gregory (2001).

3.5.1   Parameter estimation

GLMs are usually fitted using the method of maximum likelihood, which was introduced in Section 3.1.6. The parameters to be estimated are $\boldsymbol{\beta}$, the vector of coefficients in the linear predictor, along with the dispersion parameter ψ. Let $f_t(\cdot; \mathbf{x}_t, \boldsymbol{\beta}, \psi)$ denote the probability density function of Y_t (or probability mass function if Y_t is discrete), given the associated vector of covariates $\mathbf{x}_t$ and parameters $(\boldsymbol{\beta}, \psi)$. Then, if successive observations are independent, their joint density is $\prod_{t=1}^{T} f_t(y_t; \mathbf{x}_t, \boldsymbol{\beta}, \psi)$ and the log-likelihood function is

$$\ell(\boldsymbol{\beta}, \psi; \mathbf{y}) = \sum_{t=1}^{T} \log f_t(y_t; \mathbf{x}_t, \boldsymbol{\beta}, \psi). \tag{3.60}$$

Maximising this expression yields the maximum likelihood estimators (MLEs) of $\boldsymbol{\beta}$ and ψ.

For the multiple regression model (3.34), the dispersion parameter is $\psi = \sigma^2$. The results in Section 3.1.6 show that, for this model, the MLE of $\boldsymbol{\beta}$ is the same as the least squares estimator and does not depend on the value of the dispersion parameter. This is a feature of all GLMs and is due to the use of distributions in the exponential family. Indeed, it can be shown (Dobson, 2001, Section 4.3; McCulloch, Searle and Neuhaus, 2008, Section 5.4) that in any GLM the maximum likelihood estimate of $\boldsymbol{\beta}$ satisfies an equation of the form

$$\left[X'WX\right]\hat{\boldsymbol{\beta}} = X'W\mathbf{z}, \tag{3.61}$$

where $\mathbf{z}$ is a $T \times 1$ vector with typical element $z_t = \eta_t + (y_t - \hat{\mu}_t)g'(\hat{\mu}_t)$ and W is a diagonal matrix with typical element $w_{tt} = [V(\hat{\mu}_t)g'(\hat{\mu}_t)^2]^{-1}$. As usual, $\hat{\mu}_t = g^{-1}(\hat{\beta}_0 + \sum_{i=1}^{p} \hat{\beta}_i x_{it})$ is the fitted value at time t. Notice that (3.61) is almost identical to the weighted least squares equation (3.44); superficially, the only difference is that the data vector $\mathbf{y}$ has been replaced by a vector of *adjusted dependent variates*, $\mathbf{z}$. However, in (3.61) the elements of W depend on $\hat{\boldsymbol{\beta}}$, whereas in (3.44) they were known. This suggests an iterative method of fitting: start with some plausible estimates of the fitted values $\{\hat{\mu}_t\}$, solve (3.61) to obtain an estimate of $\boldsymbol{\beta}$, recalculate the fitted values using this estimate and continue until the estimate of $\boldsymbol{\beta}$ has stabilised. This *iteratively (re)weighted least squares*

(IWLS) algorithm is used by most software packages to fit GLMs; the observations $\{y_t\}$ themselves are usually used as estimates of the fitted values in the first step.

Having estimated the regression coefficients, we turn to the dispersion parameter ψ. For some distributions, such as the Poisson (see Table 3.8), this is known; in other cases it must be estimated. In principle, maximum likelihood could be used here as well. Unfortunately, however, maximum likelihood estimators of dispersion parameters can be biased in small samples and can be sensitive to, for example, rounding errors in small observations (McCullagh and Nelder, 1989, Section 8.3). It is therefore usual to estimate the dispersion parameter by a *method of moments*, equating the sample and theoretical variance of suitably standardised residuals:

$$\hat{\psi} = \frac{1}{T-k} \sum_{t=1}^{T} \frac{(y_t - \hat{\mu}_t)^2}{V(\mu_t)}.$$

In the normal linear model, this expression reduces to (3.21) since, in this case, $\eta_t = \mu_t$ and $V(\mu_t) = 1$. The maximum likelihood estimator in this case has divisor T in place of $T - k$ (this was shown in Section 3.1.6 for the linear trend model).

Once the dispersion parameter has, if necessary, been estimated, the covariance matrix of the estimated regression coefficients can be estimated as $\hat{\psi}\left[X'WX\right]^{-1}$. Tests of hypotheses about individual coefficients can then be constructed just as for multiple regression models. In this particular instance, the form of the covariance matrix follows from the weighted least squares form of the estimates (see Section 3.3). However, it is in fact a specific example of the more general result given in Section 3.1.6 that in large samples the covariance matrix of the MLE is approximately equal to the inverse of the expected information matrix J: it can be shown that $J = X'WX/\psi$ for the regression coefficients in a GLM. The approximate large-sample nature of the result is worth noting, however, as with many other models that have been discussed so far, the approximations are often very accurate, but are usually justified on the basis of assumptions that do not always hold in a trend analysis. Fortunately, alternative justifications are available in many situations of practical interest. We conclude this section with a brief outline of the issues involved.

The usual derivation of the large-sample approximation for an MLE (see, for example, Davison, 2003, Section 4.4 and Cox, 2006, Chapter 6) requires, among other things, that in a neighbourhood of the true parameter value the expected information matrix converges to a constant positive definite matrix upon suitable normalisation, usually after dividing by T. In a GLM, the (i, j)th element of the expected information $X'WX/\psi$ is $\sum_{t=1}^{T} w_{tt}x_{ti}x_{tj}/\psi$. If one of the covariates is the time index t, and others represent nontrending environmental variables, different elements of the information matrix may have different orders of magnitude (see, for example, Example 3.4 on the linear model, where the information matrix is proportional to $X'X$). In this case no simple normalisation will yield the required convergence. A resolution to the problem can be found in the results of Sweeting (1980), who provides much more general conditions under which the usual large-sample approximations can be justified.[3] These conditions usually hold when some covariates are functions of the time index, providing the expected information increases without bound as more data are collected. This justifies the use of standard likelihood based procedures in many trend analyses. There are, however, known situations in which

---

[3] There is a typographical error in Sweeting (1980): at the bottom of page 1379, his condition $\mathrm{Var}\{\mathcal{I}_t(\theta)\} \to_u 0$ should read $\mathrm{Var}\{A_t(\theta)^{-1}\mathcal{I}_t(\theta)A_t(\theta)^{-1}\} \to_u 0$.

the standard theory does not apply, even in the case of the multiple regression model where no approximations are usually involved. These situations, which are sometimes referred to as *spurious regressions*, are discussed further in Chapter 5.

3.5.2   Model comparison

In Section 3.2, we saw that sequences of nested linear models could be compared using F tests based on their residual sums of squares. A similar idea can be applied when using GLMs in large enough samples. The starting point is to note, from (3.32), the connection between the log-likelihood and the residual sum of squares in the multiple regression model. In a GLM, the role of the residual sum of squares is taken by the *deviance*, defined as

$$D = 2\psi \left[\ell_{\text{FULL}}(\mathbf{y}) - \ell(\boldsymbol{\beta}, \psi; \mathbf{y}) \right], \tag{3.62}$$

where $\ell_{\text{FULL}}(\mathbf{y})$ is the log-likelihood for a model that fits the data perfectly, i.e. in which $\mu_t = y_t$. Like the sum of squares (to which it corresponds in the case of the multiple regression model), D is always non-negative and smaller values correspond to better-fitting models. When the dispersion parameter ψ is unknown, F tests can be used to compare nested models in exactly the same way as for the multiple regression case; all that is required is to substitute deviances for sums of squares in (3.41). The procedure is referred to as *analysis of deviance*, but is implemented in R via the `anova` command.

In general, the use of F tests to compare models is necessary to account for an unknown dispersion parameter (Krzanowski, 1998, Section 5.5). In some GLMs, however, the dispersion parameter has a known value – for example, from Table 3.8, the dispersion parameters for the binomial and Poisson distributions are $1/n$ and 1 respectively. In such situations, suppose we wish to compare two models $\mathcal{M}_0$ and $\mathcal{M}_1$, say, where $\mathcal{M}_1$ is obtained by adding m extra terms to the linear predictor for $\mathcal{M}_0$. A test of the null hypothesis that the data were generated from $\mathcal{M}_0$ can be carried out by comparing the *scaled deviances* for the two models. The scaled deviance is defined as

$$D^{(s)} = D/\psi = 2 \left[\ell_{\text{FULL}}(\mathbf{y}) - \ell(\boldsymbol{\beta}, \psi; \mathbf{y}) \right]. \tag{3.63}$$

Under the null hypothesis, the difference in scaled deviances, $D_0^{(s)} - D_1^{(s)}$, say, has approximately a chi-squared distribution with m degrees of freedom (Davison, 2003, Section 10.2.1). The `anova` command in R can be used once again. Notice from (3.63) that the difference $D_0^{(s)} - D_1^{(s)}$ can equivalently be written as

$$2(\ell_1 - \ell_0), \tag{3.64}$$

where ℓ_0 and ℓ_1 are the maximised log-likelihoods for the respective models. In this form, the difference in scaled deviances is often called the *(log-)likelihood ratio statistic*. In fact, its use is not restricted to GLMs: in large samples, its null distribution remains approximately χ_m^2 when comparing nested models fitted by maximum likelihood in a very wide range of situations.

For the comparison of non-nested models, methods such as the AIC can be used for GLMs in the same way as for multiple regression models. Model comparisons made using AIC or related criteria are effectively comparisons of log-likelihoods (recall the definition of AIC in Section 3.2.3); hence there is an operational connection with tests based on the likelihood ratio statistic, even though the arguments underlying the procedures are rather different. In principle, criteria such as AIC can also be used to compare models involving different classes of distribution (for example to determine whether a normal or gamma distribution fits the data better). However, when computing log-likelihoods (and hence AIC values), some software packages may omit constant terms that are irrelevant when comparing models involving the same class of distributions. For models involving different classes, the inclusion of such terms is critical. In such situations, therefore, to be on the safe side it may be worth computing the required log-likelihoods explicitly, by summing the respective log densities.

3.5.3   Model checking

As in the normal linear model, most diagnostics for GLMs are based around residual plots, together with the analogues of influence measures such as the leverage and Cook's distance. The main differences are that the concept of a residual is not uniquely defined in a GLM and that residual plots can be uninformative for highly discrete data. For most models, the *raw residuals* $\{Y_t - \mu_t\}$ are of limited value since (for example) they all have different variances. A natural solution is to standardise each raw residual to have the same variance under the model. This is the idea behind *Pearson residuals*. The Pearson residual at time t is defined as

$$r_t^{(P)} = \frac{Y_t - \hat{\mu}_t}{\sqrt{V(\hat{\mu}_t)}}.$$

If the fitted model is correct, the Pearson residuals should appear uncorrelated, with zero mean and constant variance. Plots such as those in Section 3.1.1 can often be used to provide an informal assessment of this.

An alternative class of residuals for GLMs is based on the analogy between the deviance and the residual sum of squares. From (3.60) and (3.62), it is clear that the deviance is a sum of terms; furthermore, it can be shown that these terms are all non-negative. The *deviance residuals* are obtained by taking the square root of each term and attaching a plus or minus sign as appropriate, depending on whether the observation is greater or smaller than the fitted value (McCullagh and Nelder, 1989, Section 2.4.3). Thus the sum of squared deviance residuals is just the deviance. The idea is arguably somewhat artificial and deviance residuals are perhaps less easily interpreted than Pearson residuals; however, for diagnostic purposes they can be used in exactly the same way. For an overview of different types of residual for GLMs, see Piegorsch and Bailer (2005, Section 3.2.3) and references therein, notably Jørgensen (2002). In R, the `plot.glm` command produces a variety of diagnostics based on deviance residuals.

If the data are highly discrete, residual plots can be difficult to interpret. With binary-valued data, for example, residual plots appear as two curves, corresponding respectively to the zeros and ones in the data. In this case it may be more helpful to construct residuals based on aggregated data. Suppose, for example, that logistic regression has been used to model the monthly sightings of a species over a period of several years and that the analyst wants to check that the residuals do not show any trend. Let p_{my} be the modelled probability of a sighting in month m of year y and let Y_{my} be the corresponding 0/1 observation. Then, rather than studying individual residuals, one could aggregate to an annual timescale. Under the model, the expected number of months with sightings in year y is

$$E \sum_{m=1}^{12} Y_{my} = \sum_{m=1}^{12} p_{my}$$

and the variance of this sum is

$$\sum_{m=1}^{12} p_{my}(1 - p_{my}).$$

The quantities

$$\frac{\sum_{m=1}^{12}\left(Y_{my} - p_{my}\right)}{\sqrt{\sum_{m=1}^{12} p_{my}(1 - p_{my})}}$$

therefore have mean zero and variance 1 under the model; a time series plot of these quantities may be easier to interpret than a plot of individual monthly residuals. For more discussion of the construction and interpretation of diagnostics in models for binary data, see Cox and Snell (1989, Section 2.7).

3.5.4 Prediction with GLMs

Extrapolation using a GLM is similar to that for a multiple regression model (Section 3.1.4). Given an estimated parameter vector $\hat{\boldsymbol{\beta}}$ and a vector $\mathbf{x}_{T+\ell}$ of covariates at future time point $T + \ell$, the corresponding linear predictor can be estimated as $\hat{\eta}_{T+\ell} = \mathbf{x}'_{T+\ell}\hat{\boldsymbol{\beta}}$. With known dispersion parameter ψ, an approximate 95 % confidence interval for this linear predictor is

$$\hat{\eta}_{T+\ell} \pm 1.96\psi\sqrt{\mathbf{x}_{T+\ell}\left(X'WX\right)^{-1}\mathbf{x}'_{T+\ell}}.$$

The estimate $\hat{\eta}_{T+\ell}$, together with the endpoints of the confidence interval, can then be transformed back to the scale of the original observations to obtain an approximate 95 % confidence interval for the underlying mean at time $T + \ell$. Usually this interval will not be symmetric; this is due to the nonlinear relationship between the linear predictor $\eta_{T+\ell}$ and the mean $\mu_{T+\ell}$.

The construction of prediction (rather than confidence) intervals is more difficult for GLMs than for multiple regression models, however. This is because when $Y_{T+\ell}$ is predicted using $\hat{\mu}_{T+\ell}$, the two sources of error $\hat{\mu}_{T+\ell} - \mu_{T+\ell}$ and $Y_{T+\ell} - \mu_{T+\ell}$ have different types of distribution; the former is approximately normal whereas the latter depends on the model. In general, it is difficult to derive useful expressions for the quantiles of a sum of normal and non-normal variables (in the multiple regression case, both sources of error have normal distributions and so their sum is also normal). For this reason, the predict.glm function in R produces extrapolations of the linear predictor together with associated standard errors, but does not attempt to produce prediction intervals. If such intervals are desired, a crude approach is simply to proceed as though there is no estimation error and to calculate the appropriate quantiles of the chosen family of distributions with mean $\hat{\mu}_{T+\ell}$ and dispersion $\hat{\psi}$. If the standard error of $\hat{\eta}_{T+\ell}$ is small relative to the standard deviation of $Y_{T+\ell}$ (or if $Y_{T+\ell}$ is highly discrete so that prediction intervals have limited interpretability in any case), this may be perfectly adequate. Failing this, other methods must be used. As with the changepoint problem discussed in Section 3.4, bootstrap methods offer one solution. A simple bootstrap algorithm for obtaining prediction intervals is as follows:

1. Generate a new sample $Y_1^*, \ldots, Y_T^*$ by simulating random numbers from the appropriate family of distributions, with respective means $\hat{\mu}_1, \ldots, \hat{\mu}_T$ and dispersion parameter $\hat{\psi}$.

2. Refit the model to the new sample and use the parameter estimates from this refitted model to construct an estimate of the underlying mean at time $T + \ell$; call this estimate $\hat{\mu}_{T+\ell}^*$.

3. Simulate a large number (M, say) of observations from the chosen family of distributions, with mean $\hat{\mu}_{T+\ell}^*$ and dispersion parameter $\hat{\psi}$.

4. Repeat steps 1 to 3 N times, where N is another large number.

The result of this algorithm is a set of NM simulated observations, from which an empirical distribution can be constructed. Prediction intervals can be taken from the appropriate percentiles of this distribution. Like the bootstrap algorithm for the changepoint problem, the method works by generating artificial data sets with (hopefully) similar structure to the observations. Uncertainty in the estimation of the model parameters is accounted for by refitting the model repeatedly to N different artificial data sets, whereas the additional random error at time $T + \ell$ is accounted for by generating M realised values of $Y_{T+\ell}$ for each fitted model. Notice, however, that the success of the method depends on the artifical data being 'sufficiently similar' to the original observations. If this is not the case, a more sophisticated algorithm such as that described in Davison and Hinkley (1997, Section 7.2) may be required.

3.5.5   Extensions and refinements

We conclude this discussion of GLMs by considering some extended features that are often useful. The first follows from the observation that, according to Equation (3.61), the parameter estimates in a GLM depend on the underlying distribution only through the variance function $V(\mu)$, which enters via the weight matrix W. This suggests that the precise form of the distribution is relatively unimportant and hence that in any situation where the variance appears related to the mean, models could be fitted *as if* the data were generated from the corresponding exponential family distribution. The idea is due to Wedderburn (1974), who coined the term *quasi-likelihood* to describe it. The procedure delivers valid estimators, and confidence intervals and hypothesis tests for the regression coefficients can be interpreted in the usual way. Depending on the variance function used, quasi-deviances can also be constructed for comparing nested models; however, these are not always uniquely defined (Davison, 2003, Section 10.6). Models fitted using quasi-likelihood can also be used to extrapolate the underlying mean of a process into the future, as described in the previous subsection; however, it is of course not possible even in principle to calculate prediction intervals without being more specific about the form of the distribution.

In Section 3.3, we identified four categories of heteroscedasticity in multiple regression models. The second of these involved a direct relationship between the variance and the mean; clearly quasi-likelihood offers a way of dealing with this, as an alternative to data transformation. The `glm` command in R is able to fit models using quasi-likelihood, and offers the facility for the user to define their own variance functions if the inbuilt ones are not adequate.

The only form of heteroscedasticity that has not yet been addressed is that in which the variance of Y_t appears systematically related to other factors. GLMs can also be used to accommodate this (McCullagh and Nelder, 1989, Section 10.2), by writing the multiple regression model (3.34) as

$$Y_t = \mu_t + \varepsilon_t = \mu_t + \sigma_t Z_t, \text{ say,}$$

where Z_t is a standard normal random variable and σ_t depends on covariates. Noting that Z_t^2 has a chi-squared distribution on one degree of freedom, which is the same as a gamma distribution with both parameters equal to $1/2$ (Rice, 2006, Section 2.3), we deduce that $U_t = \sigma_t^2 Z_t^2$ is distributed as $\Gamma\left(1/2, 1/2\sigma_t^2\right)$. This gamma distribution is in the exponential family (see Table 3.8) and has the expected value σ_t^2. Given the sequence $\{U_t\}$, we could therefore fit a gamma GLM relating some function of σ_t^2 to the underlying covariates (call this the 'variance model'); the fitted values would then estimate the variances of the individual observations. The regression coefficients for the means $\{\mu_t\}$ (the 'mean model') could then be estimated via weighted least squares, using the estimated variances to construct the weights. In practice, of course, we cannot observe $\{U_t\}$, so we use the residuals from an ordinary least squares fit of the mean model as preliminary estimates. After refitting using weighted least squares, these residuals will obviously change, whence it is necessary to refit the variance model. Model fitting therefore alternates between weighted least squares fits of the mean model and gamma GLM fits of the variance component, until convergence is achieved. This procedure has been applied to monthly temperature time series by Chandler (2005) and to daily and weekly series of potential evapotranspiration by Yang et al. (2005).

As with multiple regression models, autocorrelation in GLMs can be handled either directly (by analogy with generalised least squares) or by including lagged values of the response variable as covariates in the model. The *generalised estimating equation* (GEE) framework of Liang and Zeger (1986) provides the analogue of generalised least squares; short introductions can be found in Dobson (2001, Section 11.4) and McCulloch, Searle and Neuhaus (2008, Section 9.3). Most applications of GEEs have been to clinical trials in which repeated measurements are made on each of a sample of patients; in these applications, measurements from the same patient tend to be correlated but those from different patients are usually independent. This data structure is implicit in much of the GEE literature; specifically, it is common to refer to independent 'clusters' of measurements. In the current context, when analysing a single time series we effectively have just one cluster, but the methodology can still be used providing the sample size is large enough relative to the strength of the autocorrelation. GEEs can be fitted in R using the geepack library (Yan, 2002).

When fitting models using maximum likelihood, the justification for dealing with auto-correlation by including lagged responses is that in the presence of dependence between successive observations, the joint density of the observations can always be written as a product of conditional densities:

$$\prod_{t=1}^{T} f_t\left(y_t | y_1, \ldots, y_{t-1}; \mathbf{x}_t, \boldsymbol{\beta}, \psi\right), \text{ say.}$$

This result follows from the definition of a conditional density; see Rice (2006, Section 3.5) for example. The log-likelihood for $\boldsymbol{\beta}$ and ψ can therefore be written as a sum of terms, each conditioned on all previous observations:

$$\ell\left(\boldsymbol{\beta}, \psi; \mathbf{y}\right) = \sum_{t=1}^{T} \log f_t\left(y_t | y_1, \ldots, y_{t-1}; \mathbf{x}_t, \boldsymbol{\beta}, \psi\right). \tag{3.65}$$

Compare this with the independence case (3.60). The conditioning can usually be achieved by incorporating a few previous values into the linear predictor; often, including just

Y_{t-1} will be enough to remove most of the residual autocorrelation. Consideration of measurement units suggests that previous values should be transformed to the same scale as the linear predictor where possible: our experience is that this usually brings substantial benefits in terms of model fit. For example, in a loglinear model, since the linear predictor is $\eta_t = \log \mu_t$, $\log Y_{t-1}$ is a more natural covariate than Y_{t-1}. Of course, if some of the observations are zero this cannot be done; in this case, an alternative such as $\log(1 + Y_{t-1})$ may be considered. For more details of this approach to handling autocorrelation in GLMs, see Chapter 8.

3.6   Inference with small samples

With the exception of the standard multiple regression model, the methods introduced in this chapter are justified on the basis of large-sample approximations. When these approximations are not sufficiently accurate, computer-intensive methods often provide a feasible alternative. One of the most popular modern techniques is the bootstrap. This has already been introduced in Sections 3.4 and 3.5 and is in fact a collection of related techniques, all of which involve the use of artifical data sets generated in such a way as to resemble the observations with respect to carefully chosen features. These artificial data sets all involve resampling from the observed data, or from models fitted to the observed data: the name 'bootstrap' derives from the phrase 'to pull oneself up by one's bootstrap' (Efron and Tibshirani, 1993, page 5). Hypothesis tests, confidence intervals and prediction can all be handled within the bootstrap framework. We give only a brief overview of the main ideas here; in particular, we do not discuss the construction of bootstrap confidence intervals. More details can be found in Efron and Tibshirani (1993) and Davison and Hinkley (1997). The latter authors give a very clear and readable account, supported in R by the `boot` library (Canty and Ripley, 2008).

The most easily implemented of the bootstrap techniques is the *parametric* or *model based* bootstrap, in which artificial data sets are generated from the fitted model itself. The bootstrap prediction algorithm for GLMs in Section 3.5 is an example. Hypothesis testing can also be performed using the parametric bootstrap, as follows:

1. Calculate a statistic for testing the null hypothesis in the observed data. This may be, for example, a t statistic for a coefficient in a regression model or an F statistic for comparing nested models.

2. Fit a model corresponding to the null hypothesis of interest (for example a regression model or GLM in which a particular covariate is omitted), if this has not already been done in step 1.

3. Use the fitted model from step 2 to generate a large number, N say, of artificial data sets satisfying the null hypothesis. For each of these data sets, calculate the same test statistic as in step 1. The result will be a collection of N test statistics, each obtained under the null hypothesis.

4. Compare the observed value of the test statistic with the quantiles of the distribution obtained from the N values in step 3, to obtain a p-value for testing the null hypothesis.

Strictly speaking, the p-value obtained in step 4 above is an estimate of the actual p-value for the test. In practice it is just the proportion of the bootstrap samples yielding a test statistic that is more extreme (in a sense determined by the precise testing situation) than that observed. The standard error of this proportion is $\sqrt{p(1-p)/N}$, where p is the actual p-value. This can be used to determine an appropriate value for N. For example, choosing $N = 10\,000$ will deliver a standard error of $\sqrt{p(1-p)}/100$ so that the estimate can be considered accurate to two decimal places; this should be adequate for most applications.

The bootstrap test procedure assumes that any significant differences between the observations and artificial data sets are due to violation of the null hypothesis in the observations. This is a strong assumption, which can only be made if the model provides an adequate representation of the data; for example, if the residuals from a multiple regression model are not normally distributed, the artifical data sets generated in step 3 will not resemble the original data. Further, autocorrelation is always a potential difficulty in any trend analysis; artificial data sets must be generated so as to reproduce realistically any autocorrelation in the observations. For multiple regression models, this can be achieved either by generating autocorrelated error sequences from a model such as the AR(1) (see Section 3.3.3), or by including previous Y values in the regression model and generating artificial data sets sequentially, at each stage conditioning on the values already generated. For GLMs, the notion of an error sequence is less well defined and the latter approach is easier to implement.

In cases where the parametric bootstrap is inappropriate, perhaps because of concerns over the distributional assumptions of the model, a *nonparametric bootstrap* may be used instead. The main difference between parametric and nonparametric bootstraps is the way in which the artificial data sets are generated. For regression models, the nonparametric version works by sampling artificial error sequences and adding these to the 'structural' part of a fitted model (i.e. the estimated means $\hat{\mu}_1, \ldots, \hat{\mu}_T$). The error sequences are sampled from the model residuals (possibly standardised as defined by Equation (3.27) in Section 3.1.3), with replacement; the procedure is 'nonparametric' in that no explicit form is assumed for the error distribution. The bootstrap approach to changepoint testing in Section 3.4 is an example. The justification for the procedure is essentially that if one is not prepared to make any assumptions about the distribution from which the residuals are drawn, the best estimate of this distribution is that of the residuals themselves. The idea is, however, difficult to apply to GLMs, again because of the difficulty in defining useful residuals; see Davison and Hinkley (1997, Section 7.2) for some possible approaches.

Autocorrelated residuals are also difficult to handle within the nonparametric bootstrap framework. One possible option is to generate new error sequences by sampling blocks of successive residuals, rather than sampling one at a time. The result is the *block bootstrap*. Some care is required with the use of such techniques, however. We do not discuss them further here; for a thorough discussion of bootstrap methods for time series data, see Davison and Hinkley (1997, Section 8.2). Of course, in models where autocorrelation is handled by including lagged Y values as covariates, there is no problem since the residuals from such models should be uncorrelated in any case.

The bootstrap is not the only way to proceed when standard approximations break down. An alternative procedure, at least for hypothesis testing, is based on randomly reordering one or more components of the data so as to generate artifical data sets in which the null hypothesis is known to hold because any relationship among the components has been broken by the random reordering. As with the bootstrap, any test statistic computed from the observations can then be compared with the distribution of test statistics from the artificial data sets. A brief but clear overview of such *permutation tests* can be found in Faraway (2005, Section 3.3).

References

Akaike, H. (1974) A new look at the statistical model identification. *IEEE Transactions on Automatic Control*, **AC-19**(6), 716–723.

Alliot, T. E. H., Curtis, C. J., Hall, I., Harriman, R. and Battarbee, R. W. (1995) The impact of nitrogen deposition on upland freshwaters in Great Britain: a regional assessment. *Water, Air and Soil Pollution*, **85**, 297–302.

Azzalini, A. (1996) *Statistical Inference Based on the Likelihood*. Chapman & Hall, London.

Bates, D. M. and Watts, D. G. (1988) *Nonlinear Regression Analysis and Its Applications*. John Wiley & Sons, Inc., New York.

Belitz, C., Brezger, A., Kneib, T. and Lang, S. (2009) BayesX–*Software for Bayesian Inference in Structured Additive Regression Models*. Version 2.0.1. Available from http://www.stat.uni-muenchen.de/~bayesx.

Brockwell, P. J. and Davis, R. A. (2002) *Introduction to Time Series and Forecasting*, 2nd edition. Springer-Verlag, New York.

Buckland, S. T., Anderson, D. R., Burnham, K. P., Laake, J. L., Borchers, D. L. and Thomas, L. (2004) *Advanced Distance Sampling*. Oxford University Press, Oxford.

Canty, A. and Ripley, B. (2008) boot: *Bootstrap R (S-Plus) Functions*. R package version 1.2-34.

Carter, C. K. and Kohn, R. (1997) Semiparametric Bayesian inference for time series with mixed spectra. *Journal of the Royal Statistical Society, Series B*, **59**, 255–268.

Chandler, R. E. (1998) Orthogonality. In *Encyclopedia of Biostatistics* (eds. P. Armitage and T. Colton). John Wiley & Sons, Ltd, Chichester. pp. 3203–3209.

Chandler, R. E. (2005) On the use of generalized linear models for interpreting climate variability. *Environmetrics*, **16**, 699–715.

Chatfield, C. (2000) *Time-Series Forecasting*. Chapman & Hall/CRC, Boca Raton, Florida.

Chatfield, C. (2003) *The Analysis of Time Series – An Introduction*, 6th edition. Chapman & Hall/CRC Press, Boca Raton, Florida.

Cochrane, D. and Orcutt, G. H. (1949) Application of least squares regression to relationships containing autocorrelated error terms. *Journal of the American Statistical Association*, **44**, 32–61.

Cook, R. D. and Weisberg, S. (1999) *Applied Regression Including Computing and Graphics*. John Wiley & Sons, Inc., New York.

Cox, D. R. (2006) *Principles of Statistical Inference*. Cambridge University Press, Cambridge.

Cox, D. R. and Snell E. J. (1989) *Analysis of Binary Data*, 2nd edition. Chapman & Hall, London.

Davidson, R. and Mackinnon, J. G. (1993) *Estimation and Inference in Econometrics*. Oxford University Press, New York.

Davidson, R. and Mackinnon, J. G. (2004) *Econometric Theory and Methods*. Oxford University Press, New York.

Davison, A. C. (2003) *Statistical Models*. Cambridge University Press, Cambridge.

Davison, A. C. and Hinkley, D. V. (1997) *Bootstrap Methods and Their Application*. Cambridge University Press, Cambridge.

Dobson, A. J. (2001) *An Introduction to Generalized Linear Models*, 2nd edition. Chapman & Hall, London.

Draper, N. R. and Smith, H. (1998) *Applied Regression Analysis*, 3rd edition. John Wiley & Sons, Inc., New York.

Efron, B. and Tibshirani, R. J. (1993) *An Introduction to the Bootstrap*. Chapman & Hall, London.

Faraway, J. J. (2005) *Linear Models with R*. Chapman & Hall/CRC, Boca Raton, Florida.

Fewster, R. M., Buckland, S. T., Siriwardena, G. M., Baillie, S. R. and Wilson, J. D. (2000) Analysis of population trends for farmland birds using generalized additive models. *Ecology*, **8**, 1970–1984.

Ficetola, G. F. and Denoel, M. (2009) Ecological thresholds: an assessment of methods to identify abrupt changes in species-habitat relationships. *Ecography*, **32**, 1075–1084.

Fisher, R. A. (1929) Tests of significance in harmonic analysis. *Proceedings of the Royal Society*, **A125**, 54–59.

Flury, B. (1997) *A First Course in Multivariate Statistics*. Springer, New York. xviii + 277 pp.

Fox, J. (2002) *An R and S-PLUS Companion to Applied Regression*. Sage Publications, Thousand Oaks, California.

Fox, J. (2009) *car: Companion to Applied Regression*. R package version 1.2-11.

Franses, P. H., Neele, J. and van Dijk, D. (2001) Modelling asymmetric volatility in weekly Dutch temperature data. *Environmental Modelling and Software*, **16**, 131–137.

Freeman, S. N., Baillie, S. R. and Gregory, R. D. (2001) Statistical analysis of an indicator of population trends in farmland birds. Joint Report by the British Trust for Ornithology and Royal Society for the Protection of Birds to the Ministry of Agriculture, Fisheries and Food. Also available as Research Report 251, British Trust for Ornithology, Thetford, UK.

Frei, C. and Schär, C. (2001) Detection probability of trends in rare events: theory and application to heavy precipitation in the Alpine region. *Journal of Climate*, **14**, 1568–1584.

Freidlin, M. and Pavlopoulos, H. (1997) On a stochastic model for moisture budget in an Eulerian atmospheric column. *Environmetrics*, **8**, 425–440.

Fried, R. and Gather, U. (2005) Robust trend estimation for AR(1) disturbances. *Austrian Journal of Statistics*, **34**, 139–151.

Fried, R. and Imhoff, M. (2004) On the online detection of monotonic trends in time series. *Biometrical Journal*, **46**, 90–102.

Garisch, I. and Groenewald, P. C. N. (1999) The Nile revisited: changepoint analysis with auto-correlation. In *Bayesian Statistics, vol. 6* (eds. J. M. Bernardo, J. O. Berger, A. P. Dawid and A. F. M. Smith). Oxford University Press, Oxford. pp. 753–760.

Garthwaite, P. H., Jolliffe, I. T. and Jones, B. (2002) *Statistical Inference*, 2nd edition. Oxford University Press, Oxford.

Gelman, A., Carlin, J., Stern, H. and Rubin, D. (1995) *Bayesian Data Analysis*. Chapman & Hall, London.

Greene, W. H. (2003) *Econometric Analysis*, 5th edition. Prentice-Hall, New Jersey.

Haddon, M. (2001) *Modelling and Quantitative Analysis in Fisheries*. Chapman & Hall/CRC, Boca Raton, Florida.

Hannan, E. (1973) The estimation of frequency. *Journal of applied Probability*, **10**, 510–519.

Healy, M. J. R. (2002) *Matrices for Statistics*, 2nd edition. Oxford University Press, Oxford.

Hinkley, D. V. (1970) Inference about the change-point in a sequence of random variables. *Biometrika*, **57**, 1–17.

Hinkley, D. V. (1971) Inference about the change-point from cumulative sum tests. *Biometrika*, **58**, 509–523.

Horn, R. A. and Johnson, C. R. (1985) *Matrix Analysis*. Cambridge University Press, Cambridge.

Inclán, C. and Tiao, G. C. (1994) Use of cumulative sums of squares for restrospective detection of changes in variance. *Journal of the American Statistical Association*, **89**(427), 913–923.

Jarušková, D. (1997) Some problems with application of change-point detection methods to environmental data. *Environmetrics*, **8**, 469–483.

Jørgensen, B. (2002) Generalized linear models. In *Encyclopedia of Environmetrics* (eds. A. H. El-Shaarawi and W. W. Piegorsch) John Wiley & Sons, Ltd, Chichester. pp. 873–880.

Julious, S. A. (2001) Inference and estimation in a changepoint regression problem. *The Statistician*, **50**(1), 51–61.

Kauermann, G. and Carroll, R. J. (2001) A note on the efficiency of sandwich covariance estimation. *Journal of the American Statistical Association*, **96**, 1387–1396.

Krzanowski, W. J. (1988) *Principles of Multivariate Analysis*. Oxford University Press, Oxford.

Krzanowski, W. J. (1998) *An Introduction to Statistical Modelling*. Arnold, London.

Kundzewicz, Z. W. and Robson, A. J. (2004) Change detection in hydrological records – a review of the methodology. *Hydrological Sciences Journal*, **49**(1), 7–19.

Leith, N. A. and Chandler, R. E. (2010) A framework for interpreting climate model outputs. *Journal of the Royal Statistical Society, Series C*, **59**(2), 279–296.

Liang, K. Y. and Zeger, S. (1986) Longitudinal data analysis using generalized linear models. *Biometrika*, **73**(1), 13–22.

Lund, R. and Reeves, J. (2002) Detection of undocumented changepoints – a revision of the two-phase regression model. *Journal of Climate*, **15**, 2547–2554.

Lunn, D. J., Thomas, A., Best, N. and Spiegelhalter, D. (2000) WinBUGS – a Bayesian modelling framework: concepts, structure and extensibility. *Statistics and Computing*, **10**, 325–337. Software available from http://www.mrc-bsu.cam.ac.uk/bugs/.

McCullagh, P. and Nelder, J. A. (1989) *Generalized Linear Models*, 2nd edition. Chapman & Hall, London.

McCulloch, C. E., Searle, S. R. and Neuhaus, J. M. (2008) *Generalized, Linear, and Mixed Models*. John Wiley & Sons, Inc., Hoboken, New Jersey.

Manly, B. F. J. (2001) *Statistics for Environmental Science and Management*. Chapman & Hall/CRC, Boca Raton, Florida.

Menne, M. W. and Williams, C. N. (2005) Detection of undocumented changepoints using multiple test statistics and composite reference series. *Journal of Climate* **18**, 4271–4286.

Muggeo, V. M. R. (2003) Estimating regression models with unknown breakpoints. *Statistical Medicine*, **22**, 3055–3071.

Nelder, J. A. (2000) Functional marginality and response-surface fitting. *Journal of Applied Statistics*, **27**, 109–112.

Nelder, J. and Wedderburn, R. (1972) Generalized linear models. *Journal of the Royal Statistical Society, Series A*, **135**, 370–384.

Piegorsch, W. W. and Bailer, A. J. (2005) *Analyzing Environmental Data*. John Wiley & Sons, Ltd, Chichester.

Pinheiro, J. C. and Bates, D. M. (2000) *Mixed-effects models in S and S-PLUS*. Springer-Verlag, New York.

Poole, D. (2006) *Linear Algebra: A Modern Introduction*, 2nd edition. Thomson Brooks/Cole, Belmont, California.

Prais, S. J. and Winsten, C. B. (1954) Trend estimators and serial correlation. Cowles Commission Discussion Paper 373, Chicago, Illinois.

Priestley, M. B. (1981) *Spectral Analysis and Time Series*. Academic Press, New York.

Quinn, B. (1996) Statistical problems in the analysis of underwater sound. In *Athens Conference on Applied Probability and Time Series*, *vol. II* (eds. P. Robinson and M. Rosenblatt). Springer-Verlag, New York. pp. 324–338. Lecture Notes in Statistics 115.

Rice, J. (2006) *Mathematical Statistics and Data Analysis*, 3rd edition. Duxbury Press, Belmont, California.

Santer, B. D., Wigley, T. M. L., Boyle, J. S., Gaffen, D. J., Hnilo, J. J., Nychka, D., Parker, D. E., Parker, D. E. and Taylor, K. E. (2000) Statistical significance of trends and trend differences in layer-average atmospheric temperature time series. *Journal of Geophysical Research*, **105**(D6), 7337–7356.

Schwartz, J. (1994) Air pollution and daily mortality: a review and meta analysis. *Environmental Research*, **64**, 36–52.

Sturtz, S., Ligges, U. and Gelman, A. (2005) R2WinBUGS: a package for running WinBUGS from R. *Journal of Statistical Software*, **12**, 1–16.

Sweeting, T. J. (1980) Uniform asymptotic normality of the maximum likelihood estimator. *Annals of Statistics*, **8**, 1375–1381.

Thomas, A., O'Hara, B., Ligges, U. and Sturtz, S. (2006) Making BUGS Open. R *News*, **6**, 12–17. Software available from http://mathstat.helsinki.fi/openbugs.

Tol, R. (1996) Autoregressive conditional heteroscedasticity in daily temperature measurements. *Environmetrics*, **7**, 67–75.

Tol, R. S. J. (1997) Autoregressive conditional heteroscedasticity in daily wind speed measurements. *Theoretical Applied Climatology*, **56**, 113–122.

Tomé, A. R. and Miranda, P. M. A. (2004) Piecewise linear fitting and trend changing points of climate parameters. *Geophysical Research Letters*, **31**, L02207. DOI:10.1029/2003GL019100.

Venables, W. N. and Ripley, B. D. (1999) *Modern Applied Statistics with S-Plus*. Springer-Verlag, New York.

von Storch, H. and Zwiers, F. W. (1999) *Statistical Analysis in Climate Research*. Cambridge University Press, Cambridge.

Wackerly, D., Mendenhall, W. and Scheaffer, R. (2007) *Mathematical Statistics with Applications*, 7th edition. Duxbury Press, Belmont, California.

Wedderburn, R. W. M. (1974) Quasi-likelihood functions, generalised linear models, and the Gauss–Newton method. *Biometrika*, **61**, 439–447.

Weisberg, S. (2005) *Applied Linear Regression*, 3rd edition. John Wiley & Sons, Inc., Hoboken, New Jersey.

Yan, J. (2002) geepack: Yet another package for generalized estimating equations. *R-News*, **2/3**, 12–14.

Yang, C., Chandler, R. E., Isham, V. S. and Wheater, H. S. (2006) Quality control for daily observational rainfall series in the UK. *Water and Environment Journal*, **20**(3), 185–193. DOI:10.1111/j.1747-6593.2006.00035.x.

Yang, C., Chandler, R. E., Isham, V. S., Annoni, C. and Wheater, H. S. (2005) Simulation and downscaling models for potential evaporation. *Journal of Hydrology*, **302**, 239–254.

4

Nonparametric trend estimation

The previous chapter introduced the idea of a statistical model for trend and explored a variety of parametric techniques for estimation and testing. These techniques were all based on an assumption that the underlying trend had a specified mathematical form: for example that it was linear or that it depended linearly on covariates. However, such assumptions are not always appropriate. In this chapter, therefore, we describe some simple *nonparametric* techniques, incorporating the main features of environmental time series including seasonality, cycles, trends and relationships with other measured environmental variables. The key feature of these techniques is that the modelled trends and relationships are determined from the data themselves, rather than being forced to conform to a prespecified mathematical structure. This does *not* mean, however, that the techniques are completely assumption-free. There are a number of key texts introducing this modelling framework including Hastie and Tibshirani (1990), Simonoff (1996), Bowman and Azzalini (1997) and Wood (2006).

4.1 An introduction to nonparametric regression

We have seen that most model-based approaches to trend analysis can be regarded as some form of regression, in which the covariates may include the time index along with other variables that are thought to influence changes in the quantity of interest. The development in the previous chapter started with the linear model (3.3) in which the response was considered to depend linearly upon a single covariate. To remove the linearity assumption, and hence to allow greater flexibility where necessary, one could consider the following alternative version of equation (3.3):

$$Y_t = m(x_t) + \varepsilon_t \qquad (t = 1, \ldots, T), \qquad (4.1)$$

where (ε_t) is a white noise sequence with variance σ^2 as usual, but now the parametric trend function $\beta_0 + \beta_1 x_t$ has been replaced by an arbitrary function $m(x_t)$ whose

Statistical Methods for Trend Detection and Analysis in the Environmental Sciences, First Edition.
Richard E. Chandler and E. Marian Scott.
© 2011 John Wiley & Sons, Ltd. Published 2011 by John Wiley & Sons, Ltd.

mathematical form is not specified. Equation (4.1) defines a *nonparametric regression model* with independent errors.

In practice, one probably would not want the regression function $m\,(\cdot)$ in (4.1) to be completely arbitrary. In particular, it will often be reasonable to assume that it is smooth so that small changes in x lead to small changes in $m\,(x)$, and hence are associated with small changes in Y on average. Thus, many nonparametric regression techniques are based on some form of smoothing. In the context of a trend analysis in which x_t is just the time index t, we have already encountered the use of smoothing as a tool for exploratory analysis, in Section 2.2. Here we develop the idea further. By working within a model based framework, however, we are no longer merely describing the apparent structure in a scatterplot of Y against x: instead, we are explicitly estimating the dependence of the expected value of Y upon x.

4.1.1   Linear smoothing

We start by considering techniques for which the estimated regression function $\hat{m}(\cdot)$ is obtained via linear transformations of the data $y_1, \ldots, y_T$. Specifically, let $\mathbf{y}$ denote the vector of observations as usual and denote by $\mathbf{m}$ a vector containing the values of the regression function at a specified set of covariate values $\mathbf{x}$. Then the methods considered in this section are such that the estimate of $\mathbf{m}$ can be written as

$$\hat{\mathbf{m}} = S\mathbf{y}, \qquad (4.2)$$

where S is a *smoothing matrix*. By way of justification for this, consider once again the simple moving averages examined in Section 2.2:

$$\tilde{y}_t = \sum_{j=-k}^{k} w_j y_{t+j}.$$

In their simplest form, we saw that such moving averages could be used to calculate a smoothed series $\tilde{y}_{k+1}, \tilde{y}_{k+2}, \ldots, \tilde{y}_{T-k}$. In matrix–vector notation, this smoothed series can be written as

$$
\begin{pmatrix} \tilde{y}_{k+1} \\ \tilde{y}_{k+2} \\ \vdots \\ \tilde{y}_{T-k} \end{pmatrix}
=
\begin{pmatrix}
w_{-k} & w_{1-k} & \cdots & w_k & 0 & \cdots & 0 \\
0 & w_{-k} & \cdots & w_{k-1} & w_k & \cdots & 0 \\
\vdots & \vdots & \ddots & \ddots & \ddots & \ddots & \vdots \\
0 & 0 & \cdots & w_{-k} & \cdots & w_{k-1} & w_k
\end{pmatrix}
\begin{pmatrix} y_1 \\ y_2 \\ \vdots \\ y_{T-1} \\ y_T \end{pmatrix}.
$$

Equation (4.2) has exactly the same form as this, and hence can be seen as a direct extension of the moving average principle. Indeed, smoothers generally perform local averaging by averaging the observations within the neighbourhood of each point of interest.

Notice that if $\mathbf{x}$ is equal to the observed vector of covariate values, Equation (4.2) defines the fitted values of the regression function for each of the observations. Denote the corresponding smoothing matrix by $\tilde{S}$: this has dimensions $T \times T$ and is the direct analogue of the hat matrix defined in Section 3.1.3 in the context of parametric regression models.

4.1.2   Local linear regression

To carry out smoothing in practice, it is necessary to specify a means of generating the smoothing matrix S. There are two main decisions to be made:

- How to average the response values in each neighbourhood, i.e. the type of smoothing to perform.

- What size of neighbourhood is required. The size of the neighbourhood is typically expressed in terms of a *smoothing parameter*, sometimes referred to as a *bandwidth*.

Many types of smoothing are available, including those based on kernel methods, local polynomial fitting and regression splines; see, for example, Bowman and Azzalini (1997), Hastie and Tibshirani (1990), Simonoff (1996). Kernel methods are the equivalent, in a regression framework, of the moving averages considered in Section 2.2; local polynomial fitting was also introduced there as a means of visualising trend. Although these smoothing methods are all different, the end results, in terms of estimation, are often similar. We start by discussing the local linear smoothing method, first proposed in this context by Cleveland (1979).

Local linear smoothing is computationally straightforward to implement (in R, routines are provided in the sm library of Bowman and Azzalini, 2007) and to modify, and has a number of attractive properties. Figure 2.6 provides a graphical representation of the procedure, in the case when the covariate is the time index: $x_t = t$. For a particular covariate value x, the first step is to carry out a weighted least squares linear regression of the response variable upon $x_t - x$:

$$\min_{\alpha, \beta} \sum_{t=1}^{T} \{y_t - [\alpha_x + \beta_x (x_t - x)]\}^2 \, w (x_t - x; h) , \qquad (4.3)$$

where $w(\cdot; h)$ is a *kernel weight function*, chosen to ensure that observations close to x have the greatest contribution to the weighted sum of squares and hence that the estimated regression line, say $\hat{\alpha}_x + \hat{\beta}_x (x_t - x)$, can be regarded as an approximation to the true relationship in the neighbourhood of the value x of interest. The quantity h is a smoothing parameter, controlling the 'width' of the kernel function, i.e. the speed with which it decays and hence downweights observations with covariate values very different from x. For example, one might choose to use a Gaussian density centred on zero and with standard deviation h:

$$w (x_t - x; h) \propto \exp \left[- \frac{(x_t - x)^2}{2h^2} \right].$$

Wider kernels lead to more smoothing, just as with the simple moving averages in Section 2.2.

According to the fitted local regression line, when $x_t = x$ the expected response is $\hat{\alpha}_x$. This is taken as the estimate of $m(x)$. The whole procedure is then repeated for a range of different values of x, to build up an estimate of the complete underlying regression function.

Notice from Equation (3.44) that $\hat{\alpha}_x$ is a linear function of the observations $\{y_t\}$; thus we can write $\hat{\alpha} = s_x y$, say, where s_x is some row vector (the elements of s_x can be calculated explicitly; see Bowman and Azzalini, 1997, page 50). The local nature of the regression ensures that $\mathbf{s}_x$ allocates most weight to those elements of $\mathbf{y}$ with covariate values close to x. In the context of Equation (4.2), $\mathbf{s}_x$ gives the row of the smoothing matrix S corresponding to the covariate value x. Simultaneous estimation of the regression function at a collection of values of x can therefore be carried out easily by assembling S from its constituent rows, and then using (4.2).

Fan and Gijbels (1992) and Fan (1993) show the excellent theoretical properties of the local linear estimator. In particular, its behaviour near the ends of the range of x values is superior to the simpler local mean or *Nadaraya–Watson* estimator in which, rather than fitting regression lines, one simply calculates a kernel-weighted average of the observations at each covariate value of interest. Fan and Gijbels (1992) also consider the idea of varying the bandwidth in local linear smoothers, since there may be situations where a constant bandwidth is not flexible enough for estimating curves with a complicated shape.

An attractive feature of local linear smoothers is that as the smoothing parameter becomes very large, the kernel function becomes essentially flat over the range of the data so that the weights $w(x_t - x; h)$ attached to each observation in (4.3) become effectively equal. In this case, the estimate $\hat{m}(x)$ is merely the fit of the least squares regression line at x. Hence ordinary least squares regression can be regarded as a special case of local linear smoothing.

4.1.3   Spline smoothing

The idea behind local regression is intuitive and appealing. However, it is not the only way of constructing linear smoothers. An alternative is to build an estimate of the unknown regression function from a collection, or *basis*, of simpler functions. In fact we have seen an example already, in Section 2.3, where we used the fact that any sequence can be represented using a collection of sine and cosine functions: this collection is usually referred to as the *Fourier basis* for the sequence. Here, we introduce some alternative bases that emerge naturally from considerations of smoothness.

Suppose initially that we are prepared to make no assumptions whatsoever about the regression function $m(\cdot)$, but that we want to estimate it from data $y_1, \ldots, y_T$ by least squares. Thus we wish to minimise

$$\sum_{t=1}^{T} \left[y_t - m(x_t) \right]^2 .$$

The minimum attainable value of this expression is zero, and is attained by any function $m(\cdot)$ that interpolates the observations (i.e. satisfies $m(x_t) = y_t$ for $t = 1, \ldots, T$). This clearly does not help to estimate the regression function.

The assumption of smoothness provides a way out of the problem. Smoothness implies that the second derivative $m''(\cdot)$ of the regression function is 'small', since large values of the second derivative imply that the regression function has high curvature and therefore tends to be (in the terminology of Wood, 2006) 'wiggly'. An overall measure of the (lack of) smoothness of $m(\cdot)$ over the interval (a, b) is therefore

$$\int_a^b \left[m''(x) \right]^2 dx .$$

Thus, rather than minimising the sum of squares we might choose to minimise a *penalised* sum of squares

$$\sum_{t=1}^{T} \left[y_t - m(x_t) \right]^2 + \lambda \int_{\mathbb{R}} \left[m''(x) \right]^2 dx, \tag{4.4}$$

where λ is a constant controlling the relative contributions of the two terms. When $\lambda = 0$, the original sum of squares is obtained; progressively increasing λ gives more weight to the second term and hence discourages the estimated regression function from being too wiggly. Thus the second term can be thought of as a 'roughness penalty', and λ can be interpreted as a smoothing parameter in exactly the same way as the bandwidth h for local linear smoothers: increasing λ leads to smoother estimates of $m(\cdot)$. In particular, increasing λ indefinitely will ultimately lead to an estimated regression function for which $m''(x) \equiv 0$: this is just a straight line. As with local linear smoothing, the penalised sum of squares approach contains ordinary least squares regression as a special case.

Denoting the ordered covariate values in the observations by $x_{(1)} < x_{(2)} < \cdots < x_{(T)}$, it can be shown (Wood, 2006, Section 4.1) that the estimated regression function $\hat{m}(\cdot)$ obtained by minimising (4.4) consists of $T + 1$ connected segments, defined respectively on the intervals

$$\left(-\infty, x_{(1)}\right), \left(x_{(1)}, x_{(2)}\right), \ldots, \left(x_{(T-1)}, x_{(T)}\right), \left(x_{(T)}, \infty\right).$$

The outermost segments are linear (whence their contributions to the second term in (4.4) are zero); the $T - 1$ remaining segments are cubic polynomials, constrained so that the estimated regression function is continuous throughout and has continuous first and second derivatives. Such a function is called a *cubic smoothing spline*; the individual linear and cubic components form the basis for representing the regression function in this context. Cubic smoothing splines are linear smoothers (Simonoff, 1996, Section 5.6; Davison, 2003, page 532).

Several variants on cubic splines are available for nonparametric regression. For example, in a *cubic regression spline* the segments are defined not by the individual covariate values but rather by a set of $K \ll T$ *knots*, chosen to cover the overall range of the covariate; fitting is still carried out by minimising a penalised sum of squares of the form (4.4). Thus the basis dimension (i.e. the number of 'building blocks' used to construct the estimated regression function) is reduced from $T + 1$ to $K + 1$. This reduction brings computational advantages with little loss of flexibility; thus, providing K is large enough, cubic regression splines usually provide perfectly adequate estimates of the regression function. An alternative approach is to constrain the spline coefficients in adjacent segments to be similar to each other in some sense; this leads to the *penalised splines* of Eilers and Marx (1996). Penalised splines are used in Chapter 10 of the present volume; a more extensive description is given there. Wood (2006) gives a good review of different types of spline; the methods described therein are implemented in the mgcv library for R.

4.1.4   Choice of smoothing parameter

In Section 2.3 we saw that when a time series is smoothed to visualise a trend, the results depend on the precise choice of weights. Similarly, in a nonparametric regression context the result depends on the smoothing parameter chosen. For both local linear and spline smoothers, increasing the smoothing parameter indefinitely will cause the estimated regression function to approach a straight line; in contrast, as smoothing decreases the smoother tends to interpolate the data. In Section 2.3 we saw that special filters can be designed to extract specific features of interest, and indeed similar considerations can be used to choose a smoothing parameter in the current context (see Hastie and Tibshirani, 1990, Section 3.7 for example). However, the model based framework of this chapter provides an alternative means of choosing the smoothing parameter: rather than thinking about specific features that might be of interest, we can ask how best to estimate the regression function $m(x)$ in (4.1).

It turns out that there is a trade-off between the bias and variance of the estimator $\hat{m}(x)$. For local linear smoothers, an increase in the smoothing parameter leads to more data points being given appreciable weight in (4.3), so that the 'effective sample size' used for estimation of $m(x)$ increases: this reduces the variance of the estimator. On the other hand, if the true regression function is markedly nonlinear then it will be poorly approximated by a straight line except in the immediate neighbourhood of x: in this case, choosing too large a smoothing parameter will generally lead to bias in the estimation of $m(x)$. In Chapter 7 of the present volume, Giannitrapani *et al.* give a precise formulation of this. Here we give a simple formal demonstration of the bias problem. Recall that the observations **y** are regarded as the realised values of random variables **Y**, and let **m** and $\boldsymbol{\varepsilon}$ denote the corresponding vectors of regression function values and errors. Then we have $\mathbf{Y} = \mathbf{m} + \boldsymbol{\varepsilon}$ so that

$$E[SY] = E[S(\mathbf{m} + \boldsymbol{\varepsilon})] = S\mathbf{m}.$$

Thus, for any linear smoother to be unbiased we require $S\mathbf{m} = \mathbf{m}$, i.e. that the true regression function $m(\cdot)$ is unaffected by smoothing. For local linear smoothing, it should be clear that this is the case if $m(\cdot)$ is constant or linear (if the input is a straight line, the output will be the same straight line): the method is therefore unbiased in these specific cases. In general, however, $S\mathbf{m} \neq \mathbf{m}$ and hence $\hat{m}(\cdot)$ is a biased estimator.

The bias–variance trade-off is a feature of all nonparametric regression methods, not just local linear smoothing. In general, increasing the smoothing parameter leads to reduced variance and increased bias (and vice versa). Thus it is important to choose the smoothing parameter appropriately. There are various methods available for making this choice, including cross-validation and generalised cross-validation, as discussed in Hastie and Tibshirani (1990), Simonoff (1996), Wood (2006). The idea here is, omitting each data point in turn, to fit the model using a range of different smoothing parameters and use the fitted models to predict the response variable at the omitted data points. Some measure of prediction error, aggregated over all data points, is then computed for each of the candidate smoothing parameter values, and the 'optimal' value is taken as the one yielding the smallest aggregate error. However, such techniques can be unreliable in the presence of autocorrelation (for example Bowman and Azzalini, 1997, Section 7.5; Simonoff, 1996, Section 5.5; Hart, 1991). This is essentially because it is difficult to distinguish between smooth trends and strong autocorrelation; see Section 4.4 below.

As an alternative to cross-validation, the appropriateness of a particular smoothing parameter value may be checked by calculating the *effective degrees of freedom* of the fitted model. In parametric settings, the degrees of freedom (df) of a model usually correspond to the number of regression coefficients estimated and the residual degrees of freedom correspond to the number of independent data points remaining. It is not immediately obvious that these concepts can be applied to nonparametric models; however, Hastie and Tibshirani (1990, Section 3.5) discuss some definitions for use with linear smoothers that are nowadays generally accepted. These are

$$\mathrm{df}_{\mathrm{par}} = \mathrm{tr}(\tilde{S}) \qquad \text{and} \qquad \mathrm{df}_{\mathrm{err}} = T - \mathrm{tr}(2\tilde{S} - \tilde{S}\tilde{S}'), \tag{4.5}$$

where 'tr' denotes the trace of a matrix (i.e. the sum of its diagonal elements) and $\tilde{S}$ is the analogue of the hat matrix as defined earlier. Here, $\mathrm{df}_{\mathrm{par}}$ denotes the degrees of freedom for the model (i.e. the effective number of parameters) and df_{err} denotes the residual degrees of freedom.

To give some insight into these definitions, recall that the hat matrix in a parametric regression model is defined (see Section 3.1.3) as $\tilde{H} = X \left(X'X \right)^{-1} X'$, where X is the $T \times k$ design matrix of covariates. Using the standard result that $\mathrm{tr}\,(AB) = \mathrm{tr}\,(BA)$ for any matrices A and B such that both matrix products are well defined (Healy, 2002, page 17; Schott, 1997, pages 4–5), we have

$$\mathrm{tr}(\tilde{H}) = \mathrm{tr}\left[X \left(X'X \right)^{-1} X' \right] = \mathrm{tr}\left[\left(X'X \right)^{-1} X'X \right] = \mathrm{tr}\,(I_k),$$

where I_k is the $k \times k$ identity matrix. Clearly the trace of this is just k, the number of regression coefficients estimated. So in a parametric setting, df_{par} is equal to the number of regression coefficients. Similarly, using the fact that that H is symmetric and idempotent (Section 3.1.3) it can be shown that df_{err} is equal to the residual degrees of freedom in this setting.

The analogy with parametric regression models shows that df_{par} can be thought of as a measure of the complexity of the fitted regression function. Moreover, in general there is a one-to-one relationship between the smoothing parameter and df_{par} (Hastie and Tibshirani, 1990, Section 3.5) so that there is a unique smoothing parameter associated with each value of df_{par} (although a numerical root-finding algorithm is usually required to find it). This provides the opportunity to make an informed choice of smoothing parameter if the analyst has some idea of the likely complexity of the regression function. For example, in a trend analysis it may be deemed unlikely that the trend function (i.e. a regression function with the time index as covariate) has more than two turning points over the period of record. In some sense, such a function is of equivalent complexity to a cubic polynomial. Since a cubic polynomial has four parameters, choices of smoothing parameter leading to $df_{par} > 4$ would be considered inappropriate. This is similar in spirit to some of the techniques discussed in Section 2.3, in that the analyst is required to provide some input regarding the kind of structure of interest.

For a cubic regression spline, the basis dimension (or, equivalently, the number of knots K) dictates the maximum possible degrees of freedom; this observation can be used to determine the number of knots to use in any particular application. In general K should comfortably exceed the anticipated degrees of freedom of the regression function, so that the spline has sufficient flexibility to represent the structure of interest. (Generalised) cross-validation can then be used to choose the smoothing parameter λ in (4.4) as usual; this will typically lead to an estimated regression function with fewer degrees of freedom than the theoretical maximum (if this is not the case, the implication is that the structure in the data may be too complex to be represented by a regression spline with K knots, and hence the value of K should be increased). This idea is used in the mgcv library in R.

For spline based smoothers, a completely different approach to smoothing parameter selection is based on the observation that a penalised least squares criterion such as (4.4) can also be regarded as the log-likelihood for a *mixed-effects* model in which the regression coefficients are split into two groups, some of which are considered as fixed and others (corresponding to the individual spline contributions) as the realised values of random variables (Wood, 2006, Section 6.6). This has close connections with state space modelling (Section 5.5), where underlying trends can be modelled as evolving stochastically but smoothly through time. In a mixed-effects formulation, the smoothing parameter appears as an intrinsic part of the model specification and hence can be estimated using likelihood based or Bayesian methods. Kneib and Fahrmeir discuss the Bayesian approach in more detail in Chapter 10 of the present volume.

As an alternative to these 'formal' techniques, graphical methods may be (and commonly are) used to choose an appropriate smoothing parameter: here the analyst makes a judgement based on a visual inspection of the data and estimated regression function, perhaps taking into account knowledge of the system under investigation to judge whether a specific estimate of the regression function is reasonable. Such an approach is, of course, open to accusations of subjectivity: it may be helpful, therefore, to repeat any critical analyses using a range of plausible smoothing parameter values in order to determine whether or not the conclusions are sensitive to the precise choice.

Example 4.1

To illustrate the methods described above we revisit the series of annual mean wind speeds at De Bilt, seen previously in Figures 1.1 and 2.1. In Figure 4.1(a), local linear smoothing has been used to produce two different estimates of the underlying trend in this series. For the solid curve, the smoothing parameter h in (4.3) has been chosen using cross-validation; the resulting estimate has $df_{par} = 5.0$ effective degrees of freedom. For the dashed curve, a smoothing parameter corresponding to 4 effective degrees of freedom has been chosen. With fewer degrees of freedom, the resulting structure is (formally) slightly less complex than for the cross-validated estimate, although visually the two curves appear very similar in this example.

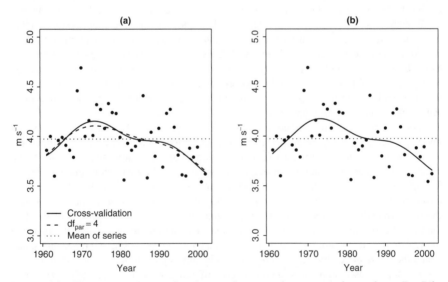

Figure 4.1   Nonparametric trend estimates for annual mean wind speeds at De Bilt, obtained using (a) local linear smoothing and (b) cubic regression splines.

Figure 4.1(b) shows the same data. Here, however, the trend is estimated using a cubic regression spline with the smoothing parameter λ in (4.4) chosen using cross-validation. This was produced using the gam() function in the R package mgcv, with the number of knots K set to the default value (which is 10, according to Wood, 2006, page 224). The result is very similar to the cross-validated local linear estimate. Unsurprisingly, the effective degrees of freedom is also similar: $df_{par} = 4.9$ here. This illustrates the point made in Section 4.1.2, that in nonparametric regression the end results are often insensitive to the precise method of smoothing used.   ∎

4.1.5   Variance estimators

After model fitting, it is necessary to provide an estimate of the error variance σ^2. In linear modelling the residual sum of squares (RSS) provides a basis for this, as discussed in Section 3.1. The analogue in the nonparametric setting is

$$\text{RSS} = \sum_{t=1}^{T} [Y_t - \hat{m}(x_t)]^2$$

and, as described above, the approximate degrees of freedom for error can be taken as $df_{err} = T - \text{tr}(2\tilde{S} - \tilde{S}\tilde{S}')$. The quantity

$$\hat{\sigma}^2 = \text{RSS}/df_{err} \tag{4.6}$$

is therefore a natural estimator of σ^2, being the direct analogue of (3.21) in the parametric case. With the definition of df_{err} given here, if $\hat{m}(\cdot)$ was unbiased as an estimator of $m(\cdot)$ then $\hat{\sigma}^2$ would be unbiased as an estimator of σ^2 (Hastie and Tibshirani, 1990, Section 3.9).

However, since $\hat{m}(\cdot)$ is usually biased, in general (4.6) is a biased estimator of σ^2 in nonparametric regression models. Alternative variance estimators have been suggested by Rice (1984) and Gasser, Sroka and Jennen-Steinmetz (1986). Rice (1984) proposed an estimator based on differences of the response variable, in order to remove the principal effects of the underlying mean curve. In contrast, the approach by Gasser, Sroka and Jennen-Steinmetz (1986) involves a form of linear interpolation. *Pseudo-residuals* are defined as

$$\tilde{e}_t = \frac{x_{t+1} - x_t}{x_{t+1} - x_{t-1}} Y_{t-1} + \frac{x_t - x_{t-1}}{x_{t+1} - x_{t-1}} Y_{t+1} - Y_t \qquad (t = 2, \ldots, T-1)$$
$$= a_t Y_{t-1} + b_t Y_{t+1} - Y_t, \text{ say,}$$

and the error variance is then estimated as

$$\tilde{\sigma}^2 = \frac{1}{T-2} \sum_{t=2}^{T-1} c_t^2 \tilde{e}_t^2, \tag{4.7}$$

where $c_t^2 = (a_t^2 + b_t^2 + 1)^{-1}$. This works because $\tilde{e}_t$ is the difference between Y_t and the line joining its two immediate neighbours. This line can be thought of as some form of minimal smoother, since it is based on just two data points. Thus it provides an estimate of $m(x_t)$ with low bias, so that (4.7) has attractive properties as a variance estimator.

4.1.6   Standard errors for the regression function

Having estimated the vector **m** of regression function values along with the error variance, it is usually of interest to quantify the uncertainty in the regression function estimate. In the first instance this is most easily achieved by calculating standard errors for each element of $\hat{\mathbf{m}}$. For a linear smoother, these standard errors are easily derived from the covariance matrix of $\hat{\mathbf{m}} = S\mathbf{Y}$:

$$\text{Var}\,(\hat{\mathbf{m}}) = \text{Var}\,(S\mathbf{Y}) = S\text{Var}\,(\mathbf{Y})\,S' = \sigma^2 SS'.$$

In practice, σ^2 must be replaced in this expression by an estimate obtained using any of the methods described above.

For presentational purposes, it can be helpful to produce a plot of the estimate $\hat{\mathbf{m}}$ along with approximate pointwise 95 % variability bands (Bowman and Azzalini, 1997, Section 4.4). If the errors $\{\varepsilon_t\}$ are normally distributed then these bands can be computed, for each element of $\hat{\mathbf{m}}$, as

$$\text{Estimate} \pm 2 \text{ standard errors.}$$

Note that, in general, the resulting intervals cannot be interpreted as confidence intervals: this is due to the potential bias of $\hat{\mathbf{m}}$ as an estimator of $\mathbf{m}$. Even in the absence of bias, care is required in the interpretation of such bands since they are not able to represent uncertainty about all values of $\mathbf{m}$ simultaneously. In general, the relation between pointwise and global uncertainty assessments is far from straightforward (Hastie and Tibshirani, 1990, Section 3.8). Nonetheless, variability bands are often useful aids to the interpretation of nonparametric regression estimates.

4.1.7   Testing for consistency with parametric models

It is often of interest to assess formally whether the data are consistent with some pre-specified form of regression function. For example, a constant regression function implies that the covariate has no effect on the response: if the covariate is the time index then this corresponds to a complete absence of trend. Alternatively, if the regression function can be regarded as linear then there is some justification for switching to a parametric modelling approach.

By analogy with classical regression theory, it is natural to use residual sums of squares to construct tests of the null hypothesis that the underlying regression function has a specified parametric form. Specifically, if RSS_0 and RSS_1 denote the residual sums of squares under the parametric and nonparametric models respectively, then a direct analogue of the usual F statistic (see Section 3.2.3) is

$$F = \frac{(\text{RSS}_0 - \text{RSS}_1)\,/(\text{df}_0 - \text{df}_1)}{\text{RSS}_1/\text{df}_1}. \tag{4.8}$$

Here, df_0 is the residual degrees of freedom for the parametric model and df_1 is its nonparametric counterpart, defined as df_{err} in (4.5). In the parametric setting, under the null hypothesis the F statistic is distributed as $F(\text{df}_0 - \text{df}_1, \text{df}_1)$. Heuristic arguments have been used to justify the continued use of this result in the nonparametric setting (Hastie and Tibshirani, 1990, Section 3.9) and experience suggests that it performs reasonably in practice. In the absence of strong theoretical justification for its use, however, an alternative may be helpful. We describe here a procedure given by Bowman and Azzalini (1997, Section 5.2), once again under the assumption that the errors $\{\varepsilon_t\}$ are normally distributed. The procedure is appropriate for use whenever the nonparametric estimator is unbiased under the null hypothesis (H_0, say), i.e. when $\tilde{S}m = m$. In Section 4.1.4, we saw that this is the case for local linear smoothing when the true regression function is constant or linear. The same is true for cubic splines, as should be clear by inspection of the penalised least squares criterion (4.4).

Notice first that (4.8) is proportional to

$$\tilde{F} = \frac{RSS_0 - RSS_1}{RSS_1} \qquad (4.9)$$

and the p-value for the test can be written as as $P\left(\tilde{F} > \tilde{F}_{obs}\right)$, where $\tilde{F}_{obs}$ is the observed value of $\tilde{F}$. Now the residual vector for the nonparametric model can be written as $e_1 = \left(I_T - \tilde{S}\right)Y$, where I_T denotes the $T \times T$ identity matrix. Note that if the nonparametric estimator is unbiased under H_0 then $\left(I_T - \tilde{S}\right)m = 0$, so the residual vector can equivalently be written as $e_1 = \left(I_T - \tilde{S}\right)(Y - m) = \left(I_T - \tilde{S}\right)\varepsilon$, where $\varepsilon = (\varepsilon_1 \cdots \varepsilon_T)'$ is the vector of (unobserved) errors in the model (4.1). Thus, under H_0, we have

$$RSS_1 = e_1'e_1 = \varepsilon'\left(I_T - \tilde{S}\right)'\left(I_T - \tilde{S}\right)\varepsilon = \varepsilon'A\varepsilon, \text{ say.}$$

Similarly, the parametric least squares estimator is unbiased under H_0. Thus, letting H denote the hat matrix for the parametric fit, we have

$$RSS_0 = \varepsilon'\left(I_T - H\right)'\left(I_T - H\right)\varepsilon = \varepsilon'\left(I_T - H\right)\varepsilon.$$

The last step here follows from the fact that the hat matrix is guaranteed to be symmetric and idempotent (see Section 3.1.3).

The numerator in (4.9) can now be written as

$$RSS_0 - RSS_1 = \varepsilon'[I_T - H - A]\varepsilon = \varepsilon'B\varepsilon, \text{ say.}$$

Therefore we can write $\tilde{F} = \varepsilon'B\varepsilon/\varepsilon'A\varepsilon$ and the p-value of the test is

$$P\left(\tilde{F} > \tilde{F}_{obs}\right) = P\left[\varepsilon'B\varepsilon > \tilde{F}_{obs}\varepsilon'A\varepsilon\right] = P\left[\varepsilon'C\varepsilon > 0\right] \text{ say,} \qquad (4.10)$$

where $C = B - \tilde{F}_{obs}A$ is a symmetric matrix. Now define $Z = \varepsilon/\sigma$ so that $\varepsilon = \sigma Z$, where σ^2 is the error variance in (4.1). Substituting this into (4.10), we find that

$$P\left(\tilde{F} > \tilde{F}_{obs}\right) = P\left[\sigma^2 Z'CZ > 0\right] = P\left[Z'CZ > 0\right],$$

which is a probability statement involving a quadratic form in standard normal random variables. This, coupled with the symmetry of C, enables (4.10) to be approximated by referring to the quantiles of a scaled and shifted chi-squared distribution. The technique is used in Chapter 7 where more details are given; see also Bowman and Azzalini (1997, pages 87–88).

This procedure completely avoids the need to define approximate degrees of freedom for the nonparametric regression estimate, and hence has a more solid theoretical foundation than the F test based on (4.8). In R, the sm.regression function in the sm library implements the procedure for testing the null hypotheses that the true regression function is constant or linear. As an alternative, however, parametric or nonparametric bootstrap methods[1] (see Section 3.6) could be used to determine the distribution of a test statistic such as (4.9) under the null hypothesis.

The result of any of these test procedures will, of course, depend on the choice of smoothing parameter. In particular, if the data themselves may have been used to select the amount of smoothing (for example via cross-validation or subjective visualisation), then the choice of smoothing parameter represents a source of uncertainty that is not accounted for by the theory above. As noted by Wood (2006, Chapter 4), the effect is generally to underestimate standard errors and to produce p-values that are too small when testing hypotheses. It is therefore worth exploring the sensitivity of the result to this choice, at least within a reasonable range. Bowman and Azzalini (1997, Chapter 5) suggest using a *significance trace* to determine this sensitivity: this is a plot of the p-value against the smoothing parameter.

To visualise the result of a test of the hypothesis that the regression function has a specified parametric form, it can be helpful to produce a graphical diagnostic by plotting a reference band (Bowman and Azzalini, 1997, Chapter 5) to indicate where the smooth curve should lie under the null hypothesis. For notational convenience, we consider the evaluation of such bands only at the observed covariate values: the difference between the corresponding vectors of parametric and nonparametric estimators is then $(H - \tilde{S})\,Y$. Note that under H_0, the parametric estimator is always unbiased. If $\tilde{S}Y$ is also unbiased, we have $E\left[(H - \tilde{S})\,Y\right] = 0$. It follows that if the observations are uncorrelated, the covariance matrix of the difference between the two estimators is

$$\sigma^2 \left(H - \tilde{S}\right)\left(H - \tilde{S}\right)',$$

from which the corresponding standard errors can be extracted. Thus, under the assumption of normality, a pointwise 95 % reference band for the smooth curve under the null hypothesis is defined by the vectors

$$HY \pm 1.96\sigma \sqrt{\operatorname{diag}\left[\left(H - \tilde{S}\right)\left(H - \tilde{S}\right)'\right]}. \tag{4.11}$$

Areas where the smooth curve lies outside the reference band highlight features of departure from the null model. In practice, of course, an estimate of σ^2 must be substituted; the resulting reference band is approximate rather than exact. Once again, the sm.regression function provides an implementation in R.

Example 4.2

Figure 4.2(a) shows the annual wind speed series from De Bilt, with a trend estimate and variability band superimposed. The trend estimate is the same as the solid line in Figure 4.1(a). However, the variability band now gives some indication of the precision of the estimate. Notice that the band widens at each end of the series, where fewer observations are available from which to estimate the regression function.

The variability band gives an indication of how much the true regression function might be expected to vary about its estimate. By contrast, a reference band indicates where the estimate might be expected to lie if the regression function is constant or

---

[1] Note that parametric and nonparametric bootstraps are distinguished by the way in which the stochastic components of the artificial bootstrap datasets are generated. By contrast, parametric and nonparametric regressions are distinguished by the way in which the systematic component of a model is represented.

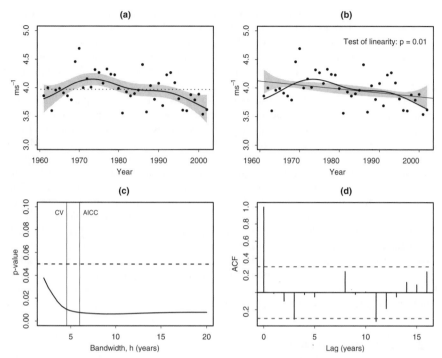

Figure 4.2   Nonparametric trend estimates for annual mean wind speeds at De Bilt using local linear smoothing (a) with approximate 95 % variability band, (b) with 95 % reference band computed under the assumption that the trend is linear. Panel (c) is a significance trace showing the p-value for the test of linearity as a function of the smoothing parameter, with 'automatic' choices indicated, and panel (d) shows the ACF of residuals from the estimated trend in (a) and (b). The smoothing parameter in panels (a) and (b) is chosen using cross-validation and corresponds to the 'CV' value in panel (c).

linear. This is illustrated in Figure 4.2(b). Here, in addition to the local linear smooth, the least squares linear trend estimate is shown along with a 95 % reference band, computed from (4.11) under the assumption that the trend is linear. The smooth estimate falls outside the reference band for the first half of the 1960s and again during the 1970s, suggesting that the data are not well modelled by a linear trend. This is confirmed by testing the null hypothesis of linearity: the p-value for this test, obtained using the scaled and shifted chi-squared approximation to (4.10) as implemented in the sm library in R, is 0.01. Notice, incidentally, that in rejecting the hypothesis of a linear trend, we also reject the hypothesis of no trend since this corresponds to a linear trend with zero slope. This therefore suggests that there is a nonlinear trend in annual mean wind speeds at De Bilt.

   To determine the sensitivity of these results to the choice of smoothing parameter, Figure 4.2(c) shows a significance trace for the test. For all smoothing parameter values in the range (2, 20) the null hypothesis of linearity is rejected comfortably at the 5 % level, giving some reassurance that the result is insensitive to the precise choice. This range of values easily encompasses any that could be considered 'reasonable', as indicated by the two vertical lines on the plot and by the fact that the smoothing parameter corresponding to four effective degrees of freedom, used to obtain the dashed trend estimate in Figure 4.1(a), is $h = 6.4$ years. The first of the vertical lines indicates the smoothing parameter obtained via cross-validation and used to obtain the results in Figures 4.2(a) and (b). The second indicates the smoothing parameter obtained using an alternative method of 'automatic' selection motivated by the AIC (see Section 3.2.3) and proposed by Hurvich, Simonoff and Tsai (1998).

Finally, as with all models, it is important to be aware of the assumptions underlying nonparametric regression techniques: (4.1) specifies that the errors $\{\varepsilon_t\}$ are uncorrelated with constant variance. As usual, this can be checked using appropriate residual plots. By way of an example, Figure 4.2(d) shows the ACF of residuals from the local linear trend estimate. Notice that the computation of residuals from a nonparametric smooth is very similar to detrending via filtering (see Section 2.3.4) and, as with any other filtering process, this induces artificial structure. Thus the approximate 95 % bands in Figure 4.2(d) are incorrect since they are computed under the assumption that the input series is white noise. For a more accurate picture, it would be necessary to consider that under model (4.1) the covariance matrix of the residuals from a linear smooth is $\sigma^2 \left(\boldsymbol{I}_T - \tilde{\boldsymbol{S}} \right) \left(\boldsymbol{I}_T - \tilde{\boldsymbol{S}} \right)'$ (this follows in exactly the same way as for the parametric case considered in Section 3.1.3). However, for present purposes it suffices to note that there is little evidence of residual autocorrelation in Figure 4.2(d). A plot of the time series of residuals (not shown) suggests that the assumption of constant variance also seems reasonable; hence we may have some confidence in the conclusions of the analyses above. ∎

4.2   Multiple covariates

4.2.1   Additive, semiparametric and bivariate models

In the presence of more than one covariate, model (4.1) can in principle be extended straightforwardly: given covariates $\{x_{1t}, \ldots, x_{pt} : t = 1, \ldots, T\}$ one could write

$$Y_t = m\left(x_{1t}, \ldots, x_{pt}\right) + \varepsilon_t \qquad (t = 1, \ldots, T).$$

Unfortunately, however, if p is large then enormous quantities of data are usually required to estimate the regression function in such a model without imposing additional assumptions. To understand this intuitively, consider what would be required to estimate $m\left(x_1, \ldots, x_p\right)$ using local linear smoothing: it would be necessary to carry out a weighted multiple linear regression of the response upon the covariates, allocating most weight to data points with covariate values 'close' to the configuration of interest. Taking the covariates individually, this is unproblematic: for the jth covariate, a specific choice of bandwidth may lead to some proportion, say π_j, of the observations being deemed close to x_j. Considering the covariates simultaneously, however, it is necessary to find observations that are close to the configuration of interest in each of p dimensions: if the covariates are roughly uncorrelated, such observations will form a proportion of about $\pi_1 \times \pi_2 \times \cdots \times \pi_p$ of the total sample. If p is large then this proportion will be very small. For example, if $p = 4$ and $\pi_1 = \pi_2 = \pi_3 = \pi_4 = 0.25$, only a proportion 0.25^4 (i.e. just under 0.5 %) of the observations will be given appreciable weight in the estimation. Clearly, any estimator constructed from such a small proportion of the data will be extremely variable unless the sample size is large. This phenomenon is referred to as the *curse of dimensionality* (Hastie and Tibshirani, 1990, Section 4.2; Simonoff, 1996, page 101).

To overcome the curse of dimensionality in practice, usually additional assumptions are required. A common assumption is that the regression function is *additive*, in which case the effects of the covariates can be considered separately in an *additive model* (Buja, Hastie and Tibshirani, 1989; Hastie and Tibshirani, 1990; Stone, 1985):

$$Y_t = \alpha + \sum_{j=1}^{p} m_j \left(x_{jt} \right) + \varepsilon_t \qquad (t = 1, \ldots, T). \qquad (4.12)$$

Here, $m_j \left(x_{jt} \right)$ is a smooth function of the jth covariate, satisfying the constraint $\sum_{t=1}^{T} m_j \left(x_{jt} \right) = 0$, and the errors $\{\varepsilon_t\}$ are assumed to be independent with mean 0 and variance σ^2. Informally, by imposing the restriction of additivity, each of the smooth functions can be estimated in turn so that to estimate $m_j(\cdot)$ one has only to find data points that are close to the value(s) of interest in the jth dimension. If the assumption of additivity is not warranted, however, bivariate terms can be included to model pairwise interactions between covariates; see McMullan, Bowman and Scott (2007) for an example in which nonparametric trend estimates are allowed to vary with spatial location. Alternatively, one might choose to impose additional assumptions on the smoothness of the regression function instead of assuming additivity; this leads to *thin plate regression splines* as a generalisation of the cubic regression splines already encountered (Wood, 2006, Section 4.1).

In a model such as (4.12), the individual smooth functions may be estimated and combined using the *backfitting algorithm*, described below. Smooths can be obtained for each of the model components separately, using different smoothing parameters and methods of smoothing for each component if necessary. The representation of seasonality in an environmental time series is one specific example where different smoothing methods may be required for different components. For example, a simple additive model for a monthly environmental or ecological time series might be

$$Y_t = \alpha + m_1 \,(\text{year}) + m_2 \,(\text{month}) + \varepsilon_t,$$

where 'month' denotes the month number within the year, from 1 to 12. In this model, the smooth functions for year and month represent trend and seasonality respectively. The trend component could be estimated using local linear or spline smoothers. However, without modification neither of these techniques would deliver a satisfactory estimate of the seasonal cycle, because by definition this should be periodic: months 12 and 1 are 'neighbours' and the estimated smooth function $\hat{m}_2 (\cdot)$ should ensure a smooth transition between them. In the context of spline smoothing, this can be dealt with via the use of cyclical splines as described in Wood (2006, Section 4.1). For local linear smoothing one could use a periodic kernel function, for example based on the von Mises rather than the Gaussian distribution (see, for example, Fisher, 1993, or Mardia and Jupp, 1999). The von Mises distribution yields local regression weights

$$w(x_t - x; h) = \exp \left[\frac{1}{h} \cos \left(2\pi \frac{x_t - x}{M} \right) \right],$$

where M is the cycle length (12 in the hypothetical example above). Even in this case, however, a local linear regression of Y_t upon the month number would be inappropriate; a local mean (Nadarya–Watson) estimator can therefore be used instead, to create a circular smoother. This technique is implemented in the `sm.regression` routine in R.

Additive models can be extended to generalised additive models (GAMs) in a similar way that linear models can be extended to generalised linear models (GLMs; see Section 3.5). GAMs are identical to GLMs, except that the linear predictors $\{\eta_t\}$ are now replaced with additive predictors. Thus μ_t, the expected response at time t, satisfies

$$\eta_t = g\left(\mu_t\right) = \alpha + \sum_{j=1}^{p} m_j\left(x_{jt}\right) \qquad (4.13)$$

for smooth functions $\{m_j\left(\cdot\right) : j = 1, \ldots, p\}$. Relevant applications in the literature include Clarke *et al.* (2003) who model trends of seabird populations, Fewster *et al.* (2000) who examine population trends in farmland birds and Bellido, Pierce and Wang (2001) who consider variation in abundance of squid in Scottish waters. Implementations in R can be found in the mgcv library already mentioned, as well as the gam library (Hastie, 2008).

Further extensions include Hastie and Tibshirani (2000), who provide a Bayesian procedure for posterior sampling from additive and generalised additive models, and Green, Jennison and Seheult (1985). The latter authors consider semiparametric models in which the linear predictor contains both parametric and smooth components.[2] In a semiparametric model, the linear predictor η_t is written as

$$\eta_t = \alpha + \sum_{j=1}^{p} m_j(x_{jt}) + \mathbf{z}_t \beta, \qquad (4.14)$$

so that the additive model structure (4.13) is augmented with a linear combination of other covariates $\mathbf{z}_t$. The fitting of such models can be speeded up by exploiting the partly parametric structure; see Hastie and Tibshirani (1990, Section 5.3.3) for example.

4.2.2 The backfitting algorithm

Additive and generalised additive models can be fitted by the backfitting algorithm: see Hastie and Tibshirani (1990, Section 4.4) and Wood (2006, Section 4.11). This involves iteratively smoothing with respect to each covariate, using as response the *partial residuals* (i.e. residuals computed from the other model components) and at each stage carrying out additional adjustment to ensure that $\sum_{t=1}^{T} \hat{m}_j\left(x_{jt}\right) = 0$. The mean of the observations provides an estimate of α: $\hat{\alpha} = \bar{y}$, say. Full details are given by Giannitrapani *et al.* in Chapter 7 of the present volume: therefore we restrict ourselves here to some relatively brief comments in the context of linear smoothers. Most of the algebraic manipulations are the same regardless of the set of evaluation points (i.e. covariate configurations at which the model components are estimated); however, for purposes such as the explicit construction of residuals, the model components must be evaluated at the observed covariate values. The development below therefore focuses mainly on this setting.

Let S_j^* denote the fundamental smoothing matrix associated with the jth model component (i.e. the matrix used to smooth the residuals at each iteration of the backfitting algorithm). Because of the iterative nature of the algorithm, in general there is

---

[2] Strictly speaking, the term *semiparametric* originally referred to a model in which just one of the covariate effects was represented nonparametrically (e.g. Hastie and Tibshirani, 1990, page 118). Our usage is consistent with more recent practice, for example Ruppert, Wand and Carroll (2003).

no simple relationship between the $\{S_j^*\}$ and the estimated components $\{\hat{\mathbf{m}}_j\}$ obtained on convergence: we do not have $\hat{\mathbf{m}}_j = S_j^* \mathbf{y}$ for example. However, since each iteration involves linear transformations of the data vector, for each $j \in \{1, \ldots, p\}$ we do have $\hat{\mathbf{m}}_j = \tilde{S}_j \mathbf{y}$ for *some* matrix $\tilde{S}_j$. The $\{\tilde{S}_j\}$ are usually referred to as *projection matrices* in the literature, since they map the observations $\mathbf{y}$ into the fitted model components. In Chapter 7, Giannitrapani *et al.* show how these projection matrices can be constructed recursively as the backfitting algorithm progresses: see Equation (7.11). The idea has also been used by Bowman, Giannitrapani and Scott (2009).

In the notation above, the vector of fitted values for the model can be obtained upon convergence as

$$\hat{\mathbf{y}} = \hat{\alpha}\mathbf{1} + \sum_{j=1}^{p} \hat{\mathbf{m}}_j = \sum_{j=0}^{p} \tilde{S}_j \mathbf{y} = \tilde{S}\mathbf{y}, \quad \text{say},$$

where $\tilde{S} = \sum_{j=0}^{p} \tilde{S}_j$ and $\tilde{S}_0$ is the $T \times T$ matrix filled with the value T^{-1}. The effective degrees of freedom $\mathrm{df}_{\mathrm{par}}$ and $\mathrm{df}_{\mathrm{err}}$ can now be defined using (4.5) exactly as before, and used to construct dispersion parameter estimates analogous to (4.6) (for more discussion of these estimates, see Chapter 7). However, analogues of the alternative dispersion parameter estimators, introduced previously in the context of regression on a single covariate, are more difficult to construct here.

Noting that the effective number of model parameters is defined as $\mathrm{df}_{\mathrm{par}} = \mathrm{tr}(\tilde{S})$ and that $\tilde{S} = \sum_{j=1}^{p} \tilde{S}_j$, it is natural to define the effective number of parameters for the jth model component as $\mathrm{tr}(\tilde{S}_j)$: thus the total number of parameters can be decomposed into contributions from the individual covariates. As before, this can be used to check the choice of smoothing parameter(s) when the analyst has some idea of the likely complexity of the individual model components (see Chapter 7 for an example of this). Notice, however, that $\mathrm{tr}(\tilde{S}_j)$ depends not only on the fundamental smoothing matrix S_j^* but also on the other components in the model: this differs from the situation in parametric regression models, where the degrees of freedom for each component is always equal to the number of additional parameters required to specify that component. In the nonparametric case, the dependence upon other model components reflects the fact that the complexity of the estimate $\hat{m}_j(\cdot)$ will itself vary depending on whether or not other components are included in the model: for example, a nonparametric model has the flexibility to try and adapt to structure arising from the omission of an important covariate.

As a more easily computed alternative to the definition given above, Hastie and Tibshirani (1990) suggest that the degrees of freedom for the estimate $\hat{m}_j(\cdot)$ can be approximated by $\mathrm{tr}(S_j^*) - 1$. The reduction by 1 here is due to the constraint $\sum_{t=1}^{T} m_j(x_{jt}) = 0$: this is automatically incorporated into the 'exact' definition $\mathrm{tr}(\tilde{S}_j)$ (again, see Chapter 7 for full details).

As described above, the backfitting algorithm provides estimates of the smooth functions $\{m_j(\cdot)\}$ only at the observed covariate values. If estimates are required at other values as well (for example to represent the model components on a regular grid for plotting purposes) then some extra work is needed. For smoothers that use basis functions, such as cubic regression splines, this work is relatively straightforward: one simply evaluates the basis functions at the covariate values of interest, as implemented in the `predict.gam` routine in the R library `mgcv` (Wood, 2006, Section 5.2.6). For local linear smoothers, more care is required: one solution is to smooth the partial residuals from the final iteration of the backfitting algorithm on to the desired evaluation points. To ensure that the results for the jth component are consistent with the constraint $\sum_{t=1}^{T} m_j(x_{jt}) = 0$, these evaluation points should include at least one of the observed values of x_{jt}: any change in the estimated component here can then be added back to all of the results.

An extension of the basic backfitting algorithm can be used to fit generalised additive models. This extension is best understood with reference to the iteratively weighted least squares (IWLS) algorithm used for fitting GLMs (see Section 3.5.1). The IWLS algorithm consists of a sequence of weighted least squares fits in which both the response variable and the weights change at each iteration. Recall from Equation (3.43) that any weighted least squares fit is equivalent to an unweighted fit after transforming both the response and covariates. Thus, to adapt the IWLS algorithm for use with GAMs, each weighted least squares fit is replaced with a complete cycle of the backfitting algorithm in which the response and covariates are all adjusted and transformed to yield an unweighted smoothing problem. For further details, see Hastie and Tibshirani (1990, Section 6.3).

The backfitting algorithm is implemented in the gam library in R. However, it is not the only way to fit GAMs. In principle at least, providing the individual smoothers are linear the smooth functions can be estimated simultaneously as described by Wood (2006, Chapter 3) and Hastie and Tibshirani (1990, Section 5.2); the former treatment exploits the special features of spline based smoothing. The price to pay for simultaneous estimation is an increase in computational cost; however, with modern computing power this is not a problem except when very large data sets are involved.

4.2.3   Inference for additive models

In nonparametric regression, as in any other modelling exercise, to answer questions of substantive interest requires the use of tools that allow us to draw inferences about the model components (e.g. to determine whether they are statistically significant or to assess the uncertainty in an estimated smooth function).

As usual, the key is the estimation of standard errors for the individual components and the formulation of test statistics for model comparison. For additive models based on linear smoothers, much of the necessary theory is a straightforward extension of that described in Section 4.1. For example, the standard errors associated with the estimated component $\hat{\mathbf{m}}_j$ can be derived by considering the covariance matrix

$$\text{Var}\left(\hat{\mathbf{m}}_j\right) = \text{Var}\left(\tilde{\mathbf{S}}_j \mathbf{Y}\right) = \tilde{\mathbf{S}}_j \text{Var}\left(\mathbf{Y}\right) \tilde{\mathbf{S}}_j' = \sigma^2 \tilde{\mathbf{S}}_j \tilde{\mathbf{S}}_j',$$

which can be estimated by replacing σ^2 with its estimate $\hat{\sigma}^2 = \text{RSS}/\text{df}_{\text{err}}$ as before. A variability band can then be constructed, as described in Section 4.1.6.

In the case of a single covariate, we have shown in Section 4.1.7 how to construct tests of hypotheses that the underlying regression function is constant or linear. Such techniques can be regarded as extensions of the classical procedures for comparing nested models (Sections 3.2 and 3.5), since a smooth regression function includes 'constant' or 'linear' functions as special cases. When dealing with multiple covariates, the range of possibilities is substantially greater: for example, one may be interested in comparing nested models to determine whether or not a particular subset of covariates influences the response (which is equivalent to testing the hypothesis that the corresponding model components are zero), to determine whether some or all of the covariate effects are linear or perhaps to determine whether a bivariate smooth term can be split into additive univariate components.

For additive models, 'F-like' statistics, of the same form as (4.8) but where now df_0 and df_1 in (4.8) correspond to the residual degrees of freedom for the reduced and extended models respectively, can be used to test hypotheses. For generalised additive models the residual sums of squares are replaced by residual deviances, by direct analogy with the usual procedure for GLMs (see Section 3.5.2). As in Section 4.1.7, for models with an unknown dispersion parameter it is common practice to proceed as though the test statistic is distributed as $F(\mathrm{df}_0 - \mathrm{df}_1, \mathrm{df}_1)$ under the null hypothesis that the data were generated from the simpler of the two models (see Wood, 2006, Section 4.10 for an accessible discussion of the issues involved here); alternatively a quadratic form test analagous to (4.10) can be used. It is worth noting, however, that the derivation of the latter test relies on the nonparametric estimators being unbiased under the null hypothesis: in general this will not be the case in additive models. Exceptions occur if the null hypothesis corresponds to a fully parametric model in which all terms are either constant or linear, at least if local linear smoothing or regression splines are used. In general, however, the presence of bias ensures that the earlier theory for computing the p-value corresponding to (4.10) is, strictly speaking, incorrect (Chapter 7 discusses this issue further). Similar comments apply to the construction of reference bands, indicating where the nonparametric estimate should lie under the corresponding null hypotheses. For hypothesis testing purposes, the problem of bias can be reduced by deliberately undersmoothing: having chosen a model on the basis of the test results, the estimates can then be refined using a more appropriate smoothing parameter. In this case, the price to pay for undersmoothing is a reduction in the power of tests to detect genuine structure, which follows from the increased variance of the estimators. As always, bootstrap methods provide an alternative approach that eliminates the need for theoretical calculations, although here it may be necessary to use different smoothing parameters when fitting the null model for the purposes of generating bootstrap samples: see Figueiras, Roca-Pardiñas and Cadarso-Suárez (2005) and Roca-Pardiñas, Cadarso-Suárez and González-Manteiga (2005) for examples involving environmental applications. Finally, in situations where one wishes to test whether a particular covariate is needed in the model at all, permutation tests (see Section 3.6) may be used: for these tests, the potential effect of bias is perhaps less severe because the resampled data sets are not generated from a possibly biased model fit. All of these options have their drawbacks, however, and the problem of model selection and comparison in a nonparametric context remains an active topic of statistical research.

There is a further feature of nested model comparisons in a nonparametric context that is, at first sight, surprising: test statistics such as (4.8) can occasionally take negative values. This indicates that of the two models considered ($\mathcal{M}_0$ and $\mathcal{M}_1$, say), the simpler one ($\mathcal{M}_0$) provides the closest fit to the data. In a parametric setting, such an outcome is impossible by definition of the least squares and maximum likelihood fitting procedures. With nonparametric methods, however, there is more than one way in which the situation can occur. The first is if smoothing parameters for both models are selected using some automatic procedure such as (generalised) cross-validation. In this case, there is a possibility that a larger smoothing parameter will be selected for $\mathcal{M}_1$ so that it is unable to capture the same amount of detail as $\mathcal{M}_0$: thus, despite appearances, the models are not necessarily nested. Wood (2006, page 206) suggests that to avoid this problem care should be taken to ensure that each smooth term in $\mathcal{M}_0$ has no more effective degrees of freedom than the same term in $\mathcal{M}_1$.

Another scenario in which the simpler model $\mathcal{M}_0$ can fit better than $\mathcal{M}_1$ is when $\mathcal{M}_0$ contains a highly variable, but nonetheless relatively parsimonious and correctly specified, parametric component that is represented nonparametrically in $\mathcal{M}_1$. A high-frequency sine wave would be an example of such a component. In this case, too large a choice of smoothing parameter in $\mathcal{M}_1$ will lead to an oversmooth estimate that cannot track the changes in the data as well as the correctly specified parametric model $\mathcal{M}_0$. In practice, this difficulty would usually be overcome by an appropriate choice of smoothing parameter; however, it is worth being aware of the potential problem.

These two examples illustrate that inference in GAMs can be complicated by the need to choose smoothing parameters; thus, as in the case of a single covariate, it is helpful to explore the sensitivity of results to the precise choice.

Example 4.3

In Section 4.1 we used nonparametric methods to model trends in annual mean wind speeds at De Bilt. Turning now to monthly instead of annual means, it becomes necessary to consider seasonality as well as trend. In Chapter 2 (specifically Figure 2.2) we saw that wind speeds tend to be lower and less variable in summer than in winter. The nonconstant variability is potentially problematic for an additive modelling approach; as discussed in Section 3.3, one way to try and allow for this is to transform the data prior to analysis. Here, we fit models to the logarithms of the monthly means. For ease of interpretation, the logarithms are to base 10.

The first model to be fitted is additive, with components corresponding to trend (i.e. a nonparametric smooth of the time index) and seasonality (a smooth of the 'month' variable). The trend is estimated using a cubic regression spline and the seasonal cycle using a cyclical spline as described above and implemented in the mgcv library in R; smoothing parameters are chosen using cross-validation. The estimated components are shown in Figures 4.3(a) and (b) respectively, along with 95 % variability bands. Note that the smooths are centred about zero because, in an additive model, they represent deviations from the overall mean. Apart from this, the trend component is very similar to that obtained from the annual means in the previous section; this is hardly surprising. The estimated seasonal cycle also seems reasonable, indicating that the highest wind speeds occur during the winter months (specifically from December to March).

Under the assumption of additivity, the trend is the same in all months of the year. It is, of course, possible that trends vary between months or, equivalently, that the seasonal cycle changes through time. In a parametric setting, this would be investigated by testing for the existence of interactions between the trend and seasonal covariates (see Section 3.2.2). The nonparametric equivalent is to compare the additive model with a bivariate smooth of the two variables. A bivariate model fit is shown in Figure 4.3(c): here, the horizontal and vertical axes represent 'year' and 'month' respectively and the bold contours show the expected $\log_{10}$ mean monthly wind speed according to the fitted model. The smoothing parameters have again been chosen using cross-validation, but in such a way that the chosen values are no larger than those for the previous additive model. This is to try and ensure that the additive model is nested within the bivariate model, as discussed above.[3] If this precaution is not taken then the phenomenon described previously occurs for these data, with the 'simpler' additive model yielding the smaller residual sum of squares.

The dashed contours in Figure 4.3(c) indicate the standard errors of the fit: as before, these are largest at the ends of the series. Inspection of Figure 4.3(c) reveals that there does indeed seem to be seasonal variation in the trends: the characteristic increase around 1970 that was seen in the analyses of annual wind speeds and in Figure 4.3(a) seems to be confined mainly to the winter months in the bivariate smooth, with a more or less monotonic decrease between May and September.

---

[3] In practice, because of the way the mgcv library works, it was necessary first to determine the smoothing parameters for the bivariate model, and then to set these as lower bounds in the search for optimal smoothing parameters in the additive model.

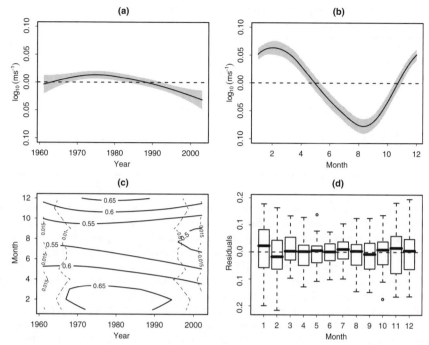

Figure 4.3   Additive and bivariate models for the logarithms (to base 10) of monthly mean wind speeds at De Bilt, using cubic regression splines for trend and cyclic splines for seasonality. Smoothing parameters are chosen using generalised cross-validation. Panels (a) and (b) are estimated trend and seasonal components for the additive model, with 95% variability bands. Panel (c) shows the estimated bivariate smooth (including the overall mean); thin lines and small labels are contours of the associated standard errors. Panel (d) shows monthly boxplots of residuals from the bivariate smooth.

Before reading too much into the patterns shown in Figure 4.3(c), however, it is worth checking to see if the fit is significantly better than that of the additive model. The simplest way to do this is using the approximate F test based on (4.8). The mgcv library in R provides routines for this purpose; the output is given in Table 4.1. This shows that the bivariate model reduces the residual sum of squares (labelled 'deviance' in the output) by 0.034, at the cost of an extra 3.4 effective degrees of freedom. The F statistic is 2.207, which yields a p-value of 0.08 when compared with an $F(3.4, 493.2)$ distribution. Taken at face value, this suggests accepting the hypothesis of additivity at the 5% level. However, given the approximate nature of the F test, it may be worth cross-checking the result using different techniques. This is particularly true since the log transformation has not entirely succeeded in removing the seasonal structure in the variance: this is illustrated by the monthly residual distributions in Figure 4.3(d). Alternative transformations were explored, but none was found that overcame the problem. The residual ACF (not shown) also indicates that there is a small amount of residual autocorrelation (at lag 1, the coefficient is 0.120); again, this may affect the inference.

As suggested previously, bootstrap methods provide an alternative means of establishing the distribution of the test statistic under the null hypothesis. In fact, a nonparametric

Table 4.1   Edited R output showing a comparison of additive and bivariate models for monthly wind speed, using an approximate F test.

```
Analysis of Deviance Table

Model 1: logy ~ s(Year) + s(Month, bs ="cc")
Model 2: logy ~ te(Year, Month, bs = c("cr", "cc"))
  Resid. Df Resid. Dev      Df Deviance      F  Pr(>F)
1  496.6694    2.27720
2  493.2357    2.24275  3.4337  0.03445 2.2067 0.07748 .
---
Signif. codes:  0 *** 0.001 ** 0.01 * 0.05 . 0.1   1
```

block bootstrap (see Section 3.6) opens up the possibility to account both for the seasonally changing variance and for the residual autocorrelation. This has been done here, with the error series for each bootstrap sample generated by resampling entire years (with replacement) from the residual time series for the bivariate model. By resampling entire years we preserve most of the residual autocorrelation, and also ensure that the different monthly residual distributions are represented in each bootstrap sample; 10 000 bootstrap samples were generated to ensure that the estimated p-value is accurate to two decimal places (see Section 3.6). For each bootstrap sample, exactly the same procedure for choosing smoothing parameters and fitting models was followed as for the original data. The resulting bootstrap p-value (i.e. the proportion of bootstrap-generated test statistics exceeding the observed value of 2.207) was 0.272. This is very different from the value of 0.08 obtained from the approximate F test. Neither p-value is completely accurate: the performance of the F approximation is likely to be degraded by the seasonal variance structure in the residuals, and the bootstrap may be affected by the resampling of residuals from a biased estimate of the null model. Nonetheless, both procedures suggest that the data structure is described reasonably by an additive model: hence there is little evidence for seasonal variation in the trends, and the apparent structure in Figure 4.3(c) is primarily due to sampling variation.   ∎

4.2.4   Handling autocorrelation

The effect of autocorrelation upon parametric regression models has been discussed extensively in Chapter 3. Many of the issues involved (see Section 3.1.1) are essentially the same in the nonparametric case: if autocorrelation is ignored when fitting models, the estimates are valid but subject to some loss of precision, standard errors are usually too small and formal methods of model comparison are likely to be liberal. Opsomer, Wang and Yang (2001), McMullan, Bowman and Scott (2007) and Bowman, Giannitrapani and Scott (2009) (see also Chapter 7) review the effects of correlation on standard forms of nonparametric regression.

As in the parametric setting, a simple way of dealing with autocorrelation is to fit models in the first instance as though observations are independent, and subsequently to adjust standard errors and model comparison statistics for the dependence. For example, the adjustment (3.47) of standard errors for parametric regressions generalises straightforwardly when linear smoothers are used in a nonparametric context: if S_j is the smoothing matrix for the jth component of an additive model so that the component estimator is

$\hat{\mathbf{m}}_j = S_j \mathbf{Y}$, then the covariance matrix of $\hat{\mathbf{m}}_j$ is

$$S_j \mathbf{\Sigma}^{-1} S'_j, \tag{4.15}$$

where $\mathbf{\Sigma}$ is the covariance matrix of the observations. Standard errors for the individual smoothed values are given by the square roots of the diagonal elements of (4.15). In practice, again as in the parametric setting, the covariance matrix $\mathbf{\Sigma}$ can be estimated by fitting an appropriate time series model to the regression residuals. For an example of such an approach, see Giannitrapani *et al.* (2006). In addition, these authors describe dependence-adjusted versions of F statistics for comparing nested models; see also Ferguson *et al.* (2007) and McMullan, Bowman and Scott (2007).

As in Section 3.3.3, an alternative to adjusting for autocorrelation is to allow for it explicitly in the model-fitting process. For local linear smoothers, one way of achieving this starts from the observation that the local least squares criterion (4.3) can be written in vector–matrix form as

$$\left(\mathbf{y} - \left[\alpha\mathbf{1} + \mathbf{x}^*\beta\right]\right)' W \left(\mathbf{y} - \left[\alpha\mathbf{1} + \mathbf{x}^*\beta\right]\right),$$

where $\mathbf{1}$ is a vector of T ones, $\mathbf{x}^*$ is a vector with tth element $x_t - x$ and W is a $T \times T$ diagonal matrix with (t, t)th element $w(x_t - x; h)$. An analogy with generalised least squares (see Section 3.3.3) suggests that autocorrelation can be accounted for by replacing W in this expression with $\mathbf{\Gamma}'W\mathbf{\Gamma}$, where $\mathbf{\Gamma}'\mathbf{\Gamma} \propto \mathbf{\Sigma}^{-1}$. This approach is described further in Chapter 7. It is not the only way to proceed, however; for alternatives, see Cai (2007), Welsh, Lin and Carroll (2002) and Bowman and Azzalini (1997, Section 7.5). The latter reference forms the basis for the `sm.regression.autocor` function in the `sm` library for R.

To accommodate the effects of autocorrelation when using spline based smoothers, a completely different approach can be used. This exploits the mixed model representation of spline smoothers mentioned in Section 4.1.4. Once the necessary machinery is in place to fit models within this framework, the addition of an additional layer of correlated random variables to account for residual dependence is relatively straightforward. The 'necessary machinery' referred to is, however, rather technical: see Ruppert, Wand and Carroll (2003, Section 4.9) and Wood (2006, Section 6.6) for example. The methods described in the latter reference are implemented in R via the `gamm` ('generalised additive mixed model') routine in the `mgcv` library. State space models (Section 5.5) provide an alternative viewpoint that is arguably slightly more transparent. Bayesian approaches are also possible; see Fahrmeir and Lang (2001), Fahrmeir, Kneib and Lang (2004) and Chapter 10 of the present volume.

The discussion above has centred mainly on the effects of autocorrelation upon the calculation of standard errors and model comparison statistics. In a nonparametric setting, an additional problem arises if one wants to use the data themselves to choose an appropriate smoothing parameter. Techniques such as cross-validation, for example, tend to undersmooth because they aim to minimise some measure of prediction error: in the presence of autocorrelation, the information available for predicting the value of y_t is concentrated in its immediate neighbours. The issue is discussed in more detail by Bowman and Azzalini (1997, Section 7.5).

Example 4.4

In the previous example, we fitted a GAM to the monthly mean wind speed series from De Bilt and noted that some residual autocorrelation was present. In that example, where the main aim was to discriminate between additive and bivariate model structures, autocorrelation and heteroscedasticity were handled using a block bootstrap to determine the distribution of the F statistic under the null hypothesis of additivity. Here, we use the same data to illustrate the effect of autocorrelation upon 'automatic' procedures for smoothing parameter selection.

Figure 4.4(a) shows a nonparametric estimate of the trend component in an additive model similar to that shown in Figure 4.3(a): in fact, the only difference between the two plots is that in Figure 4.3(a) the cross-validation search was constrained to ensure that the chosen smoothing parameter was no less than that for the bivariate model illustrated in Figure 4.3(c): by contrast, in Figure 4.4(a) the search was not so constrained. Notice that the estimated trend here is less smooth, with a 'plateau' in the 1980s.

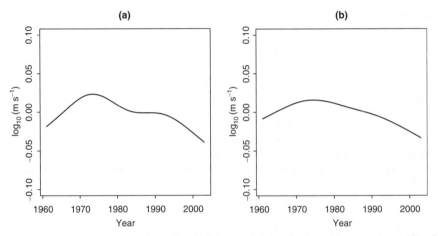

Figure 4.4   Trend estimates from the additive model for the logarithms (to base 10) of monthly mean wind speeds at De Bilt, with smoothing parameters chosen by cross-validation (a) without accounting for autocorrelation, (b) accounting for autocorrelation using the gamm *routine in* R.

As noted above, autocorrelation can lead to undersmoothing if cross-validation is used to choose a smoothing parameter and, to avoid this, when autocorrelation is present it is necessary to accommodate it in the fitting process. In panel (b) of Figure 4.4, the same additive model has been fitted as in panel (a) but now using the gamm routine with the error sequence (ε_t) in (4.12) taken as an AR(1) process (see Section 3.3.3). As expected, the estimated trend is now smoother: the effective degrees of freedom are reduced from 4.4 in panel (a) to 2.7 in panel (b). The parameter ϕ of the AR(1) process is estimated as 0.14; this agrees reasonably with the lag 1 residual autocorrelation coefficient of 0.12 in the previous example. The plateau in the 1980s is no longer present in panel (b). Although for many applications this difference would be relatively unimportant, it nonetheless illustrates that if automatic methods are used for smoothing parameter selection, a failure to account for even weak autocorrelation can lead to visible changes in estimated model components.   ∎

4.3   Other nonparametric estimation techniques

The nonparametric methods discussed so far in this chapter have all been based on local polynomial and spline based smoothers. Although these are appealing and intuitive, there are many other techniques that may be of use in the context of nonparametric trend estimation. In this section we briefly review some of these.

4.3.1   Lowess smoothing

One of the most widely used of all nonparametric regression techniques is the *locally weighted scatterplot smoother* (lowess) introduced by Cleveland (1979). In R, this is implemented via the `lowess()` and `loess()` functions; the form is a precursor to the latter, although the two implementations behave slightly differently. Originally designed for use with a single covariate, lowess has many features in common with local linear smoothing: the regression function at each covariate value of interest is from a weighted straight-line fit to the data, with most weight given to 'neighbouring' data points. The main differences between lowess and local linear smoothing are as follows:

- The straight lines are not fitted using weighted least squares, but rather using a robust estimation procedure (see Section 3.1.5) that aims to reduce the influence of a moderate number of outlying data points.

- The kernel weight function used to define the 'neighbourhood' of each covariate value of interest is taken to have finite rather than infinite support. The original proposal in Cleveland (1979), which has been followed in most subsequent implementations of the method, is to use the 'tricube' kernel

$$w\left(x_t - x; h\right) = \max\left(\left[1 - \left|\frac{x_t - x}{h}\right|^3\right]^3, 0\right);$$

the notation here is the same as for Equation (4.3).

- The kernel does not have a fixed width, but instead is chosen to include a fixed proportion of the observations; thus the bandwidth h in the tricube kernel above should strictly be written as $h(x)$. The kernel will therefore tend to be quite narrow in regions with a high density of observations, and the estimated regression function will have a reduced bias in these regions. By contrast, in areas with a low density of observations the kernel will tend to be rather wide. The bias in these areas is inevitably increased, but this is arguably a price worth paying since the data contain little information about the regression function here in any case. In the context of nonparametric trend estimation, this 'nearest neighbour' approach may be particularly useful for irregularly spaced series.

As a result of the robust estimation procedure used in lowess, formal statistical inference (for example to construct standard errors and to test hypotheses about the regression function) is more difficult than for the linear smoothers studied earlier in the chapter. Perhaps for this reason, neither of the two implementations in R provides standard errors or variability bands for the estimated regression function. These implementations are therefore perhaps best viewed as sophisticated tools to assist in the informal interpretation of graphical displays. The most obvious role for such tools is at the exploratory stage of an analysis; however, they can also be helpful when examining residuals from a fitted model, as seen already in Chapter 3 (see, for example, Figures 3.3 and 3.4).

Although lowess was originally designed as a tool for estimating the effect of a single covariate upon a response, the technique can also be used with multiple covariates, for example to estimate one of the components of an additive model (in this setting, lowess estimation would be carried out at the appropriate point in the backfitting algorithm). For the analysis of time series data, the *seasonal–trend lowess* (STL) decomposition (Cleveland *et al.*, 1990) is another example where lowess smoothing has been used to provide robust estimates of both trend and seasonality; this can be regarded as a modern development of classical procedures for splitting a series into 'trend', 'seasonal' and 'irregular' components (see Section 2.3.4).

The STL procedure is designed for use with regularly spaced time series where the seasonal period (i.e. the number of observations, say S, making up a complete seasonal cycle) is fixed. This allows S seasonal subseries to be defined: for monthly data, these subseries would consist of all the Januaries, all the Februaries and so on. Given an initial estimate of the trend, the original data are detrended. The seasonal component is then estimated by applying a lowess smoother separately to each of the subseries, reassembling the results into a single series and then high-pass filtering to remove any remaining trend. Finally, the trend component is updated by subtracting the estimated seasonal components from the original series and lowess smoothing the result. The process is iterated to convergence. Notice that by computing the seasonal component separately for each subseries, the resulting seasonal cycle is not guaranteed to be smooth: Cleveland *et al.* (1990) suggest that if this is seen as a problem, the final seasonal component can itself be smoothed using lowess.

In order to compute seasonal subseries for the STL procedure, clearly the observations must form a regularly spaced time series. In principle, missing data can be accommodated in the procedure, although the `stl` routine in R does not currently allow this. For an alternative procedure that achieves essentially the same goal and is capable of handling potentially large numbers of missing observations, see Section 5.5.5.

To use the STL procedure in practice, it is necessary to choose two smoothing parameters: one for the seasonal component and one for the trend. Cleveland *et al.* (1990) provide some suggestions for how to choose these, based on frequency considerations such as those outlined in Section 2.3. These suggestions form the basis for the default settings in the `stl` routine in R. As a lowess based technique, however, STL is often used as an essentially exploratory tool: for this purpose a subjective choice of smoothing parameter may be perfectly adequate. In this context it may be helpful to define the 'effective span' of a kernel as the range of observations receiving appreciable weight in the smoothing. For a regularly spaced series the tricube weight function used by lowess allocates around 90 % (respectively 95 %) of the total weight to observations within $0.6h$ (respectively $0.7h$) of the time point of interest, so the procedure may be thought of as averaging over windows of roughly this length.

Example 4.5

Figure 4.5 shows an STL decomposition of the monthly wind speed series from De Bilt. The top panel shows the original time series and the remaining three panels give the seasonal, trend and irregular components respectively. The smoothing parameters for both the trend and seasonal components have been chosen subjectively, in each case such that the lowess smoothing window spans 10 years on either side of the time point of interest (i.e. 21 years in total). As noted above, the procedure thereby gives appreciable weight only to observations within 6 or 7 years of each time point; a crude interpretation is therefore that we are averaging roughly over 15-year windows here. This choice ensures a smooth estimate of the trend, similar to those obtained earlier in this chapter using techniques such as cross-validation (note that some differences are due to the use of logged data in the earlier analyses, whereas Figure 4.5 is a decomposition of the untransformed wind speeds). A similarly large smoothing parameter is chosen to define the seasonal component; again, this is partly informed here by the earlier analyses, which suggested that there are no significant changes in the seasonal cycle through time.

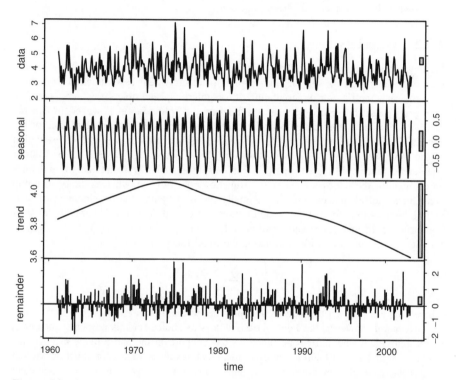

Figure 4.5   STL decomposition for monthly mean wind speeds from De Bilt. Units on all plots are $m\,s^{-1}$.

The grey bars at the right-hand end of each plot are all the same height and therefore give an impression of the relative magnitudes of the different components: for example, the trend accounts for a very small proportion of the overall variation.

Overall, given our earlier analyses of these data, the STL decomposition reveals little that is new. Note, however, the suggestion of two winter peaks in the estimated seasonal cycle throughout a large part of the record. Of course, formal methods would be required to determine whether this feature represents anything more than sampling variation. In particular, its apparent persistence over a period of more than 20 years should not be overinterpreted, because the seasonal component estimates in successive years are constructed from overlapping windows. However, if the feature is genuine then it probably would not have been picked up by our earlier analyses because they were based on an implicit assumption that the seasonal cycle in (log) mean wind speeds is smooth. The STL analysis provides the opportunity to detect such features because the seasonal component is computed separately for each seasonal subseries, as described above.   ■

Although the STL technique has been presented here primarily as an exploratory or descriptive tool, the literature indicates that it nonetheless can be extremely useful as a means of revealing structure in environmental time series. Applications include the assessment of trends in water quality (Esterby, 1993), turtle abundance and egg production (Balazs and Chaloupka, 2004; Chaloupka, 2001), atmospheric ozone (Carslaw, 2005) and groundwater levels (Shamsudduha *et al.*, 2009).

4.3.2 Wavelets

We have seen that spline smoothers can be regarded as representing an arbitrary smooth regression function as a sum of contributions from individual basis functions; a similar idea underlies the Fourier representation of a regularly spaced time series in Section 2.3. There is a slight difference between the Fourier and spline bases, however, in that the latter emerges as a representation of the underlying regression function by virtue of the penalised least squares criterion (4.4), whereas the former provides a direct representation of the observations $y_1, \ldots, y_T$ via Equation (2.11). Given this property of the Fourier coefficients, one could ask whether they can be split in such a way that one group defines an estimate of the regression (i.e. trend) function $m(\cdot)$ in (4.1), whereas the other group constitutes 'noise': indeed, this is essentially the idea behind the development of low- and high-pass filters for estimating and removing trend in Section 2.3. Formally, given a regularly spaced time series and set of basis functions $\{\psi_j(\cdot)\}$ such that $y_t = \sum_j c_j \psi_j(t)$ for each t, one might consider estimating the trend function as

$$\hat{m}(t) = \sum_{j \in \mathcal{C}} b_j \psi_j(t),$$

where $\mathcal{C}$ denotes the set of indices corresponding to the trend.

In general, however, the Fourier basis is a poor choice for this purpose, since each of the individual sine and cosine functions has support on the entire range of covariate values: this makes it difficult to represent local features in the regression function without using a large number of Fourier coefficients. This is undesirable since each coefficient itself will be influenced by the noise (ε_t) as well as the regression function: thus the need for a large number of coefficients will in general lead to an estimator with high variance.

Of course, the Fourier basis provides a parsimonious description of a sinusoidal or periodic regression function. Similarly, the polynomials of degree 0 to degree $T - 1$ provide a basis for the observations $y_1, \ldots, y_T$: this basis provides a parsimonious description of trends that are well approximated by low-order polynomials, since in this case one needs to extract just the first few coefficients in the basis representation. In general, however, in a nonparametric regression context one is unable or unwilling to specify a particular form for the underlying regression. The question thus arises: Is it possible to find bases that provide relatively parsimonious descriptions of a wide range of regression functions?

At the risk of offending mathematical readers by failing to define exactly what is meant by 'a wide range of regression functions' (for details, see Donoho *et al.* 1995), the answer to this question is 'yes'. The bases in question are known as *wavelet* bases, since the constituent functions can be regarded in some sense as 'local' versions of the sine and cosine waves used in a Fourier representation. We here give a very heuristic summary of the main ideas. For a more detailed introductory account, see Sadiku, Akujuobi and Garcia (2005) or Nason and Silverman (1994); the latter also serves as a reference for the R package wavethresh (Nason, Kovac and Maechler, 2009), which can be used to implement the methods described below. Those wishing to go beyond mere dilettantism should consult book-length treatments such as Nievergelt (1999), Percival and Walden (2000) or Nason (2008).

Each constituent function in a wavelet basis is defined so as to correspond roughly to fluctuations at a specific time point and temporal scale or 'resolution'. To emphasise this, they are usually written with a double subscript: $\{\psi_{jk}\}$, say, where j and k are the 'resolution' and 'time' indices respectively. The basis function representation of the original series is therefore

$$y_t = \sum_j \sum_k c_{jk} \psi_{jk}(t). \tag{4.16}$$

To guarantee that this representation is valid and unique, the wavelet bases discussed here are constrained to be orthonormal; i.e. the functions $\{\psi_{jk}(\cdot)\}$ satisfy $\int \psi_{jk}^2(t)dt = 1$ and $\int \psi_{jk}\psi_{j'k'}(t)dt = 0$ for $(j, k) \neq (j', k')$. A further feature is that each function is a scaled and shifted version of a 'mother wavelet' $\psi(\cdot)$: specifically, $\psi_{jk}(t) = 2^{j/2}\psi(2^j t - k)$; some examples are shown in Figure 4.6. The mother wavelet has compact support (i.e. is nonzero only over a finite range), in an attempt to ensure that the individual $\{\psi_{jk}(\cdot)\}$ are associated with specific time intervals. Specifically, $\psi_{jk}(\cdot)$ can be thought of as corresponding to an interval of length $T/2^j$ ending at time $kT/2^j$, whence a large value of $|c_{jk}|$ in (4.16) suggests that there is some important structure in the vicinity. The correspondence is often rather approximate, however, because individual wavelets tend to extend for some distance outside this nominal interval. This is illustrated in Figure 4.6, where the grey band indicates the location of the interval relative to the specific wavelets illustrated there.

Perhaps surprisingly, it is not necessary to evaluate the wavelets themselves to calculate the coefficients $\{c_{jk}\}$: indeed, many of the wavelets in current use do not have an analytical representation and are defined only indirectly. The *discrete wavelet transform* (DWT) of a time series is usually computed by a succession of filtering operations in which, starting with the finest resolution, a high-pass filter is used to extract $T/2$ wavelet coefficients and a companion *scaling filter* is used to define what is effectively an aggregated series of $T/2$ *scaling coefficients*. The choice of wavelet determines the filter weights for both the wavelet and scaling filters; the length of these filters (i.e. the number of nonzero weights), L, say, is always an even number. Taking the scaling coefficients as data, the operation is then repeated to obtain wavelet and scaling coefficients at the next resolution. At each resolution, the total number of wavelet and scaling coefficients is always T.

Clearly, if T is a power of two then the successive aggregation into coarser-resolution series will ultimately yield a single number (this is interpretable as an overall mean level). In fact, this constraint on the series length is required by most wavelet software since this enables the use of a fast algorithm, due to Mallat (1989), for computing the DWT. Of course, this requirement is rarely met in practice: to use wavelet methods, therefore, it is usually required to pad out the ends of the series to the required length, for example

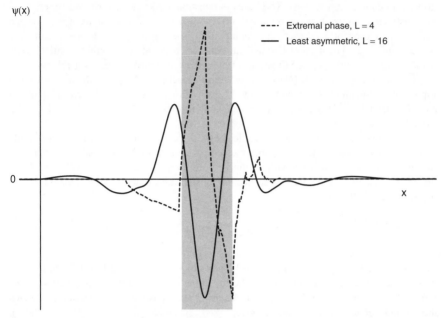

ψ(x)

Figure 4.6   Examples of mother wavelets from two of the families proposed by Daubechies (1988). L denotes the length of the associated filter. The grey band indicates the nominal support of each wavelet.

using some of the schemes suggested in Section 2.3.3. Further modification is required to ensure that the filters can be applied at the ends of the series: this is usually done by extending the data either periodically or by reflection (for example by appending the values $y_{T-1}, y_{T-2}, \ldots$ to the series as necessary). Clearly, this will affect the interpretation of the wavelet coefficients at both ends of the series and will have a greater effect for long filters than for short ones. At the coarsest resolutions where the filtering operations are carried out on short, highly aggregated series, the end effects can be substantial.

The requirements of orthonormality, compact support and high-pass filtering impose constraints on the choice of function to act as a mother wavelet $\psi(\cdot)$. However, some freedom remains, perhaps most obviously in the filter length L. As just discussed, longer filters are more likely to suffer from edge effects; however, they also have more flexibility and hence can in principle be designed to handle specific structures of interest more accurately. For example, Figure 4.6 illustrates members of the 'extremal phase' and 'least asymmetric' wavelet families of Daubechies (1988). These families have the property that for a filter of length L, information on polynomial trends of degree up to $L/2$ is concentrated entirely in the wavelet coefficients that are unaffected by end effects (Craigmile, Guttorp and Percival, 2004). In practice, this means that polynomial trends would manifest themselves primarily as large wavelet coefficients at the coarsest resolutions. This is entirely consistent with our view of trends as corresponding to low-frequency features.

Since the operations required to compute the wavelet coefficients are all linear filters, the coefficients themselves can be regarded as a linear transformation of the data: $\mathbf{c} = \mathbf{W}\mathbf{y}$, say, where $\mathbf{c}$ denotes the vector of all the wavelet coefficients (together with the single scaling coefficient representing the overall mean) and $\mathbf{W}$ is a $T \times T$ matrix. The orthonormality of wavelet bases ensures that $W'W = I_T$, the $T \times T$ identity matrix: thus W' is the inverse of W and the original data can be reconstructed from the wavelet coefficients as $y = W'\tilde{c}$.

We can now return to the problem of trend estimation. Noting that information on trends is likely to be concentrated in the coarse-resolution wavelet coefficients, one possibility would be to modify the coefficient vector c by setting to zero the coefficients at all but the coarsest resolutions, and then carrying out the inverse DWT: $\hat{m} = W'\tilde{c}$, say, where $\tilde{c}$ denotes the modified coefficient vector and $\hat{m}$ is the estimate of the trend function. However, such an approach would fail to pick up features at finer resolutions that may also be of interest: for example, a changepoint or discontinuity (see Sections 3.4.3 and 4.3.4) would be expected to give rise to large wavelet coefficients at several resolutions, near the time point at which the discontinuity occurred. As an alternative, therefore, we may choose to set to zero all wavelet coefficients with magnitudes below some threshold and to reconstruct the trend estimate from the remaining coefficients. A further refinement, essentially acknowledging that the noise (ε_t) affects all wavelet coefficients including those influenced by the trend, is to shrink the remaining coefficients towards zero as well. In this case, the kth modified coefficient at resolution j is

$$\tilde{c}_{jk} = \text{sign}\left(c_{jk}\right) \times \max\left(\left|c_{jk}\right| - c_{\text{crit}}, 0\right), \qquad (4.17)$$

where c_{crit} is the chosen threshold. This idea is known as *wavelet shrinkage*, and the scheme (4.17) is referred to as *soft thresholding*.

To use wavelet shrinkage in practice, it is necessary to choose the threshold c_{crit}: this plays exactly the same role as the smoothing parameter or bandwidth in other nonparametric estimation procedures. Based on ideas of minimising some measure of expected estimation error for a wide variety of regression functions, Donoho and Johnstone (1994) proposed the use of a *universal threshold*: $c_{\text{crit}} = \hat{\sigma}\sqrt{2\log T}$, where $\hat{\sigma}$ is an estimate of the error standard deviation in (4.1), which is assumed constant. Nason (1996) suggested a cross-validation procedure (the usual 'leave-one-out' cross-validation cannot be applied directly in a wavelet analysis, because after omitting an observation the sample size is no longer a power of two), although as usual this does not perform well in the presence of autocorrelation; threshold selection for correlated sequences is considered by Johnstone and Silverman (1997). Alternatively, one may use different thresholds at each resolution level j (Donoho *et al.*, 1995). In the context of a trend analysis, it may be appropriate to set a threshold of zero for all coarse-resolution coefficients so that any low-frequency components are automatically included in the trend estimate, and to apply wavelet shrinkage only at the finer resolutions. In R, wavelet shrinkage is implemented in the `wavethresh` package already mentioned, as well as in the `waveslim` library (Whitcher, 2007).

Example 4.6

To illustrate the application of wavelet shrinkage in a trend analysis context, we consider once again the monthly mean wind speed series from De Bilt. As in Section 4.2, we work with logarithms to stabilise the variance. The series runs from 1961 to 2002 and hence contains $42 \times 12 = 504$ monthly observations. This is fortuitously close to a power of two: for the purpose of the wavelet analysis, therefore, four additional values have been appended at each end to yield a series of length $T = 512 = 2^9$. In view of the strong seasonality in the series, the observations for September to December 1961 have been appended to the start and those for January to April 2002 to the end. This should avoid undue distortion in the results.

Figure 4.7(a) shows the wavelet coefficients for the padded series; each 'row' should be interpreted as a time series plot of the coefficients at the corresponding resolution level. The wavelet used here is of the Daubechies 'least asymmetric' class with filter length $L = 16$ (solid curve in Figure 4.6). The finest-resolution wavelet coefficients (level $j = 8$) are shown at the bottom of the plot; there are $T/2 = 256$ of these, each corresponding roughly to a 2-month period. There is little obvious structure at this resolution, except possibly for a single large negative coefficient towards the end: this presumably indicates a large change in wind speed in two successive months around the corresponding time. In general, the coefficients at most other levels are similarly uninteresting. At levels $j = 6$ and $j = 5$, however, visually at least the coefficients display some regularity. This is due to the seasonality in the data: these levels correspond to structures on scales of roughly 8 and 16 months respectively. It is a slight disadvantage of the wavelet approach that the seasonal cycle cannot be resolved explicitly here because the cycle length is not a power of two.

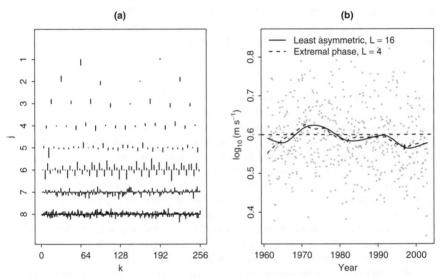

Figure 4.7   Wavelet analysis of monthly wind speeds at De Bilt: (a) wavelet coefficients based on a Daubechies least asymmetric wavelet of length $L = 16$, (b) trend estimates using wavelet shrinkage for two different wavelets. Grey points indicate data values.

The seasonality in the series creates potential problems for trend estimation using wavelet shrinkage if the same threshold is used at all resolution levels. The many large coefficients at level $j = 6$ are associated with seasonal variation rather than trend, and these should not be allowed to influence the results unduly. In this situation it may be preferable to use resolution-dependent thresholds, so that a coefficient at level j will only be retained if it is large relative to the typical values of other coefficients at that level. Moreover, as indicated above, it is probably worth keeping all of the coarse-resolution coefficients. The solid line in Figure 4.7(b) is the estimated trend obtained by soft-thresholding, with a separate threshold for each of levels 3 to 8 and leaving levels 1 and 2 unchanged; the analysis has been carried out using the `wavethresh` library in R. The dashed line shows an estimate obtained in the same manner but using a different wavelet. Apart from end effects, the two estimates are broadly similar, although this second estimate is not as smooth as the first: this reflects the differences in the mother wavelet shapes shown in Figure 4.6. The solid curve displays a local irregularity in the mid 1990s: this is due to the single large coefficient at resolution level 8 that was noted in Figure 4.7(a). Although this irregularity is of limited interest in the current context (it can probably be attributed either to sampling variability or to an outlier in the data), it does illustrate that wavelet shrinkage provides the opportunity to detect local features that could be smoothed away using other approaches. If a trend estimate is required that does not include this feature, this can be achieved by setting to zero all of the finest-resolution wavelet coefficients.

Apart from the irregularity in the 1990s, the trend estimates in Figure 4.7(b) are perhaps visually less appealing than the spline and lowess based versions seen earlier (Figures 4.3 and 4.5): although qualitatively similar (increasing until around 1970 and declining thereafter), they are less smooth and there is a possibility that end effects may be causing some bias. It is worth bearing in mind, however, that wavelet shrinkage is designed to cope with a worst-case scenario: if the analyst suspects that the data may conceal nasty but genuine structures of unknown form, wavelets provide a robust and relatively safe way of estimating them. Generality comes at a price however: methods such as splines, local linear smoothing and lowess are tailored to the estimation of smooth trends and hence will tend to do a better job if the assumption of smoothness is justified.   ■

As in this example, trend estimates obtained from different choices of wavelet will usually be qualitatively similar, at least so long as the wavelet filter length L is small compared with the sample size T (otherwise, information on trends may 'leak' into the finer-resolution coefficients as a result of end effects, making it difficult to associate specific coefficients with the structure of interest). Nonetheless, it may be helpful to assess whether there are situations in which some wavelets are likely to perform better than others. Horgan (1999) reports the results of some simulation studies involving different trend functions and different wavelets, and derives some recommendations regarding the choice of wavelet and threshold in different settings.

Apart from the use of wavelet shrinkage to estimate trends, an attractive feature of the DWT is its ability to resolve and identify features in the data at different scales. However, except at the finest resolutions it is difficult to line up features from a plot such as Figure 4.7(a) with the corresponding features in the original series. This is partly because the number of wavelet coefficients halves with each decrease in resolution and partly because each wavelet also picks up structure falling outside its nominal range of support, as illustrated in Figure 4.6. If it is of interest to locate specific features as accurately as possible, it may be preferable to carry out a modified version of the DWT in which wavelet coefficients are calculated for every time point $1, \ldots, T$ at each resolution level. The resulting *maximal overlap discrete wavelet transform* (MODWT) enables features in the wavelet decomposition to be lined up unambiguously with those in the original time series; it has the additional advantage that the series length is not restricted to a power of two. For further discussion of the MODWT, with an application to the analysis of fluctuations in sea level, see Percival and Mofjeld (1997). The technique has also been used to analyse relationships between different atmospheric time series (Whitcher, Guttorp and Percival, 2000b), in which context it can be seen as an extension to the cross-correlation analysis discussed in Section 2.1.5. It is implemented in the `waveslim` library in R, as well as in the `wavelets` library (Aldrich, 2009).

4.3.3 Varying coefficient models

In environmental systems, relationships between variables may change over time and throughout the year. To capture such effects in a statistical model, it is necessary to allow the model parameters to vary over time. This can be achieved using state space methods, as discussed in Section 5.5.2. Alternatively, *varying coefficient models* (VCMs) can be used. In a VCM the covariate effects are modelled linearly but with coefficients that change smoothly with the values of other variables (Hastie and Tibshirani, 1993). Thus the linear predictor η_t has the form

$$\eta_t = \beta_0 \left(r_{0t} \right) + \beta_1 \left(r_{1t} \right) x_{1t} + \cdots + \beta_p \left(r_{pt} \right) x_{pt}. \tag{4.18}$$

Here, for each $j \in \{0, \ldots, p\}$, r_{jt} is the value of some variable (which may be one of the other covariates in the model) that affects the regression coefficient of x_{jt} via the smooth function $\beta_j(\cdot)$–this can be seen as a form of interaction, as noted by Ruppert, Wand and Carroll (2003, Section 12.4). The model can, of course, be simplified if required by setting some of the regression coefficients to be constant; this leads to the class of semivarying coefficient models (Xia, Zhang and Tong, 2004).

In the present context, the most relevant application of VCMs is to allow the regression coefficients to depend on the time index t. In this case, the model (4.18) becomes

$$\eta_t = \beta_0 \left(t \right) + \beta_1 \left(t \right) x_{1t} + \cdots + \beta_p \left(t \right) x_{pt},$$

which has a simple and direct interpretation.

Example 4.7

To represent the classic environmental time series structure of 'trend plus seasonality', one could start by writing down a model of the form

$$Y_t = \beta_0 \left(t \right) + \beta_1 \left(t \right) \cos \left[2\pi t - \tau \left(t \right) \right] + \varepsilon_t, \tag{4.19}$$

where here the time index t is measured in years. In this model the smooth term $\beta_0(\cdot)$ represents the trend and $\beta_1(\cdot)$ is a varying coefficient seasonal term, modelled parametrically via the use of a cosine function. The phase $\tau(\cdot)$ of the seasonal cycle is also allowed to vary smoothly over time.

At first sight, the presence of the time-varying phase in (4.19) suggests that the model is not a VCM as defined previously. However, using trigonometric identities in the same way as for the parametric regression models in Section 3.2.1, it is straightforward to show that (4.19) can equivalently be written as

$$Y_t = \beta_0 \left(t \right) + \gamma_1 \left(t \right) \cos 2\pi t + \gamma_2 \left(t \right) \sin 2\pi t + \varepsilon_t,$$

where $\gamma_1(t) = \beta_1(t) \cos \left[\tau \left(t \right) \right]$ and $\gamma_2(t) = \beta_1(t) \sin \left[\tau \left(t \right) \right]$. This is a VCM with two varying coefficients; having fitted this model, the time-varying phase can be reconstructed as $\tau(t) = \arctan \left[\gamma_2(t) / \gamma_2(t) \right]$.

Notice that, by contrast with the additive models of Section 4.2, in this VCM the seasonal cycle is allowed to vary over time. The STL method, discussed in Section 4.3.1, is another way of achieving this. ∎

Estimation and inference for VCMs was initially discussed by Hastie and Tibshirani (1993), with subsequent developments by Fan and Zhang (1999), Fan and Zhang (2000) and Cai (2002). Smoothing parameter selection via cross-validation is considered by Hoover *et al.* (1998).

In one sense, VCMs can be viewed as a level in a hierarchy of increasingly complex model structures. At the bottom of this hierarchy is the additive model (4.13), followed by the additive semiparametric model (4.14) in which parametric and nonparametric components are combined. The VCM represents the next level of complexity, allowing smooth variation in the parameter vector β. Eilers and Marx (2002) describe a practical modelling approach that encompasses all of these structures, with a spline based representation of smooth functions that enables simultaneous estimation of the model components using penalised likelihood.

4.3.4   Discontinuity detection

In Section 3.4.3 we reviewed a variety of methods for detecting changepoints in environmental time series. Those methods were based on parametric models, where the changepoints corresponded to times at which one or more of the parameters experienced an abrupt change. In a nonparametric setting, clearly it is not meaningful to talk about changes in parameters. One may nonetheless be interested in the possibility of discontinuities in regression functions that are otherwise smooth.

There have been several interesting recent developments attempting to tackle this problem. Whitcher, Guttorp and Percival (2000a) use wavelet methods (see Section 4.3.2) to study changes in variance in the presence of autocorrelation, whereas Cheng and Raimondo (2008) develop kernels that are designed to detect changepoints when used as smoothers. Many other approaches (Bowman, Pope and Ismail, 2006; Hall and Titterington, 1992; Loader, 1996; Müller, 1992; Qiu and Yandell, 1998) are based on a simple idea: linear smooths obtained from data to the left and right of a particular location (i.e. covariate value) will yield similar results if the regression function is continuous there, but will differ in the presence of a discontinuity. This suggests a formal procedure to test the null hypothesis of no discontinuities: calculate left and right smooths at a range of locations and compute some overall measure of the difference between them. Large values of such a measure indicate the presence of at least one discontinuity. We now describe a version of this idea due to Bowman, Pope and Ismail (2006); the procedures are implemented in the sm.discontinuity function of the sm package in R.

Let $\hat{m}_L(x_t)$ and $\hat{m}_R(x_t)$ denote the left and right smooths obtained at covariate value x_t; then a plausible statistic for testing the null hypothesis is

$$T = \sum_t [\hat{m}_L(x_t) - \hat{m}_R(x_t)]^2 / V_t,$$

where the sum is over all locations at which the left and right smooths are computed and V_t denotes the variance of $\hat{m}_L(x_t) - \hat{m}_R(x_t)$ (this can be estimated using techniques similar to those described in Section 4.1.6). In the presence of autocorrelation the test statistic may be modified, using ideas such as those discussed in Section 4.2.4. The null distribution can be approximated using a ratio of quadratic forms, whence a p-value can be obtained. As usual, it is worth using a significance trace to explore the sensitivity of results to the choice of smoothing parameter. For an example of this in an environmental context, see Bates *et al.* (2011).

The test just described is global, in the sense that rejection of the null hypothesis indicates merely that at least one discontinuity is present somewhere in the regression function. To indicate the discontinuity location(s), it can be helpful to plot the left and right smooths at each location and to overlay a reference band, centred on the average of the two smooths and with width $2\sqrt{V_t}$. Both smooths will fall outside this band where they differ by more than 2 standard errors: these points can be considered as candidate discontinuity locations. In practice, if the null hypothesis is rejected then one might consider that there is a discontinuity at the point with the largest standardised difference, and then repeat the procedure separately for the observations on either side of this location to search for further discontinuities. For more details and discussion, see Bowman, Pope and Ismail (2006).

4.3.5   Quantile regression

So far we have dealt exclusively with models for the conditional expectation of the response variable Y given a covariate vector $\mathbf{x}$. If we are prepared to make specific assumptions about the distributional form of the response (for example that it is Gaussian with a constant variance), then this could be used to make inferences about other aspects of the distribution, such as the way in which covariates affect specific percentiles. Examples of applications in which this might be of interest include air pollution (where the higher percentiles may have a direct impact on human health) or the study of rainfall (where high percentiles may be associated with the risk of flood defence failure).

If, however, we are unwilling to make specific assumptions about the distribution of the response variable, then the techniques studied so far provide little information about aspects other than the mean. In this case, one may consider specifying a regression style model directly for the quantile(s) of interest. This can be achieved using the technique of *quantile regression* (Koenker, 2005; Koenker and Bassett, 1978), which was first developed by econometricians as an extension to the linear model. Appealing features of quantile regression include the flexibility to deal with heterogeneous conditional distributions and robustness to outliers. Median regression, for example, has often been proposed as a robust regression method (this is demonstrated below); this is an example of quantile regression in which the median (i.e. 0.5 quantile) of the response distribution is modelled. More general quantile regression is a natural extension to this.

A *linear quantile regression model* takes the form

$$Q_Y(\theta|\mathbf{x}) = \beta_0^{(\theta)} x_0 + \beta_1^{(\theta)} x_1 + \cdots + \beta_p^{(\theta)} x_p \quad \text{for } \theta \in (0, 1), \tag{4.20}$$

where $Q_Y(\theta|\mathbf{x})$ is the θth quantile (i.e. the 100θth percentile) of the distribution of Y conditional on $\mathbf{x}$. In (4.20), the coefficients $\{\beta_j^{(\theta)}\}$ also depend on the quantile θ.

Models such as (4.20) cannot be fitted using methods that require the distribution of the response variable to be specified in parametric form: quantile regression deliberately avoids any such specification. Thus neither maximum likelihood nor Bayesian methods can be used. Furthermore, methods such as least squares are clearly inappropriate. It is therefore not immediately clear how to estimate the coefficients $\{\beta_j^{(\theta)}\}$. The key is to start by considering the quantiles of a set of data $y_1, \ldots, y_T$ in the absence of covariates. For simplicity of exposition, we also assume that there are no ties and denote the ordered data values by $y_{(1)} < y_{(2)} < \cdots < y_{(T)}$. For notational convenience, define $y_{(T+1)} = \infty$. Providing θT is not an integer, the θth quantile (q_θ, say) is given by $y_{(\lceil \theta T \rceil)}$: here, $\lceil x \rceil$ is the value obtained by rounding x up to the next integer (thus $\lceil 4.1 \rceil = 5$ for example). If θT is an integer, q_θ is not uniquely defined: all values in the semi-open interval $[y_{(\theta T)}, y_{(\theta T+1)})$ are candidates.

It is not immediately obvious, but q_θ can be defined equivalently as the minimiser of the function

$$L_\theta(q) = \sum_{t=1}^{T} (y_t - q) \left[\theta I (y_t \geq q) - (1 - \theta) I (y_t < q) \right],$$ (4.21)

where $I(\cdot)$ is an indicator function taking the value 1 if its argument is true and 0 otherwise. To see this, note that for $q \in \left(y_{(i)}, y_{(i+1)} \right]$ we have

$$L_\theta(q) = \theta \sum_{j=i+1}^{T} \left(y_{(j)} - q \right) - (1 - \theta) \sum_{j=1}^{i} \left(y_{(j)} - q \right),$$ (4.22)

with the understanding that the first and last terms are zero for $i = T$ and $i = 1$ respectively. This is linear in q; thus $L_\theta(q)$ is piecewise linear and the overall minimum must occur at the end of one of the intervals $\left\{ \left(y_{(i)}, y_{(i+1)} \right] \right\}$. Furthermore, (4.22) can be used to show that

$$L_\theta(y_{(i+1)}) - L_\theta(y_{(i)}) = [\theta T - i] \left(y_{(i)} - y_{(i+1)} \right).$$

Since $y_{(i)} < y_{(i+1)}$ by definition, it follows that $L(q)$ decreases over intervals where $i < \theta T$, remains constant where $i = \theta T$ and increases where $i > \theta T$. The minimum therefore occurs at the right-hand end of the last interval with $i < \theta T$, i.e. at $y_{(\lceil \theta T \rceil)}$, which is the required quantile.

By analogy with (4.21), Koenker and Bassett (1978) proposed to estimate the coefficients $\{\beta_j^{(\theta)}\}$ in (4.20) by minimising the expression

$$L_\theta(\boldsymbol{\beta}) = \sum_{t=1}^{T} (y_t - q_t(\boldsymbol{\beta})) \left[\theta I (y_t \geq q_t(\boldsymbol{\beta})) - (1 - \theta) I (y_t < q_t(\boldsymbol{\beta})) \right],$$ (4.23)

where now $q_t(\boldsymbol{\beta}) = \sum_{j=0}^{p} \beta_j x_{jt}$. If the model is specified correctly then, under assumptions similar to those underlying the standard theory of maximum likelihood estimation, it may be shown that the resulting estimates converge to their true values as the sample size increases. As in Section 3.5.1, however (and for the same reason), some additional work is required to demonstrate this in the context of a trend analysis where one of the covariates is the time index.

Notice that if $\theta = 1/2$, corresponding to median regression, (4.23) can be written as

$$L_{1/2}(\boldsymbol{\beta}) = \frac{1}{2} \sum_{t=1}^{T} |y_t - q_t(\boldsymbol{\beta})|.$$

The estimates resulting from median regression are therefore the same as those from least absolute deviations (LAD) regression (see Section 3.1.5). Thus, in the context of a trend analysis, LAD regression can be regarded as estimating trends in the median of the response variable. This connection also explains why median regression can be considered as robust to outliers.

Standard gradient based numerical optimisation methods cannot be used to minimise (4.23), because $L_\theta(\boldsymbol{\beta})$ is not differentiable everywhere (indeed, as with $L_\theta(q)$ in (4.21), it is not even continuous everywhere). However, the minimisation can be formulated as a linear programming problem for which well-understood algorithms are available.

Although standard linear programming algorithms can be computationally demanding for large data sets containing several thousands of observations, alternatives are available for use in such situations (see Portnoy and Koenker, 1997, for example).

As usual, bootstrap and other resampling methods may be used to obtain confidence intervals for quantile regression estimates; theoretical alternatives are also available. For a variety of different approaches to the construction of confidence intervals, see Hao and Naiman (2007), He and Hu (2002), Koenker and D'Orey (1993), Koenker and Hallock (2001), Koenker and Machado (1999). Note, however, that as usual most of these methods assume that the observations are conditionally independent given the covariates, so caution is required when analysing time series data. Methods for handling autocorrelated series are so far relatively undeveloped.

As described so far, quantile regression is nonparametric in the sense that it makes no explicit assumptions about the distribution of the response variable; however, the linear dependence of the quantiles on the covariates is restrictive. It is also slightly unsatisfactory if any of the coefficients $\{\beta_j^{(\theta)}\}$ varies with the quantile θ, since in this case there will be covariate configurations for which the fitted quantile functions at different thresholds can cross: this could lead to nonsensical situations such as a modelled lower quartile exceeding the corresponding median. In practice, such problems are unlikely to occur within the range of the data used for model fitting, but it is worth being aware of the potential problem if extrapolation is required.

Recent developments have sought to overcome the restriction of linearity by allowing a nonparametric specification of the quantile functions themselves; this is a current research topic in statistics and econometrics. For example, an additive quantile regression model can be specified, analogously to the usual GAM (see Section 4.2.1), as

$$Q_Y(\theta|\mathbf{x}) = \alpha^{(\theta)} + \sum_{j=1}^{p} g_j^{(\theta)}(x_j),$$

where the $\{g_j^{(\theta)}(\cdot)\}$ are assumed to be continuous smooth functions. Of course, semiparametric versions can also be defined. Such models can be fitted by minimising a penalised version of the usual objective function (4.23), the penalty being added to encourage smoothness in exactly the same way as for spline smoothers (see Section 4.1.3). Koenker, Ng and Portnoy (1994) propose a slightly different penalty term from that usually used for smoothing splines; the resulting fitted functions are piecewise linear with knots at the observed covariate values. As with all nonparametric regression techniques, the results depend on the choice of smoothing parameter.

The methods described above are all implemented in the quantreg package in R (Koenker, 2009).

Example 4.8

So far in this chapter, we have used nonparametric regression methods to study trends in monthly and annual mean wind speeds at De Bilt. However, as noted in Section 1.3.1, the mean wind speed is not always the most relevant quantity in applications: in determining the feasibility of a wind energy scheme, for example, potential changes in both the upper and lower tails of the wind speed distribution are of interest, at a daily (or finer) time scale. The boxplots in Figure 2.2 provide some indication that, for example, there is substantial seasonal variation in the upper tail of the daily wind speed distribution at De Bilt but little variation in the lower tail. Quantile regression provides a useful way of exploring this further.

Figure 4.8 shows the results of fitting regression models for the 5th and 95th percentiles (quantiles $\theta = 0.05$ and 0.95) of the daily wind speed distributions at De Bilt, using the rqss routine in the quantreg library in R. The fitted models are semiparametric, taking the form

$$Q_Y(\theta) = \alpha^{(\theta)} + g^{(\theta)}(t)$$

$$+ \beta_1^{(\theta)} \cos\left(\frac{2\pi \times \text{day of year}}{365}\right) + \beta_2^{(\theta)} \sin\left(\frac{2\pi \times \text{day of year}}{365}\right),$$

where 'day of year' runs from 1 to 365 (ignoring leap years) and t denotes the time index. The function $g^{(\theta)}(t)$ is a smooth term representing trend, and the sine and cosine terms provide a simple but plausible parametric representation of seasonality, as discussed in Section 3.2.1. This parametric representation is adopted partly to ease the computation. Further computational savings have been achieved by fitting to just 10 % of the data: every 10th observation has been used here. This still leaves 1534 observations for model fitting, regularly distributed through time and through the year so that accurate estimates of trend and seasonality can be obtained. To fit each model to these data takes less than a second on a modern PC. Apart from computational considerations, subsampling the data also provides a crude means of reducing autocorrelation: in Figure 2.3(d) we saw that there is very little autocorrelation in daily wind speeds at lags greater than 10 days (although in any case, autocorrelation is not too much of a problem for the analysis below because we do not attempt to compute confidence intervals or standard errors, or to allow the data to determine the smoothing parameter).

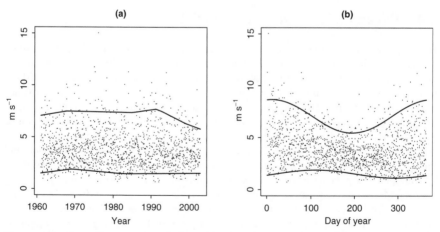

Figure 4.8   Quantile regression fit for 5th and 95th percentiles of daily wind speed distributions at De Bilt: (a) trend and (b) seasonal component. The points represent the observations. The trend component is estimated nonparametrically, with the smoothing parameter chosen by visual inspection.

The smoothing parameter for the trend estimates in Figure 4.8(a) has been chosen on the basis of visual inspection of the results from several different trial values. These estimates indicate little change in the 5th percentile of the daily wind speed distribution over the 40-year period of record; however, there is a striking decline in the 95th percentile over the final decade. Clearly, a full investigation of this would require some assessment of the uncertainty in the estimate, but it is consistent with the recent declining trend in mean wind speeds that was demonstrated in Sections 4.1.7 and 4.2. Taken together, these results suggest that the decline in mean wind speeds is at least partly associated with a reduction in strong winds.

Figure 4.8(b) shows the estimated seasonal cycles in the two chosen quantiles. This confirms the message from the boxplots in Figure 2.2: seasonality is much more pronounced in the upper tail of of the distribution. However, it also shows that the phase of the seasonal cycle seems to be different for the two quantiles. Strong winds are seen to occur more frequently in the first and fourth quarters of the year, which is unsurprising, but the lowest winds tend to occur towards the end of the year rather than (as one might naïvely expect, at least without a good knowledge of the meteorology of the area) in the summer months. This demonstrates the capacity for quantile regression to provide additional insights into the structure of a data set.   ■

In this example, although quantiles $\theta = 0.05$ and $\theta = 0.95$ are in the tails of the wind speed distribution, the corresponding wind speeds are not atypical: a wind speed above the 0.95 quantile of the distribution occurs on average every 20 days. In some applications it is necessary to study changes in quantiles that are much more extreme than this. In principle, quantile regression could be used for this purpose; however, by definition there are typically few data points available to estimate extreme quantiles and the resulting estimates are therefore likely to be very imprecise. If extreme quantiles are of interest, alternative approaches may therefore be preferable. These are discussed in Section 6.4.

4.4   Parametric or nonparametric?

Having reviewed a wide variety of parametric and nonparametric representations of trend, it is natural to ask which is preferable in any given situation. As always, there is a trade-off. Parametric models make stronger assumptions and hence, when these are satisfied, potentially lead to more powerful analyses. They also provide the opportunity for extrapolation beyond the range of the data where this appears justified: it is not obvious how to extrapolate meaningfully in a nonparametric setting (see, however, the discussion in Section 5.5). They yield more precise answers to well-defined scientific questions, if the questions themselves can be related to the values of parameters in such a model, and finally the risk of spurious residual structure is reduced (see Section 3.1). However, if the assumptions of a parametric model break down there is increased risk of misinterpreting the data. Moreover, nonparametric models offer increased flexibility in the range of structures that can be represented relatively straightforwardly. Users of both parametric and nonparametric models can protect against misinterpretation by checking their assumptions carefully, using techniques such as those introduced in Section 3.1.

With the availability of semiparametric models, it is tempting to suggest using parametric representations of relationships where possible (perhaps using techniques such as those described in Section 4.1.7 to help determine this), and nonparametric models otherwise. This suggestion is not as helpful as it sounds, however. For example, there are many time series in which the data cannot be used to distinguish between linear trends with strongly autocorrelated residuals and highly nonlinear trends that are totally unsuitable for parametric modelling (see Buckland et al., 2004, Section 5.3 for an example). This is because persistent departures from a linear trend could be viewed either as a failure of the linear assumption or as a general tendency for successive residuals to be very similar. In such situations, the choice between parametric and nonparametric approaches depends on the scientific context and in particular on whether the analyst chooses to regard the fluctuations as an interesting part of the trend or simply as a nuisance. If the former, nonparametric methods will be necessary; otherwise, it may be more appropriate to use a linear trend model with autocorrelated error structure. This suggests that the distinction between parametric and nonparametric trend modelling may not be so clear as it first appears. We return to this in Section 5.5.

References

Aldrich, E. (2009) wavelets: *A Package of Functions for Computing Wavelet Filters, Wavelet Transforms and Multiresolution Analyses*. R package version 0.2-4.

Balazs, G. H. and Chaloupka, M. (2004) Thirty-year recovery trend in the once depleted Hawaiian green sea turtle stock. *Biological Conservation*, **117**, 491–498.

Bates, B. C., Chandler, R. E., Charles, S. P. and Campbell, E. P. (2011) Assessment of apparent non-stationarity in time series of annual inflow, daily precipitation and atmospheric circulation indices: a case study from southwest Western Australia. *Water Resources Research* (in press).

Bellido, J., Pierce, G. and Wang, J. (2001) Modelling intra-annual variation in abundance of squid. *Fisheries Research*, **52**, 23–29.

Bowman, A. W. and Azzalini, A. (1997) *Applied Smoothing Techniques for Data Analysis – The Kernel Approach with S-Plus Illustrations*, vol. 18, Oxford Statistical Science Series. Oxford University Press, Oxford.

Bowman, A. W. and Azzalini, A. (2007) *R Package sm: Nonparametric Smoothing Methods (Version 2.2)*. University of Glasgow, UK and Università di Padova, Italia.

Bowman, A. W., Giannitrapani, M. and Scott, E. M. (2009) Spatiotemporal smoothing and sulphure dioxide trends over Europe. *Journal of the Royal Statistical Society, Series C*, **58**, 737–752.

Bowman, A. W., Pope, A. and Ismail, B. (2006) Detecting discontinuities in nonparametric regression curves and surfaces. *Statistical Computing*, **16**, 377–390.

Buckland, S. T., Anderson, D. R., Burnham, K. P., Laake, J. L., Borchers, D. L. and Thomas, L. (2004) *Advanced Distance Sampling*. Oxford University Press, Oxford.

Buja, A., Hastie, T. and Tibshirani, R. (1989) Linear smoothers and additive models. *Annals of Statistics*, **17**(2), 453–510.

Cai, Z. (2002) A two-stage approach to additive time series models. *Statistica Neerlandica*, **56**(4), 415–433.

Cai, Z. (2007) Trending time-varying coefficient time series models with serially correlated errors. *Journal of Econometrics*, **136**(1), 163–188.

Carslaw, D. C. (2005) On the changing seasonal cycles and trends of ozone at Mace Head, Ireland. *Atmosphere Chemistry and Physics*, **5**, 3441–3450.

Chaloupka, M. (2001) Historical trends, seasonality and spatial synchrony in green sea turtle egg production. *Biological Conservation*, **101**, 263–279.

Cheng, M. Y. and Raimondo, M. (2008) Kernel methods for optimal change-points estimation in derivatives. *Journal of Computational and Graphical Statistics*, **17**, 56–75. DOI:10.1198/106186008X289164.

Clarke, E., Spear, B., McCracken, M., Marques, F., Borchers, D., Buckland, S. and Ainley, D. (2003) Validating the use of generalised additive models and at-sea surveys to estimate size and temporal trends of seabird populations. *Journal of Applied Ecology*, **40**, 278–292.

Cleveland, W. S. (1979) Robust locally weighted regression and smoothing scatterplots. *Journal of the American Statistical Association*, **74**(368), 829–836.

Cleveland, R. B., Cleveland, W. S., McRae, J. E. and Terpenning, I. (1990) STL: a seasonal-trend decomposition procedure based on loess (with discussion). *Journal of Official Statistics*, **6**, 3–73.

Craigmile, P. F., Guttorp, P. and Percival, D. B. (2004) Trend assessment in a long memory dependence model using the discrete wavelet transform. *Environmetrics*, **15**, 313–335. DOI:10.1002/env.642.

Daubechies, I. (1988) Orthonormal bases of compactly supported wavelets. *Communications on Pure and Applied Mathematics*, **41**, 909–996.

Davison, A. C. (2003) *Statistical Models*. Cambridge University Press, Cambridge.

Donoho, D. L. and Johnstone, I. M. (1994) Ideal spatial adaptation by wavelet shrinkage. *Biometrika*, **81**, 425–455.

Donoho, D. L., Johnstone, I. M., Kerkyacharian, G. and Picard, D. (1995) Wavelet shrinkage: asymptopia? (with discussion). *Journal of the Royal Statistical Society, Series B*, **57**, 301–369.

Eilers, P. H. C. and Marx, B. D. (1996) Flexible smoothing with B-splines and penalties. *Statistical Science*, **11**(2), 89–121.

Eilers, P. H. C. and Marx, B. D. (2002) Generalized linear additive smooth structures. *Journal of Computational and Graphical Statistics*, **11**(4), 758–783.

Esterby, S. T. (1993) Trend analysis methods for environmental data. *Environmetrics* **4**, 459–481.

Fahrmeir, L. and Lang, S. (2001) Bayesian inference for generalized additive mixed models based on Markov Random Field priors. *Journal of the Royal Statistical Society, Series C*, **50**, 201–220.

Fahrmeir, L., Kneib, T. and Lang, S. (2004) Penalized structured additive regression for space–time data: a Bayesian perspective. *Statistica Sinica*, **14**, 731–761.

Fan, J. (1993) Local linear regression smoothers and their minimax efficiencies. *Annals of Statistics*, **21**(1), 196–216.

Fan, J. and Gijbels, I. (1992) Variable bandwidth and local linear regression smoothers. *Annals of Statistists*, **20**(4), 2008–2036.

Fan, J. and Zhang, W. (1999) Statistical estimation in varying coefficient models. *Annals Statistics*, **27**(5), 1491–1518.

Fan, J. and Zhang, W. (2000) Simultaneous confidence bands and hypothesis testing in varying-coefficient models. *Scandinanian Journal of Statistics*, **27**, 715–731.

Ferguson, C. A., Bowman, A. W., Scott, E. M. and Carvalho, L. (2007) Model comparison for a complex ecological system. *Journal of the Royal Statistical Society, Series A*, **170**, 691–711.

Fewster, R. M., Buckland, S. T., Siriwardena, G. M., Baillie, S. R. and Wilson, J. D. (2000) Analysis of population trends for farmland birds using generalized additive models. *Ecology*, **8**, 1970–1984.

Figueiras, A., Roca-Pardiñas, J. and Cadarso-Suárez, C. (2005) A bootstrap method to avoid the effect of concurvity in generalised additive models in time series studies of air pollution. *Journal of Epidemiology and Community Health*, **59**, 881–884.

Fisher, N. I. (1993) *Statistical Analysis of Circular Data*. Cambridge University Press, Cambridge. xviii + 277 pp.

Gasser, T., Sroka, L. and Jennen-Steinmetz, C. (1986) Residual variance and residual pattern in nonlinear regression. *Biometrika*, **73**(3), 625–633.

Giannitrapani, M., Bowman, A., Scott, M. and Smith, R. (2006) Sulphur dioxide in Europe: statistical relationships between emissions and measured concentrations. *Atmospheric Environment*, **40**, 2524–2532.

Green, P., Jennison, C. and Seheult, A. (1985) Analysis of field experiments by least squares smoothing. *Journal of the Royal Statististical Society, Series B*, **47**(2), 299–315.

Hall, P. and Titterington, D. M. (1992) Edge-preserving and peak preserving smoothing. *Technometrics*, **34**, 429–440.

Hao, L. and Naiman, D. Q. (2007) *Quantile Regression*. Sage Publications, Thousand Oaks, California. 136 pp.

Hart, J. D. (1991) Kernel regression estimation with time series errors. *Journal of the Royal Statistical Society, Series B*, **53**, 173–187.

Hastie, T. (2008) *gam: Generalized Additive Models*. R package version 1.0.

Hastie, T. and Tibshirani, R. (1990) *Generalized Additive Models*. Chapman & Hall, London.

Hastie, T. and Tibshirani, R. (1993) Varying coefficient models. *Journal of the Royal Statistical Society, Series B*, **55**(4), 757–796.

Hastie, T. and Tibshirani, R. (2000) Bayesian backfitting. *Statistical Science*, **15**(3), 196–213.

He, X. and Hu, F. (2002) Markov chain marginal boostrap. *Journal of the American Statistical Association*, **97**, 783–795.

Healy, M. J. R. (2002) *Matrices for Statistics*, 2nd edition. Oxford University Press, Oxford.

Hoover, D., Rice, J., Wu, C. and Yang, L. (1998) Nonparametric smoothing estimates of time-varying coefficient models with longitudinal data. *Biometrika*, **85**(4), 809–822.

Horgan, G. W. (1999) Using wavelets for data smoothing: a simulation study. *Journal of Applied Statistics*, **26**, 923–932.

Hurvich, C. M., Simonoff, J. S. and Tsai, C. L. (1998) Smoothing parameter selection in nonparametric regression using an improved Akaike information criterion. *Journal of the Royal Statistical Society, Series B*, **60**, 271–293.

Johnstone, I. M. and Silverman, B. W. (1997) Wavelet threshold estimators for data with correlated noise. *Journal of the Royal Statistics Society, Series B*, **59**, 319–351.

Koenker, R. (2005) *Quantile Regression*. Cambridge University Press, New York.

Koenker, R. (2009) quantreg: *Quantile Regression*. R package version 4.30.

Koenker, R. and Bassett, G. (1978) Regression quantiles. *Econometrica*, **46**, 33–50.

Koenker, R. and D'Orey, V. (1993) Computing regression quantiles. *Journal of the Royal Statistical Society, Series C*, **43**, 410–414.

Koenker, R. and Hallock, K. F. (2001) Quantile regression. *Journal of Economic Perspectives*, **15**(4), 143–156.

Koenker, R. and Machado, A. F. (1999) Goodness of fit and related inference processes for quantile regression. *Journal of the American Statistical Association*, **94**, 1296–1310.

Koenker, R., Ng, P. and Portnoy, S. (1994) Quantile smoothing splines. *Biometrika*, **81**, 673–680.

Loader, C. (1996) Change point estimation using non parametric regression. *Annals of Statistics*, **24**, 1667–1678.

McMullan, A., Bowman, A. W. and Scott, E. M. (2007) Water quality in the River Clyde: a case study of additive and interaction models. *Environmetrics*, **18**, 527–539.

Mallat, S. (1989) A theory for multiresolution signal decomposition: the wavelet representation. *IEEE Transactions on Pattern Analysis and Machine Intelligence*, **11**, 674–693.

Mardia, K. V. and Jupp, P. E. (1999) *Statistics of Directional Data*. John Wiley & Sons, Ltd, London.

Müller, H. G. (1992) Change points in non-parametric regression analysis. *Ann. Statist.* **20**, 737–761.

Nason, G. P. (1996) Wavelet shrinkage using cross-validation. *J. R. Statist. Soc., Series B* **58**, 463–479.

Nason, G. P. (2008) *Wavelet methods in Statistics with* R. Springer, Berlin.

Nason, G. P., Kovac, A. and Maechler, M., (2009) wavethresh: *Software to Perform Wavelet Statistics and Transforms*. R package version 2.2-11.

Nason, G. P. and Silverman, B. W. (1994) The discrete wavelet transform in S. *J. Computational and Graphical Statistics* **3**, 163–191.

Nievergelt, Y. (1999) *Wavelets Made Easy*. Birkhäuser, Boston.

Opsomer, J., Wang, Y. and Yang, Y. (2001) Nonparametric regression with correlated errors. *Statistical Science*, **16**(2), 134–153.

Percival, D. B. and Mofjeld, H. O. (1997) Analysis of subtidal coastal sea level fluctuations using wavelets. *Journal of the American Statistical Association*, **92**(439), 868–880.

Percival, D. B. and Walden, A. T. (2000) *Wavelet Methods for Time Series Analysis*. Cambridge University Press, Cambridge.

Portnoy, S. and Koenker, R. (1997) The Gaussian hare and the Laplacean tortoise: computability of squared-error vs absolute error estimators (with discussion). *Statistical Science*, **12**, 279–300.

Qiu, P. and Yandell, B. (1998) A local polynomial jump detection algorithm in non-parametric regression. *Technometrics*, **40**, 141–152.

Rice, J. (1984) Bandwidth choice for nonparametric regression. *Annals of Statistics*, **12**(4), 1215–1230.

Roca-Pardiñas, J., Cadarso-Suárez, C. and González-Manteiga, W. (2005) Testing for interactions in generalized additive models: application to SO_2 pollution data. *Statistics and Computing*, **15**, 289–299.

Ruppert, D., Wand, M. P. and Carroll, R. J. (2003) *Semiparametric Regression*. Cambridge University Press, Cambridge.

Sadiku, M. N. O., Akujuobi, C. M. and Garcia, R. C. (2005) An introduction to wavelets in electromagnetism. *IEEE Microwave Magazine*, **6**, 63–72.

Schott, J. R. (1997) *Matrix Analysis for Statistics*. John Wiley & Sons, Inc., New York.

Shamsudduha, M., Chandler, R. E., Taylor, R. G. and Ahmed, K. M. (2009) Seasonality and long-term trends in shallow groundwater levels in Bangladesh. *Hydrology and Earth System Sciences*, **13**, 2373–2385.

Simonoff, J. (1996) *Smoothing Methods in Statistics*. Springer-Verlag, New York.

Stone, C. J. (1985) Additive regression and other nonparametric models. *Annals of Statistics*, **13**(2), 689–705.

Welsh, A., Lin, X. and Carroll, R. (2002) Marginal longitudinal nonparametric regression: locality and efficiency of spline and kernel methods. *Journal of the American Statistical Association*, **97**(458), 482–493.

Whitcher, B., (2007) `waveslim`: *Basic Wavelet Routines for One-, Two- and Three-Dimensional Signal Processing*. R package version 1.6.1.

Whitcher, B., Guttorp, P. and Percival, D. (2000a) Multiscale detection and location of multiple variance changes in the presence of long memory. *Journal of Statistics Computerized Simulation*, **68**, 65–87.

Whitcher, B., Guttorp, P. and Percival, D. (2000b) Wavelet analysis of covariance with application to atmospheric time series. *Journal of Geophysical Research*, **105**(D11), 14941–14962.

Wood, S. N. (2006) *Generalized Additive Models: An Introduction with* R. Chapman & Hall/CRC, Boca Raton, Florida.

Xia, Y., Zhang, W. and Tong, H. (2004) Efficient estimation for semivarying-coefficient models. *Biometrika* **91**(3), 661–681.

5

Stochastic trends

Our discussion of trends has so far centred around Equation (3.1), which postulates the existence of a fixed underlying trend $\{\mu_t\}$ that is observed imperfectly due to a 'noise' component $\{\delta_t\}$. This is, however, not the only way in which long-term variation can be represented. Consider, for example, the series shown by the solid line in Figure 5.1. This shows a regular oscillating pattern and (possibly) a general upward trend, albeit with an apparent break in the structure between times 25 and 30. On the basis of the discussion so far, it may be tempting to model this using a linear trend with a periodic component (see Section 3.4) superimposed. However, the data were in fact generated by simulating from the model

$$Y_t = 0.9959Y_{t-1} - 0.5836Y_{t-2} + \delta_t, \qquad (5.1)$$

where $\{\delta_t\}$ is a sequence of uncorrelated standard normal random variables. This differs fundamentally from any model based on (3.1), in that the variation is controlled solely through dependence between successive observations rather than through a fixed underlying trend. A second simulation of the same model (shown as the dashed line in Figure 5.1) shows similar features, but the timings of peaks and breaks in the structure are different. The apparent trend can therefore be regarded as stochastic, since it varies between realisations of the process. In this chapter, we present a variety of models for representing such 'stochastic trends'. Potentially, these models can be used either in their own right to represent the structure in a time series or in combination with (3.1) to represent structure that is not accounted for by the deterministic trend $\{\mu_t\}$. Our discussion is mostly confined to regularly spaced time series; methods for irregular series are covered briefly in Section 5.4.

5.1 Stationary time series models and their properties

5.1.1 Autoregressive processes

Model (5.1) is an example of a *second-order autoregressive process* or *AR(2)*. We have already encountered the AR(1) as a convenient model for residual dependence

Statistical Methods for Trend Detection and Analysis in the Environmental Sciences, First Edition.
Richard E. Chandler and E. Marian Scott.
© 2011 John Wiley & Sons, Ltd. Published 2011 by John Wiley & Sons, Ltd.

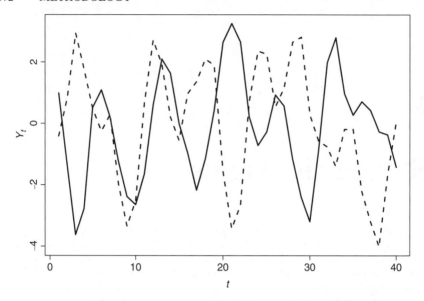

Figure 5.1   Two simulated time series, each generated from model (5.1).

in Chapter 3. More generally, we could incorporate p previous values in the model equation to define a pth-order autoregression, or AR(p):

$$Y_t = \phi_1 Y_{t-1} + \cdots + \phi_p Y_{t-p} + \delta_t, \tag{5.2}$$

where $\phi_1, \ldots, \phi_p$ are coefficients and (δ_t) is a white noise sequence with variance σ_δ^2, say.

Figure 5.1 shows that an AR(2) model can produce quasi-cyclical behaviour, but with the timing of the cycles varying between realisations. Indeed, if we were to generate a large number of realisations from (5.1), the average values at each time point would be equal (to zero, for this particular model); furthermore, the variances at each time point would be equal and the correlations between pairs of values at time points separated by the same interval would be equal. This is therefore a (weakly) stationary process (see Section 2.1.3). If, and only if, a process is stationary then it has a theoretical autocorrelation function, defined at Equation (2.2); it also has a theoretical *autocovariance function*, defined for all lags k as $\gamma(k) = \mathrm{Cov}(Y_t, Y_{t-k}) = \sigma_Y^2 \rho(k)$, where $\rho(\cdot)$ is the autocorrelation function and $\sigma_Y^2 = \gamma(0)$ is the process variance. For a stationary process, the sample ACF $r(\cdot)$ estimates the theoretical ACF $\rho(\cdot)$. This can be used, in principle, to suggest plausible models for observed time series.

Not all autoregressive processes are stationary. For example, it can be shown (Chatfield, 2003, Section 3.4) that the AR(1) process

$$Y_t = \phi Y_{t-1} + \delta_t$$

is stationary only if $|\phi| < 1$. This is closely connected with the failure of certain procedures, for example the Prais–Winsten transformation (3.51), when $|\phi| \geq 1$. In the stationary case, however, the theoretical ACF for the AR(1) process is

$$\rho(k) = \phi^{|k|}.$$

If $0 < \phi < 1$, the theoretical autocorrelations decay exponentially towards zero with increasing lag. If $-1 < \phi < 0$, a similar exponential decay is obtained, but successive coefficients have opposite signs.

The AR(2) process

$$Y_t = \phi_1 Y_{t-1} + \phi_2 Y_{t-2} + \delta_t$$

is stationary when $\phi_1 + \phi_2 < 1$, $\phi_1 - \phi_2 > -1$ and $\phi_2 > -1$. Clearly, then, (5.1) describes a stationary process. The theoretical ACF for an AR(2) process can take one of two forms, depending on the values of ϕ_1 and ϕ_2. The first is a mixture of two exponentially decaying functions:

$$\rho(k) = AB^{|k|} + (1 - A)C^{|k|},$$

for constants A, B and C depending on ϕ_1 and ϕ_2 with $|B| < 1$ and $|C| < 1$. The second is a damped sine wave:

$$\rho(k) = P^{|k|} \sin(Q|k| + R)/\sin(R),$$

for constants P, Q and R with $|P| < 1$ (Priestley, 1981, Section 3.5). The latter form corresponds to a quasi-cyclical process: the AR(2) is therefore often a good starting point for modelling processes that show apparent periodicity. This was first noted by Yule (1927), who showed that the AR(2) emerges as a natural model for the motion of a pendulum subject to random shocks (supplied, in his original example, by a boy with a peashooter). For this reason, the AR(2) process is sometimes referred to as *Yule's pendulum*.

The conditions for stationarity of the general autoregressive process (5.2) are not easily stated without introducing concepts that need not concern us here; Chatfield (2003, Section 3.4) gives the details. When fitting autoregressive models, most software packages automatically impose constraints on the parameters to ensure that the fitted models are stationary. The ACF of a stationary autoregression is a sum of components, which may include both exponentially decaying terms and damped sine waves. For this reason, it is often difficult to determine the order of an autoregressive process based solely upon inspection of its ACF. We return to this point below.

If the process (5.2) is stationary, it can be shown that its mean is zero. To obtain a process, (Y_t^*), say, with mean $\mu \neq 0$, we simply have to add μ to every observation from (5.2): $Y_t^* = Y_t + \mu$, so that $Y_t = Y_t^* - \mu$. We therefore find that an AR(p) process (Y_t^*) with mean μ can be written as

$$Y_t^* - \mu = \phi_1 \left(Y_{t-1}^* - \mu\right) + \cdots + \phi_p \left(Y_{t-p}^* - \mu\right) + \delta_t \tag{5.3}$$

or, rearranging,

$$Y_t^* = c + \phi_1 Y_{t-1}^* + \cdots + \phi_p Y_{t-p}^* + \delta_t$$

with $c = \mu(1 - \phi_1 - \cdots - \phi_p)$. The ACF of a process is unaffected by such a shift in the mean, as should be clear from the definition (2.2).

5.1.2 Moving average processes

We have seen that stationary autoregressions can produce 'trend-like' behaviour, despite merely describing the dependence between successive observations in a time series. Another class of models for such dependence is that of *moving averages*. A zero-mean moving average process of order q, or MA(q), can be written as

$$Y_t = \delta_t + \theta_1 \delta_{t-1} + \cdots + \theta_q \delta_{t-q} \tag{5.4}$$

where, as usual, (δ_t) is a white noise process with variance σ_δ^2. The terminology reflects the fact that in (5.4) Y_t is effectively a weighted average of present and past values of the underlying white noise sequence. Some authors (and software packages) reverse the signs of the coefficients when defining a moving average process, writing $Y_t = \delta_t - \theta_1 \delta_{t-1} - \cdots - \theta_q \delta_{t-q}$. There are good reasons for this; however, we use (5.4) for compatibility with R. Users of other packages should check which definition is used when fitting models.

All moving average models are stationary (Chatfield, 2003, Section 3.4). A key feature of any MA(q) process is that its autocorrelations are exactly zero for lags greater than q. Under ideal conditions, data from such a process should therefore yield a sample ACF with a (possibly unstructured) sequence of 'large' coefficients, followed by a sequence of small coefficients. In this situation, the cut-off point could be taken as an estimate of the order q of the process. This contrasts with the behaviour for autoregressive models, where the autocorrelations approach zero smoothly as the lag tends to infinity.

The mean of the process (5.4) is zero. As before, a process with mean μ can be obtained by adding μ to every observation:

$$Y_t^* = \mu + \delta_t + \theta_1 \delta_{t-1} + \cdots + \theta_q \delta_{t-q}.$$

5.1.3 Mixed ARMA processes

Having introduced autoregressive and moving average processes, it is natural to combine them to form an extended class of *autoregressive/moving average* (ARMA) processes. A zero-mean ARMA(p, q) process (Y_t) has the model equation

$$Y_t = \phi_1 Y_{t-1} + \cdots + \phi_p Y_{t-p} + \delta_t + \theta_1 \delta_{t-1} + \cdots + \theta_q \delta_{t-q}. \tag{5.5}$$

Stationarity of such a process is determined by the values of the autoregressive coefficients $\phi_1, \ldots, \phi_p$. The required conditions are the same as for an ordinary autoregression: for example, if $p = 1$, (5.5) defines a stationary process if and only if $|\phi_1| < 1$. For a stationary ARMA process (Y_t^*) with mean μ, by analogy with (5.3) the model equation is

$$Y_t^* - \mu = \phi_1 \left(Y_{t-1}^* - \mu \right) + \cdots + \phi_p \left(Y_{t-p}^* - \mu \right) + \delta_t + \theta_1 \delta_{t-1} + \cdots + \theta_q.$$

As might be expected, the ACF of a stationary ARMA(p,q) process behaves as a combination of the two separate components. In particular, the MA component has no effect after lag q, so after this stage the ACF behaves exactly as that of an AR(p). For example, the ACF of the ARMA(1,1) process

$$Y_t = \phi Y_{t-1} + \delta_t + \theta \delta_{t-1}$$

has the form

$$\rho(k) = \begin{cases} 1, & k = 0, \\ A\phi^{|k|-1}, & \text{otherwise,} \end{cases}$$

providing $|\phi| < 1$ (Brockwell and Davis 2002, Section 3.2). This shows exponential decay as for the AR(1), but starting at lag 1 rather than lag zero.

Extensions of the ARMA family of models are available for series with seasonal structure; see, for example, Chatfield (2003, Section 4.6), Brockwell and Davis (2002, Section 6.5) and Kendall and Ord (1990, Chapter 5). However, seasonal cycles in environmental series tend not to vary much over extended time periods, whence they are perhaps best thought of as essentially fixed rather than stochastic. Therefore, the seasonal ARMA models (at least in their stationary versions) are perhaps not best suited to the modelling of environmental data. We elaborate on this point in Section 5.1.7.

5.1.4   Model identification

We have briefly introduced some classes of stationary process that may be appropriate for the modelling of environmental time series and have discussed in qualitative terms the behaviour of their autocorrelation functions. In principle, therefore, if a sequence of observations appears roughly stationary then its sample ACF may be used to guide the choice of an appropriate model from this class. Unfortunately, this is not always easy. One problem is that, except in very large samples, the sample ACF often resembles the underlying ACF rather loosely. Firstly, the sample autocorrelations tend to underestimate the underlying values, especially at high lags, and, secondly, neighbouring coefficients in the sample ACF can be moderately dependent, giving a smooth appearance (often appearing as weak oscillation) when such structure is not present in the true ACF. A good discussion of these issues may be found in Kendall and Ord (1990, Chapter 6).

Another problem is that, as noted above, many different autoregressive processes have qualitatively similar ACFs. If an AR process is suspected therefore, the ACF may give few clues as to its order. In such situations, one could contemplate fitting AR processes of successively higher order to the data, at each stage noting the value of the final coefficient. Suppose the true model is AR(p) and an AR(k) model is fitted: then, when $k \leq p$, one would expect to obtain a nonzero coefficient of Y_{t-k}, but when $k > p$ the coefficient should be close to zero. A plot of the coefficients against the lag k could therefore be used to identify the order of the process. In fact, it can be shown (Kendall and Ord, 1990, Section 5.11) that the lag k coefficient defined in this way represents the correlation between Y_t and Y_{t+k}, adjusted for their mutual dependence on the intervening variables $Y_{t+1}, \ldots, Y_{t+k-1}$. The collection of these coefficients is called the sample *partial autocorrelation function* (PACF) and is usually plotted in a similar way to the ACF, with approximate confidence limits to help judge which coefficients differ 'significantly' from zero. If in fact the data are generated from a moving average process rather than an autoregression, the PACF will typically show smooth decay or damped oscillation (Kendall and Ord, 1990, Section 5.22). In short, the ACF of a moving average behaves like the PACF of an autoregression, and vice versa.

Although the repeated fitting of autoregressions sounds cumbersome, it is perfectly feasible with modern computing power and is used, for example, by the pacf routine in R. However, the PACF can also be computed directly from the ACF using the *Durbin–Levinson algorithm* (Kendall and Ord, 1990, Section 5.19).

For example, given sample autocorrelation coefficients $r(1)$ and $r(2)$, the first two partial autocorrelations are

$$r(1) \qquad \text{and} \qquad \frac{r(2) - r(1)^2}{1 - r(1)^2}$$

respectively. This shows, firstly, that the PACF contains no new information (it merely reorganises the ACF in a way that is more convenient for some purposes) and, secondly, that the sample PACF will inherit the propensity of the ACF to behave poorly on occasion.

Example 5.1

Figure 5.2 shows the theoretical autocorrelation and partial autocorrelation functions for some simple time series models, along with sample versions computed, in each case, from a simulated series of 40 observations. The top row is for an AR(1) process with parameter $\phi = 0.7$. For this process the theoretical ACF decays exponentially, and the theoretical PACF is zero after lag 1. The sample PACF has a spike at lag 1, which is consistent with the underlying AR(1) mechanism; however, both sample ACF and PACF show quite pronounced oscillation. If we did not know the true model for this series, the oscillation in the sample ACF may lead us to conclude that an AR(2) component is present; similarly,

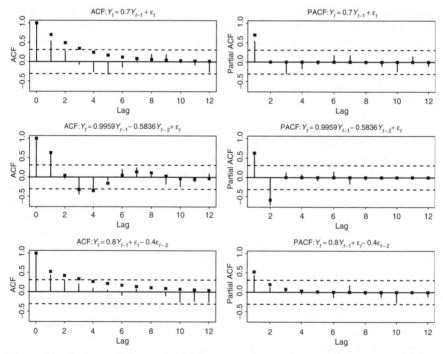

Figure 5.2   Autocorrelation and partial autocorrelation functions for (top to bottom) an AR(1), an AR(2) and an ARMA(1,1) process. The points on each plot show the true underlying function for each process; the bars show the sample version computed from a simulation of length $T = 40$.

the oscillation in the PACF may suggest an MA(2) component as well. This shows clearly the potential for spurious structure in the sample (partial) autocorrelations.

The sample functions in the remaining rows of Figure 5.2 are rather better behaved. The second row is for the AR(2) model defined in Equation (5.1) (in fact, the sample functions here correspond to the series shown as a solid line in Figure 5.1). Both theoretical and sample ACFs show clear damped oscillation, corresponding to the quasi-periodicity in this process. The theoretical PACF is zero after lag 2, and the sample version follows this quite closely.

The final row of Figure 5.2 is for an ARMA(1,1) process. The sample and theoretical correlation functions are once again in reasonable agreement, although the sample ACF shows clearly the tendency, mentioned above, for underestimation at high lags. Notice the difference between the theoretical functions for this model and those for the AR(1) in the top row. In the ARMA case, the MA(1) component manifests itself firstly via the smooth decay of the PACF and secondly via the fact that although the ACF decays exponentially from lag 1 onwards, it shows a discontinuity between lags 0 and 1.   ∎

Given the difficulties in choosing an appropriate ARMA model for a series, a pragmatic approach is to work initially with the simplest model suggested by the sample ACF, and to expand this later if this seems necessary based on a residual analysis. If, for example, the sample ACF shows a large significant coefficient at lag 1 and not much else, an MA(1) model may be appropriate; if it shows clear oscillation, then an AR(2) is a natural starting point; and if it shows smooth decay from lag 1 onwards, an ARMA(1,1) might be used.

5.1.5   Parameter estimation

Having identified, possibly tentatively, an appropriate model for a stationary series, we turn to the question of parameter estimation. As ever, the large-sample optimality properties of maximum likelihood make it the preferred estimation method in many applications. However, the routine fitting of time series models by maximum likelihood has only become feasible since the 1980s, with the development of efficient algorithms to compute the likelihood and the availability of sufficient computing power to maximise it numerically. Prior to this, fitting was often done using least squares, which has close connections with maximum likelihood in any case. A couple of general issues arise when fitting time series models using either method. These are perhaps best appreciated in the context of least squares; accordingly, we deal with this first.

Consider fitting the zero-mean autoregression (5.2) to data $y_1, \ldots, y_T$. The regression-like form of the model suggests estimating the parameter vector $\boldsymbol{\phi} = (\phi_1 \ldots \phi_p)'$ by minimising the sum of squares

$$S(\boldsymbol{\phi}) = \sum_{t=p+1}^{T} \left(y_t - \sum_{j=1}^{p} \phi_j y_{t-j} \right)^2 = \sum_{t=p+1}^{T} e_t^2(\boldsymbol{\phi}) \text{ , say.} \qquad (5.6)$$

The sum here starts at time $t = p + 1$, because contributions from earlier time points depend on unobserved values prior to time 1. Minimising (5.6) is in fact the same as maximising a log-likelihood obtained by considering the first p observations as fixed (Chatfield, 2003, Section 4.2; Priestley, 1981, Section 5.4); for this reason, (5.6) is sometimes referred to as a *conditional sum of squares*. The resulting fitted model is not guaranteed to be stationary, although it usually will be if T is large and the data are indeed from a stationary process.

Least squares estimation is less straightforward in models containing moving average components. We illustrate this using the MA(1) model

$$Y_t = \delta_t + \theta \delta_{t-1}. \tag{5.7}$$

For data $y_1, \ldots, y_T$, the least squares estimate of θ should minimise the sum of squares

$$S(\theta) = \sum e_t^2(\theta) \qquad \text{where } e_t(\theta) = y_t - \theta e_{t-1}(\theta). \tag{5.8}$$

However, there is a problem. Since $e_0(\theta)$ is unobserved, the quantity $e_1(\theta)$ cannot be calculated. The same is true for $e_2(\theta)$, and for all the other contributions to $S(\theta)$. A possible solution is to set $e_0(\theta)$ to a fixed value, say zero (which is the expected value of δ_0). If this is done, $S(\theta)$ can be calculated for any value of θ. Notice that, due to the recursive calculation involved, $e_t(\theta)$ is a polynomial of degree $t - 1$ in θ and $S(\theta)$ is a polynomial of degree $2(T - 1)$. In general, therefore, minimising $S(\theta)$ must be done numerically (contrast this with the autoregressive case (5.6), where $S(\phi)$ is quadratic in ϕ and can be minimised analytically just as in a standard multiple regression model).

It is interesting to examine the implications of fixing $e_0(\theta)$ when fitting an MA(1). Let d_t denote the realised value of δ_t corresponding to the data at hand. Then, from (5.7), we have

$$d_t = y_t - \theta d_{t-1} = y_t - \theta (y_{t-1} - \theta d_{t-2}) = y_t - \theta y_{t-1} + \theta^2 d_{t-2}$$

$$= \cdots = y_t + \sum_{j=1}^{t-1} (-1)^j \theta^j y_{t-j} + (-1)^t \theta^t d_0. \tag{5.9}$$

The same procedure can be applied to $e_t(\theta)$ as defined in (5.8):

$$e_t(\theta) = y_t + \sum_{j=1}^{t-1} (-1)^j \theta^j y_{t-j} + (-1)^t \theta^t e_0(\theta). \tag{5.10}$$

Notice next that at the true value of θ, we should have $e_t(\theta) = d_t$ for all t. From (5.9) and (5.10), the difference between $e_t(\theta)$ and d_t is

$$e_t(\theta) - d_t = (-\theta)^t [e_0(\theta) - d_0].$$

If $|\theta| < 1$, this difference tends to zero as t increases. If $|\theta| = 1$, the magnitude of the difference is always the same, and if $|\theta| > 1$, the difference increases exponentially with t. The upshot is that if $|\theta| < 1$, only the first few time points are affected to any appreciable extent by fixing $e_0(\theta)$. In this case, providing T is large, the effect on estimation of θ is negligible and the sum of squares (5.8) can be justified as a measure of model fit. However, if $|\theta| \geq 1$, then, unless $e_0(\theta) = d_0$, (5.8) has no meaningful interpretation.

Equation (5.9) gives a clue as to what is happening here. It shows that if $|\theta| < 1$ and t is large, d_t can be approximated accurately using a weighted average of the observations $\{y_{t-j} : j = 1, \ldots, t - 1\}$, with decreasing weights attached to the more distant observations. Hence it is, to all intents and purposes, observable. Correspondingly, the random variable δ_t can be approximated in terms of $Y_1, \ldots, Y_{t-1}$. Notionally at least, this approximation can be made exact by continuing the expansion into the infinite past using the random variables $Y_0, Y_{-1}, \ldots$. If this is done, and the result applied to δ_{t-1} instead of δ_t, we can substitute into the original model equation (5.7) to obtain

$$Y_t = \delta_t + \theta \left[Y_{t-1} + \sum_{j=1}^{\infty} (-1)^j \theta^j Y_{t-1-j} \right] = \sum_{j=1}^{\infty} (-1)^{j+1} \theta^j Y_{t-j} + \delta_t.$$

In doing this, we have effectively 'inverted' the moving average model specification into an autoregression, albeit of infinite order.

In general, for any model containing a moving average component of order q, the least squares estimation is hampered by not knowing the values of $\delta_0, \ldots, \delta_{-(q-1)}$. The general principle is the same as for the MA(1) process, however: under certain constraints on the moving average parameters, any such model can be written as an infinite-order autoregression with negligible weight attached to observations in the distant past. The model is then said to be *invertible*, and only the first few contributions to the sum of squares are affected to any substantial degree by not knowing the initial errors. Invertibility plays a key role in model fitting and checking, as well as in prediction (see below): it is therefore important to ensure that fitted models are invertible. In practice this restriction is not serious, since it can be shown (Priestley, 1981, Section 3.5) that for every noninvertible model there is an invertible model with the same ACF. It is at first sight remarkable that the constraints required to ensure invertibility are essentially the same as the corresponding stationarity conditions for autoregressive processes, although for moving average models defined as in (5.4) the signs of the coefficients must be reversed in the constraints. For example, the invertibility condition for the MA(1) process is $|\theta| < 1$; the conditions for the MA(2) process are $\theta_1 + \theta_2 > -1$, $\theta_1 - \theta_2 < 1$ and $\theta_2 < 1$. For more details, see Chatfield (2003, Section 3.4) and Brockwell and Davis (2002, Chapter 3).

Having dealt with least squares estimation, we can tackle maximum likelihood. Strictly speaking, we should refer to 'Gaussian' maximum likelihood, since the standard techniques for likelihood based fitting of time series models all assume that the error process (δ_t) is normally distributed. In this case, providing the process is stationary, the $\{Y_t\}$ are also normally distributed, and indeed the entire observation vector $\mathbf{Y}$ has a multivariate normal distribution. This follows from the representation of (Y_t) as a *general linear process*, in which each observation can be written as a linear combination of terms in the error process (Chatfield, 2003, Section 3.4).

Recall from Section 3.5 that a likelihood is obtained by writing down the joint probability density function of the data, and treating this as a function of the model parameters. The density of a T-variate normal distribution with mean vector $\boldsymbol{\mu}$ and covariance matrix $\boldsymbol{\Sigma}$ is

$$f(\mathbf{y}) = (2\pi)^{-T/2} |\boldsymbol{\Sigma}|^{-1/2} \exp \left[-\frac{1}{2} (\mathbf{y} - \boldsymbol{\mu})' \boldsymbol{\Sigma}^{-1} (\mathbf{y} - \boldsymbol{\mu}) \right], \tag{5.11}$$

where $|\boldsymbol{\Sigma}|$ is the determinant of $\boldsymbol{\Sigma}$. For the stationary models considered here, $\boldsymbol{\Sigma}$ contains the autocovariances (see Section 5.1.1) and has a Toeplitz structure (see Section 3.3.3). The autoregressive and moving average parameters determine the autocovariance structure and hence affect the covariance matrix $\boldsymbol{\Sigma}$, but not the mean vector $\boldsymbol{\mu}$. For these models, by definition the elements of $\boldsymbol{\mu}$ are all equal to the process mean μ. In applications, however, it can be useful to allow the elements of $\boldsymbol{\mu}$ to vary, for example to allow dependence of Y_t upon covariates as in the regression models of Chapter 3. Of course, in this case the process (Y_t) is no longer stationary; however, if we define $\tilde{Y}_t = Y_t - \mu_t$ then the process $(\tilde{Y}_t)$ is. The interpretation is that after removing any deterministic component μ, we are left with a zero-mean process that can be represented using a stationary time series model. Such series are sometimes referred to as *trend-stationary*. If we allow μ to vary and to depend on a parameter vector β then the likelihood is a function of β, along with the autoregressive parameter vector ϕ, the moving average parameter vector θ and the error variance σ_δ^2. From (5.11), we see that the log-likelihood given data $\mathbf{y}$ is

$$\ell\left(\beta, \phi, \theta, \sigma_\delta^2; \mathbf{y}\right) = -\frac{T}{2}\log(2\pi) + \log\left(|\Sigma|^{-1/2}\right) - \frac{1}{2}(\mathbf{y} - \mu)'\,\Sigma^{-1}\,(\mathbf{y} - \mu). \quad (5.12)$$

Notice that β enters here only through the mean vector μ. Therefore, for fixed values of the other parameters, maximising (5.12) with respect to β is equivalent to minimising $(\mathbf{y} - \mu)'\,\Sigma^{-1}\,(\mathbf{y} - \mu) = \tilde{\mathbf{y}}'\Sigma^{-1}\tilde{\mathbf{y}}$, say. However, this has exactly the same form as the generalised least squares criterion (3.50). This justifies our earlier claim that GLS delivers maximum likelihood estimates of regression coefficients under the assumption of normality.

In general, (5.12) cannot be maximised analytically and numerical methods must be used. However, given reasonable starting values (which may be obtained from a least squares fit, for example) and a good modern optimisation algorithm, this is not a problem. The only cause for concern is the amount of computational effort required to calculate the log-likelihood at (possibly many) different parameter values, since the number of floating point operations required to evaluate the determinant and inverse of a $T \times T$ matrix is usually roughly proportional to T^3 (Monahan, 2001, Chapter 3). Fortunately, however, for the models considered here the cost of evaluating $|\Sigma|$ and Σ^{-1} can be reduced substantially. The means of achieving this can be regarded as a generalisation of the Prais–Winsten transformation (see Section 3.3). We proceed by defining a series of *innovations* $h_1, \ldots, h_T$:

$$h_t = \begin{cases} \tilde{Y}_t, & t = 1, \\ \tilde{Y}_t - \sum_{j=1}^{t-1} w_{j,t} h_j, & \text{otherwise,} \end{cases} \quad (5.13)$$

where the 'weights' $\{w_{j,t}\}$, which depend on the parameter vectors ϕ and θ, are chosen in such a way that the innovations are mutually uncorrelated. Specifically, we have $w_{j,t} = \text{Cov}(\tilde{Y}_t, h_j)/\text{Var}(h_j)$ (Eubank, 2006, Section 1.2.2). Although the details are rather complicated, for the class of models considered here the *innovations algorithm* provides an efficient means of computing the (h_t) and their variances for any parameter values (Brockwell and Davis, 2002, Sections 2.5.2 and 3.1 to 3.3): the total number of floating point operations required is roughly proportional to T. This algorithm does require, however, that no observations are missing. The innovations can alternatively be calculated by writing the models in state space form (see Section 5.5) and using the algorithm of Gardner, Harvey and Phillips (1980). This approach, which is used in the R routine `arima`, handles missing observations exactly and the computational cost is again proportional to T. Either way, routine evaluation of the innovations, and their variances, at many different parameter values is feasible for long time series with modern computing power. We now show how this relates to (5.12).

According to (5.13), h_t is a linear combination of the mean-corrected variables $\tilde{Y}_1, \ldots, \tilde{Y}_t$, with a unit coefficient of $\tilde{Y}_t$. Therefore, we can write the vector of innovations as $\mathbf{h} = A\tilde{\mathbf{Y}}$ for some lower triangular matrix A with diagonal elements all unity. Since $\tilde{\mathbf{Y}}$ is multivariate normal (MVN) with zero mean, $\mathbf{h}$ is too: $\mathbf{h} \sim \text{MVN}(\mathbf{0}, V)$, say, where V is a diagonal matrix since the innovations are uncorrelated. Conversely, we can write $\tilde{\mathbf{Y}} = A^{-1}\mathbf{h}$ (the structure of A guarantees the existence of its inverse). Using standard results for the multivariate normal distribution (e.g. Flury, 1997, Section 3.2; Krzanowski, 1988, Section 7.2), we find that $\tilde{\mathbf{Y}} \sim \text{MVN}(\mathbf{0}, A^{-1}V[A^{-1}]')$. However, we already know that $\tilde{\mathbf{Y}} \sim \text{MVN}(\mathbf{0}, \Sigma)$. We must therefore have $\Sigma = A^{-1}V[A^{-1}]'$ and $\Sigma^{-1} = A'V^{-1}A$.

With this factorisation, the quadratic form in (5.12) can be written as

$$\tilde{\mathbf{Y}}'\Sigma^{-1}\tilde{\mathbf{Y}} = \left[A\tilde{\mathbf{Y}}\right]'V^{-1}A\tilde{\mathbf{Y}} = \mathbf{h}'V^{-1}\mathbf{h} = \sum_{t=1}^{T} h_t^2/v_t,$$

where $v_t = \text{Var}(h_j)$ is the tth diagonal element of V. Moreover, the determinant $|\Sigma|^{-1/2}$ is given by $|A|/\sqrt{|V|} = 1/\sqrt{|V|}$ (A is lower triangular, so its determinant is just the product of its diagonal elements, which are all unity) and V is diagonal so $|V| = \prod_{t=1}^{T} v_t$. Putting all this together, (5.12) becomes

$$\ell\left(\boldsymbol{\beta}, \boldsymbol{\phi}, \boldsymbol{\theta}, \sigma_\delta^2; \mathbf{y}\right) = -\frac{1}{2}\sum_{t=1}^{T}\left[\frac{h_t^2}{v_t} + \log v_t + \log(2\pi)\right], \tag{5.14}$$

where, in a slight abuse of notation, h_t is the 'observed' innovation calculated from the data $\mathbf{y}$. We have therefore succeeded in writing the log-likelihood in terms of the innovations and their variances, all of which can be computed relatively cheaply.

We have already hinted at a connection with the treatment of GLS estimation in Section 3.3. Indeed, the matrix Γ used there is proportional to $V^{-1/2}A$ in the current context. The elements of $V^{-1/2}$ are therefore proportional to the diagonal elements of Γ, and A is obtained by dividing each row of Γ by its diagonal element. For the AR(1) process, Γ is given by the Prais–Winsten matrix (Equation (3.51)). All of its diagonal elements except the first are already unity; the matrix A for this model is therefore identical with Γ except in the first row. This shows that for $t > 1$ in this model, the innovation h_t is simply the residual $e_t = \tilde{Y}_t - \phi\tilde{Y}_{t-1}$; for $t = 1$ the innovation is $h_1 = \tilde{Y}_1$. For $t = 2, \ldots, T$, the variances $\{v_t\}$ are all equal; v_1 is scaled by a factor $1/(1 - \phi^2)$. Maximising the log-likelihood (5.14) is therefore almost identical with minimising the sum of squares (5.6): the only difference is due to the $t = 1$ term in (5.14). However, including this term has an important consequence since, with v_1 proportional to $1/(1 - \phi^2)$, $\log v_1$ tends to infinity as ϕ^2 tends to 1, and therefore the log-likelihood tends to $-\infty$. Furthermore, $\log v_1$, and hence the log-likelihood, is undefined when $\phi^2 > 1$. Therefore, the maximum likelihood estimator must lie between -1 and 1, and automatically satisfies the stationarity conditions for the model.

These conclusions hold quite generally (Brockwell and Davis, 2002, Section 3.3). For an AR(p) process, when $t > p$ the innovation h_t is the residual $e_t = \tilde{Y}_t - \sum_{j=1}^{p} \phi_j \tilde{Y}_{t-j}$ and the variances v_t are all constant; the first p contributions to the log-likelihood are different and ensure that the maximum likelihood estimators automatically correspond to a stationary process. This can be checked for the AR(2) process using the form of the Γ matrix given by Barry, Burney and Bhatti (1997). For models including a moving average component, apart from the first few time points the innovations can be interpreted as 'approximate residuals' (see the next section) providing the model is invertible; in this case, the innovation variances $\{v_t\}$ tend to a constant limit as t increases.

When using maximum likelihood, all of the usual methods are available for examining and comparing models. For example, approximate standard errors can be computed for the parameter estimates, which in principle enables the significance of each term to be assessed using a t statistic. Notice, however, that the time series models considered here are effectively regression models with correlated covariates; the usual warnings apply, therefore, regarding the interpretation of t statistics in such situations (see Section 3.2). For this reason, in time series perhaps more than in other fields, the use of other likelihood based model selection criteria such as AIC is widespread.

We conclude this section by pointing out a connection between the time series log-likelihood (5.14) and the general expression (3.65) for the log-likelihood of a dependent sequence of observations. The latter is a sum of terms, with the contribution at time t corresponding to a conditional distribution given all previous observations. This is exactly the form of (5.14): here, the contribution at time t is the logarithm of a normal density that represents a conditional distribution. The variance of this normal density is v_t; the mean is zero for $t = 1$ and $\sum_{j=1}^{t-1} w_{j,t} h_j$ as defined in (5.13) otherwise: this is the expected value of $\tilde{Y}_t$ given the previous observations.

5.1.6   Model checking

As with the regression based models of earlier chapters, diagnostics for time series models are usually based on residuals. The construction of these residuals is, however, slightly more complicated in the present setting.

In the previous subsection we saw that, under the assumed model, the innovation vector $\mathbf{h}$ has zero mean and a diagonal covariance matrix. Furthermore, it follows from the discussion above that for a pth-order autoregression, if the parameter values $\phi_1, \ldots, \phi_p$ are known then the innovation h_t is simply the white noise term δ_t for $t > p$. This suggests that, in general, the innovations from a fitted model, which emerge naturally from the fitting algorithm, are closely connected with the sequence (δ_t) and will form a useful basis for diagnostics. For models involving a moving average component, this connection relies additionally upon the invertibility of the model, as might be expected from the discussion following Equations (5.9) and (5.10). As far as diagnostics are concerned, one problem with the innovations is that their variance is not constant: this is why they were described previously as 'approximate residuals'. However, since the variances $\{v_t\}$ are also calculated during the fitting of the model, it is straightforward to standardise each innovation according to its own standard deviation: $e_t^* = h_t/\sqrt{v_t}$, say, to generate a sequence of uncorrelated residuals with zero mean and unit variance, or $e_t = \sigma_\delta e_t^*$ for a sequence of uncorrelated residuals on the same scale as (δ_t). We refer to $\{e_t\}$ and $\{e_t^*\}$ as *residuals* and *standardised residuals* respectively.

In practice, of course, the model parameters are not known, so the residuals are not exactly equal to the underlying white noise sequence, even in the case of a correctly specified autoregression. Essentially the same issue arose when discussing diagnostics for the linear trend model in Section 3.1. However, as in the linear trend case, it seems intuitively reasonable that for a well-chosen model the residuals should behave more or less like a white noise sequence, and hence that all of the diagnostic techniques introduced in Chapter 3 can be applied to check for unexplained structure. A formal justification for this assertion is possible (Brockwell and Davis, 2002, Section 5.3). Therefore, for example, a time series plot of residuals can be used to check for unexplained trends and a residual correlogram can be used to check for autocorrelation that is not captured by the model.

When any correlogram is calculated, the usual 95 % limits defined in Section 2.1.3 provide a rough guide as to whether an individual coefficient may be considered significantly different from zero. However, in general this tells only part of the story. For example, if the residual autocorrelations all lie within the limits but several of them, at consecutive lags, are 'almost significant', one might suspect the presence of genuine (albeit weak) residual autocorrelation that has not been accounted for. It can therefore be useful to try and summarise the entire residual correlogram, as an alternative to focusing on individual coefficients. One way to proceed is via the *Ljung–Box–Pierce statistic*, defined for any lag m as

$$Q(m) = T(T + 2) \sum_{k=1}^{m} \frac{r_e^2(k)}{T - k},$$

where $r_e(k)$ is the residual autocorrelation at lag k. If a fitted ARMA(p, q) model with $p, q \geq 0$ is correctly specified, the distribution of $Q(m)$ can be approximated by a chi-squared distribution with $m - p - q$ degrees of freedom (Kendall and Ord, 1990, pages 111–112). This can be used as the basis for a test of the null hypothesis that the underlying residual ACF is zero at lags 1 through m. Such a test is sometimes called a *portmanteau test*. In R, the `tsdiag` command plots the p-values from this test for different values of m, as well as producing a time series plot of the standardised residuals and their ACF.

As with the linear regression models discussed in Chapter 3, the normality assumption is not critical to the estimation of parameters in time series models: maximising the Gaussian likelihood of the previous subsection is often a reasonable procedure even when this assumption does not hold. However, normality becomes much more important if different models are to be compared using criteria based on the likelihood function itself. Such criteria include the AIC (see the definition in Section 3.2.3). When fitting time series models many software packages, R included, calculate AIC values based on the assumption of normality. A normal quantile–quantile plot of the residuals (see Section 3.1.1) can provide a guide to the appropriateness of such calculations.

Example 5.2

To illustrate the application of the ideas discussed so far, we return to the haddock biomass example of Section 1.3.2. In Section 3.1, we fitted a linear trend model to the logarithms of the biomass series and found substantial autocorrelation, with a possible cyclical structure, in the residuals from this model. The periodogram analysis in Section 3.4 suggested that the residual structure could not be represented adequately using a simple periodic function. It may therefore be of interest to see if the residuals can be represented using one of the time series models introduced here.

To identify an appropriate model for the residuals, their sample ACF and PACF can be used. The ACF was shown in Figure 3.1(c). The PACF is not shown here; it looks very similar, however, to the sample PACF of the AR(2) process in Figure 5.2, with significant coefficients at the first two lags and little else thereafter. This structure in the PACF, together with the pronounced oscillation in the ACF and in the residual time series itself, suggests that in the first instance we might try to model the residuals as an AR(2) process. In principle, this could be achieved by taking the residuals as data in their own right and fitting an AR(2) model to them. However, this would fail to account for the fact that the slope and intercept of the trend line have already been estimated and that the calculation of residuals is a form of preprocessing that should ideally be accounted for (see the discussion of this in the context of multiple regression models in Section 3.2). As an alternative, therefore, we note that if the residuals form an AR(2) process then a model for the log biomass data (Y_t^*) themselves can be written as

$$Y_t^* = \beta_0 + \beta_1 x_t + \varepsilon_t,$$
$$\text{where } \varepsilon_t = \phi_1 \varepsilon_{t-1} + \phi_2 \varepsilon_{t-2} + \delta_t, \tag{5.15}$$

and (δ_t) is a white noise sequence with variance σ_δ^2. This is an example of a trend–stationary model, where the trend is $\mu_t = \beta_0 + \beta_1 x_t$; hence maximum likelihood estimates of the parameters can be obtained using the method described in the previous subsection. This method is implemented in the `arima` command in R; the results are summarised in Table 5.1 and the diagnostics from the `tsdiag` command are shown in Figure 5.3.

Table 5.1   Edited R output for model (5.15) fitted to haddock biomass data.

```
Coefficients:
             ar1       ar2   intercept       Time
          0.9959   -0.5836      5.9768    -0.0341
s.e.      0.1551    0.1492      0.1762     0.0078

sigma^2 estimated as 0.08824
log likelihood = -8.46,   aic = 26.92
```

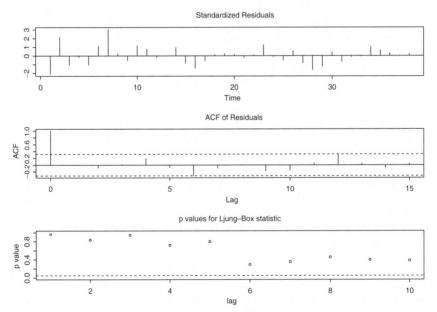

Figure 5.3   Diagnostic plots, produced in R using the `tsdiag` command, for model (5.15) fitted to haddock biomass data. Top to bottom: time series plot of standardised residuals, ACF of residuals and p-values for the Ljung–Box–Pierce portmanteau test at lags up to 10.

Focusing first on Figure 5.3, the standardised residuals in the top plot show no obvious trends (and, by contrast with the residuals from the original model in Section 3.1, the first observation no longer appears as an outlier). More interestingly, the residual ACF now shows virtually no autocorrelation. This impression is confirmed by the p-values for the Ljung–Box–Pierce portmanteau test: since all of these are well above the critical value of 0.05, there is no evidence to reject the null hypothesis that the underlying residual correlations are all zero, at least up to lag 10. It therefore seems that we have finally found a model that accounts for the structure in these data. A final check is to carry out a formal comparison of the new model with the original. The AIC for the new model is given in Table 5.1 as 26.92. That for the original can be computed, using the `AIC` command in R, as 49.79. This confirms the superiority of the new model, if such confirmation were needed.

Having established that the model appears to describe the data adequately, we can proceed to interpret the results. Comparing each of the parameter estimates in Table 5.1 with their standard errors, all of the estimates differ significantly from zero at any reasonable level of significance. The estimated trend line is very similar to that from the original model in Table 3.2 (the estimated intercept has reduced from 6.058 to 5.977 and the slope has increased from −0.037 to −0.034); however, the standard errors associated with the regression coefficients have both increased by around 20 %. These results are as expected from the general discussion in Section 3.1.1.

Table 5.1 also shows the estimates of the autoregressive parameters: $\hat{\phi}_1 = 0.9959$ and $\hat{\phi}_2 = -0.5836$, and the residual variance is estimated as 0.08824. The fitted model for the deviations from the trend line is therefore

$$\varepsilon_t = 0.9959\varepsilon_{t-1} - 0.5836\varepsilon_{t-2} + \delta_t$$

with $\mathrm{Var}(\delta_t) = 0.08824$. Apart from the residual variance, we have seen this model before: it was introduced in Equation (5.1) to motivate the use of stationary time series models for 'trend-like' behaviour. The simulated sequences in Figure 5.1 show that the model tends to generate quasi-cyclical sequences, which is consistent with the observed patterns in the residuals from the original model. ∎

This example illustrates the use of stochastic models as a means of capturing behaviour that is not easily modelled using a parametric trend function. Of course, an alternative approach would have been to use a nonparametric trend function to track the oscillation in the series: the difference between the two approaches is that in (5.15) the oscillations are regarded as essentially a distraction from the underlying linear trend, whereas a nonparametric approach would treat them as an important feature to be estimated (see the discussion at the end of Section 4.4). Subject-matter considerations should determine which approach is preferable.

In this section we have introduced the concept of time series residuals as a means of checking a fitted model. They can also be used as 'pre-whitened' series prior to computing cross-correlation functions in an exploratory analysis. Recall from Section 2.1.5 that the sample CCF between two autocorrelated series is difficult to interpret. To get around this, one possibility is to fit time series models separately to each of the individual series and to calculate cross-correlations between the resulting residual series; for further details, see Brockwell and Davis (2002, Section 7.3). The transfer function modelling framework of Box, Jenkins and Reinsel (1994, Chapters 10 and 11) extends this idea, enabling the identification and fitting of models where one time series is considered to respond to changes in the other.

5.1.7   Forecasting

In the ARMA model (5.5), Y_t is represented in terms of preceding quantities $\{Y_{t-j} : j = 1, \ldots, p\}$ and $\{\delta_{t-j} : j = 1, \ldots, q\}$. This provides the opportunity to forecast future values of the process. Suppose for the moment that the series (Y_t) follows a known ARMA process, that the values of both Y_t and δ_t have been observed for $t \leq T$ and that it is desired to forecast the future value $Y_{T+\ell}$ for some *lead time* $\ell > 0$. For many purposes, the natural predictor is the conditional expectation

$$\hat{Y}_{T+\ell} = \mathrm{E}\left(Y_{T+\ell} | \{(Y_t, \delta_t) : t \leq T\}\right),$$

which has the smallest expected squared prediction error among all functions of the observed quantities (Priestley, 1981, Section 10.1). For an ARMA model, we can write down the model equation for $Y_{T+\ell}$ and take conditional expectations of all quantities to obtain

$$\hat{Y}_{T+\ell} = \phi_1 \hat{Y}_{T+\ell-1} + \cdots + \phi_p \hat{Y}_{T+\ell-p} + \hat{\delta}_{T+\ell} + \theta_1 \hat{\delta}_{T+\ell-1} + \cdots + \theta_q \hat{\delta}_{T+\ell-q}. \quad (5.16)$$

The 'hats' here all denote expectation conditional on $\{(Y_t, \delta_t) : t \leq T\}$. Therefore, for $j \leq 0$, $\hat{Y}_{T+j}$ and $\hat{\delta}_{T+j}$ are just equal to their observed values: $\hat{Y}_{T+j} = y_{T+j}$ and $\hat{\delta}_{T+j} = d_{T+j}$, say. Further, for $j > 0$, $\hat{\delta}_{T+j} = 0$ since δ_{T+j} is uncorrelated with anything prior to time T. This provides a recursive means of calculating forecasts. Equation (5.16) is sometimes called the *difference equation* form for the forecasts. For illustration, consider the zero mean ARMA(1,1) process

$$Y_t = \phi Y_{t-1} + \delta_t + \theta \delta_{t-1}.$$

Given pairs $(Y_t = y_t, \delta_t = d_t)$ for $t = 1, \ldots, T$, the one-step-ahead forecast is just

$$\hat{Y}_{T+1} = \phi \hat{Y}_T + \hat{\delta}_{T+1} + \theta \hat{\delta}_T = \phi y_T + \theta d_T.$$

A two-step-ahead forecast can now be computed as

$$\hat{Y}_{T+2} = \phi \hat{Y}_{T+1} + \hat{\delta}_{T+2} + \theta \hat{\delta}_{T+1} = \phi \hat{Y}_{T+1},$$

and so on.

The development above is equally applicable to trend–stationary models (see Section 5.1.5): all that is required in this case is to add the trend function (μ_t) to the forecasts and prediction intervals. This effectively provides a means of prediction for regression models fitted using GLS (Section 3.3). From now on, it is convenient to employ the notation used in Section 5.1.5 for trend–stationary processes and to denote by $\tilde{Y}_t = Y_t - \mu_t$ the centred process obtained by subtracting the trend function (which, in the case of a truly stationary process, is just a constant).

In practice, it is hardly realistic to assume either that the model equation for (Y_t) is known exactly or (if moving average components are present) that the values of δ_t have been observed for $t \leq T$. The former assumption is discussed below. To deal with the latter, notice that if the underlying model is known, in principle we could use the residuals $e_1, \ldots, e_T$ to estimate the $\{\delta_t\}$. These should provide reasonable estimates, and hence a reliable basis for the construction of forecasts, if the model is invertible and T is large enough (see the discussion of invertibility in Section 5.1.5). However, rather than working directly with the observations (Y_t) and residuals (e_t), it is more accurate and illuminating to work with the innovations (h_t) defined in (5.13). We have

$$Y_t = \tilde{Y}_t + \mu_t = \mu_t + h_t + \sum_{j=1}^{t-1} w_{j,t} h_j$$

so that the conditional expectation of $Y_{T+\ell}$ is

$$\hat{Y}_{T+\ell} = \mu_{T+\ell} + \hat{h}_{T+\ell} + \sum_{j=1}^{T+\ell-1} w_{j,T+\ell} \hat{h}_j.$$

Recall, however, that the innovations $h_1, \ldots, h_T$ are just linear combinations of the variables $Y_1, \ldots, Y_T$, whose values have been observed: hence the (δ_t) are no longer needed to calculate $\hat{Y}_{T+\ell}$. It is clear that $\hat{h}_t = h_t$ for $1 \le t \le T$. Also, since the innovations are mutually uncorrelated with zero mean, $\hat{h}_t = 0$ for $t > T$. The forecast is therefore

$$\hat{Y}_{T+\ell} = \mu_{T+\ell} + \sum_{j=1}^{T} w_{j,T+\ell} \hat{h}_j. \tag{5.17}$$

As well as sidestepping the issue of estimating the (δ_t), representation (5.17) enables us to quantify the uncertainty in the forecast. The error in the forecast is $Y_{T+\ell} - \hat{Y}_{T+\ell}$, which can be written in terms of the innovations as

$$\left(\mu_{T+\ell} + h_{T+\ell} + \sum_{j=1}^{T+\ell-1} w_{j,T+\ell} h_j \right) - \left(\mu_{T+\ell} + \sum_{j=1}^{T} w_{j,T+\ell} \hat{h}_j \right)$$

$$= h_{T+\ell} + \sum_{j=T+1}^{T+\ell-1} w_{j,T+\ell} h_j. \tag{5.18}$$

This clearly has zero mean. Furthermore, (5.18) is a sum of uncorrelated terms. Its variance is therefore

$$\text{Var}\left(h_{T+\ell}\right) + \sum_{j=T+1}^{T+\ell-1} w_{j,T+\ell}^2 \text{Var}\left(h_j\right) = v_{T+\ell} + \sum_{j=T+1}^{T+\ell-1} w_{j,T+\ell}^2 v_j. \tag{5.19}$$

Under the assumption of normality, this variance can be used to construct a prediction interval for $Y_{T+\ell}$: for example, a 95 % interval is

$$\hat{Y}_{T+\ell} \pm 1.96 \sqrt{v_{T+\ell} + \sum_{j=T+1}^{T+\ell-1} w_{j,T+\ell}^2 v_j}. \tag{5.20}$$

The above development is not the only way to construct forecasts and prediction error variances. The `predict.Arima` routine in R uses an algorithm based on state space methods (see Section 5.5), although the results are numerically identical to those obtained from (5.17) and (5.19). The presentation above does, however, yield some useful insights. From the definition of the innovation weights $\{w_{j,t}\}$ after Equation (5.13), we have

$$w_{j,T+\ell} = \text{Cov}(\tilde{Y}_{T+\ell}, h_j)/\text{Var}(h_j).$$

If $\tilde{Y}_{T+\ell}$ is a stationary ARMA process it can be shown, as seems intuitively reasonable, that $\text{Cov}(\tilde{Y}_{T+\ell}, h_j)$ tends to zero as the lead time ℓ increases with $j \leq T$ fixed – after a while, the process 'forgets' about the innovation h_j. In (5.17), therefore, all contributions to the second term tend to zero as ℓ increases and, for large ℓ, the forecast is approximately equal to $\mu_{T+\ell}$ (for an MA(q) process in fact, for $\ell > q$ the equality is exact, as is easily seen from the difference equation form (5.16) of the forecasts). It follows that the forecast error is approximately $Y_{T+\ell} - \mu_{T+\ell}$, the variance of which is simply the unconditional variance of $Y_{T+\ell}$. Notice also, from (5.19), that the forecast error variance cannot decrease as ℓ increases. To summarise, if (Y_t) is trend–stationary and $(\tilde{Y}_t)$ is an ARMA process:

- Long-range forecasts converge to the underlying trend and do not depend to any appreciable degree on the available observations.

- The forecast error variance (and hence the width of prediction intervals) increases with lead time and tends to a constant value, which is the unconditional variance of $\tilde{Y}_t$.

These conclusions may seem obvious. However, as we will see shortly, they do not hold for all classes of model, and consideration of long-range forecast behaviour can provide useful guidance to the analyst. For example, as noted in Section 5.1.3, seasonal extensions of the ARMA family are available. However, the long-range forecasts from a stationary seasonal ARMA model will tend to a constant value, which is unrealistic in many environmental series showing a strong seasonal cycle that is expected to persist into the future. In such situations, it may be more appropriate to adopt a trend–stationary model in which seasonality is incorporated into the mean function (μ_t) using techniques such as those discussed in Section 3.2.

The discussion above assumes that the model is known. In practice, of course, there will usually be uncertainty regarding the precise form of the model itself, as well as the values of the parameters. The net effect is that prediction intervals, calculated as described above, tend to be too narrow. Parameter uncertainty can in principle be accounted for using bootstrap techniques (the bootstrap prediction algorithm for GLMs in Section 3.5 can be adapted for this purpose – the only change required is to simulate from the fitted time series model in the first step); however, unless the sample size is small this will usually have a relatively minor effect. Model uncertainty is much more difficult to deal with. One possibility is *Bayesian model averaging*, in which predictions from a set of plausible candidate models are combined; see Draper (1995) for example. For a more extensive discussion of these issues, see Chatfield (2000, Chapters 7 and 8). Regardless of the method chosen to compute prediction intervals, however, users could be forgiven for treating them with some scepticism unless their credibility has been demonstrated. Perhaps the simplest way to achieve this is by forecasting some observations that were not used to select or fit the model: if all is well, the mean squared error of these 'out-of-sample' forecasts should be similar to the forecast error variance calculated from the model.

Example 5.3

For the haddock biomass model, the `predict.Arima` command has been used to produce forecasts of log biomass, with associated error variances, for the period 2001–2010 (recall that the model was fitted to data up to the year 2000). The error variances have been converted to 95 % prediction intervals using Equation (5.20). The forecasts and prediction intervals are shown in Figure 5.4(a): the y axis here is labelled with actual biomass values to aid interpretation. The actual values for 2001–2004, which are also available to us but have not been used so far in any of the analyses, are also shown. Setting aside, for the moment, the complete lack of agreement between the actual values and predictions, we focus on the latter. As expected, the forecasts themselves continue the linear downward trend, and the width of the prediction intervals stabilises from around 2005 onwards (notice, however, that if the prediction intervals are converted back to the original scale as in Figure 1.2(c), they become narrower as the level decreases).

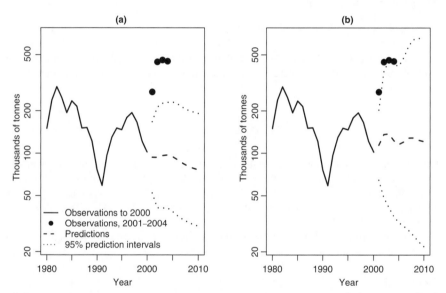

Figure 5.4   Predictions of North Sea haddock spawning stock biomass for 2001–2010, made on the basis of data to 2000 (a) using the trend–stationary AR(2) model summarised in Table 5.1, (b) using the difference–stationary ARIMA(2,1,0) model summarised in Table 5.2 (see Example 5.7). Actual biomass values for 2001–2004 are also given. The y axis scaling is logarithmic.

This example demonstrates particularly well the need for out-of-sample verification of predictive performance. The diagnostic checks in the previous section indicated an excellent fit to the data used for fitting, yet the model is completely unable to account for the sudden and rapid increase in log biomass in 2001 and 2002. This increase seems to be associated primarily with an extremely good breeding season for North Sea haddock in 1999 (ICES, 2001, Section 4.5.1). This is not apparent in the biomass data for either 1999 or 2000, presumably because juvenile fish are not part of the spawning stock. ∎

5.1.8   The backshift operator

It is often useful to seek alternative representations of time series models, in order to understand more clearly the structures that they represent and, perhaps, to appreciate the differences between apparently similar model formulations. In this respect, a very useful notational device is the *backshift operator*, denoted by B. This is used to denote the operation of moving backwards through a series: $BY_t = Y_{t-1}$, $B^2Y_t = Y_{t-2}$ and, in general, $B^kY_t = Y_{t-k}$. With this convention, the ARMA(p, q) model equation (5.5) can be written as

$$Y_t - \phi_1 BY_t - \cdots - \phi_p B^p Y_t = \delta_t + \theta_1 B\delta_t + \cdots + \theta_q B^q \delta_t$$

(5.21)

$$\text{or} \qquad \left(1 - \phi_1 B - \cdots - \phi_p B^p\right) Y_t = \left(1 + \theta_1 B + \cdots + \theta_q B^q\right) \delta_t.$$

We can also write, for example,

$$B^j B^k Y_t = B^j Y_{t-k} = Y_{t-(j+k)} = B^{j+k} Y_t$$

so that $B^j B^k = B^{j+k}$. In this respect and in many others, B behaves essentially like a number. When first encountered, perhaps its most remarkable property is that series expansions such as

$$(1 - aB)^{-1} = \sum_{j=0}^{\infty}(aB)^j \qquad \text{if } |a| < 1$$

(5.22)

are legitimate and meaningful. The right-hand side here looks like the sum of a geometric series with first term 1 and common ratio aB. The left-hand side looks like $1/(1 - aB)$, which is the usual formula for the sum of such a series. The left-hand side should be interpreted as an 'inverse operation': if $Z_t = (1 - aB)Y_t = Y_t - aY_{t-1}$, then Y_t can be thought of as $(1 - aB)^{-1}Z_t$.

Some examples will help to illustrate how the backshift operator can be used to obtain alternative representations of time series models.

Example 5.4
In Section 5.1.5, we obtained an 'infinite autoregressive' representation of the MA(1) model $Y_t = \delta_t + \theta\delta_{t-1}$. This model can be written as $Y_t = (1 + \theta B)\delta_t$, whence we can write $\delta_t = (1 + \theta B)^{-1}Y_t$. From (5.22), if $|\theta| < 1$ we have

$$(1 + \theta B)^{-1}Y_t = \sum_{j=0}^{\infty}(-\theta)^j B^j Y_t = \sum_{j=0}^{\infty}(-\theta)^j Y_{t-j}.$$

We therefore have $\delta_t = \sum_{j=0}^{\infty}(-\theta)^j Y_{t-j}$, so that $\delta_{t-1} = \sum_{j=1}^{\infty}(-\theta)^{j-1}Y_{t-j}$. Hence

$$\delta_t + \theta\delta_{t-1} = \delta_t - \sum_{j=1}^{\infty}(-\theta)^j Y_{t-j},$$

in agreement with the earlier result.   ■

Example 5.5

In Section 3.3.3 we considered the 'lagged response' regression model (Equation (3.52))

$$Y_t = \beta_0 + \sum_{i=1}^{p} \beta_i x_{it} + \phi Y_{t-1} + \delta_t.$$

This can equivalently be written as

$$Y_t - \phi Y_{t-1} = (1 - \phi B) Y_t = \beta_0 + \sum_{i=1}^{p} \beta_i x_{it} + \delta_t,$$

so that

$$Y_t = (1 - \phi B)^{-1} \left(\beta_0 + \sum_{i=1}^{p} \beta_i x_{it} + \delta_t \right).$$

Now, using (5.22), we can replace $(1 - \phi B)^{-1}$ with $\sum_{j=0}^{\infty} \phi^j B^j$ providing $|\phi| < 1$. This yields

$$Y_t = \sum_{j=0}^{\infty} \phi^j \beta_0 + \sum_{j=0}^{\infty} \phi^j B^j \sum_{i=1}^{p} \beta_i x_{it} + \sum_{j=0}^{\infty} \phi^j B^j \delta_t$$

$$= \frac{\beta_0}{1 - \phi} + \sum_{i=1}^{p} \beta_i \sum_{j=0}^{\infty} \phi^j x_{i(t-j)} + \sum_{j=0}^{\infty} \phi^j \delta_{t-j},$$

as given after Equation (3.52). However, notice that the final term above was obtained as $(1 - \phi B)^{-1} \delta_t$. Writing this final term as ε_t, we therefore have

$$\varepsilon_t = (1 - \phi B)^{-1} \delta_t \Rightarrow (1 - \phi B) \varepsilon_t = \delta_t \Rightarrow \varepsilon_t = \phi \varepsilon_{t-1} + \delta_t,$$

and (ε_t) is an AR(1) process as claimed in the earlier chapter.   ■

Example 5.6

Consider a simple ecosystem in which there are two dominant species. For the sake of argument, we will refer to these as 'predators' and 'prey'. Let X_t and Y_t be the population sizes of predators and prey respectively, in year t. Typically, the predator population in year t will depend both on the size of the breeding population and on the availability of food in year $t - 1$. Similarly, the prey population will depend on the numbers of both predators and prey during the previous year. These considerations suggest, as a very simple model of the system, the pair of equations

$$X_t = c_X + \phi_1 X_{t-1} + \psi_1 Y_{t-1} + \delta_t,$$

$$Y_t = c_Y + \phi_2 Y_{t-1} + \psi_2 X_{t-1} + \varepsilon_t,$$

where (δ_t) and (ε_t) are both white noise sequences. The parameters ϕ_1 and ϕ_2 represent dependence on the previous year's breeding population for each species; ψ_1 represents the predators' dependence on the food supply and ψ_2 (which is expected to be negative) represents prey losses from predation. The parameters c_X and c_Y control the mean population sizes.

This simple predator–prey model looks similar to a first-order autoregression, since the populations of both predators and prey depend only on the previous years' values. We can, however, use the backshift operator to investigate the properties of either time series taken individually. For the sake of illustration we consider the predator population series (X_t). To eliminate Y_{t-1} from the first equation above we can apply the operator $(1 - \phi_2 B)$ to both sides, noting (from the second equation applied to Y_{t-1} rather than Y_t) that

$$(1 - \phi_2 B)Y_{t-1} = c_Y + \psi_2 X_{t-2} + \varepsilon_{t-1}.$$

We therefore have

$$(1 - \phi_2 B)X_t = (1 - \phi_2 B)c_X + \phi_1(1 - \phi_2 B)X_{t-1}$$
$$+ \psi_1 (c_Y + \psi_2 X_{t-2} + \varepsilon_{t-1}) + (1 - \phi_2 B)\delta_t.$$

This simplifies to yield

$$X_t = [c_X (1 - \phi_2) + \psi_1 c_Y] + (\phi_1 + \phi_2)X_{t-1}$$
$$+ (\psi_1 \psi_2 - \phi_1 \phi_2)X_{t-2} + \delta_t + \psi_1 \varepsilon_{t-1} - \phi_2 \delta_{t-1}$$
$$= c_0 + \gamma_1 X_{t-1} + \gamma_2 X_{t-2} + \delta_t + \xi_{t-1}, \text{ say,}$$

where $\xi_t = \psi_1 \varepsilon_t - \phi_2 \delta_t$. This looks very similar to the model equation for an ARMA(2,1) process with nonzero mean; the only difference is the presence of the term ξ_{t-1} in place of $\theta \delta_{t-1}$. However, it can be shown that the properties of the process are identical to those of an ARMA(2,1) and hence, for all practical purposes, this can be taken as an exact model for (X_t). An identical argument holds for the prey population sizes (Y_t). This may be useful in suggesting appropriate models for time series of individual species population sizes, at least in the simple situation considered here.

The ARMA(2,1) structure for (X_t) may appear surprising, given the first-order dependence of the original model equations. However, the AR(2) component makes sense in terms of the population dynamics; in fact, it offers exactly the same types of behaviour that are demonstrated by more complex discrete-generation predator–prey models (e.g. Krebs, 2001, Chapter 13; Kot, 2001, Chapter 11), although the treatment of such models in the literature is usually deterministic, with the error series (δ_t) and (ε_t) omitted. Specifically, the options are that one or both series are nonstationary (which will result in either a population explosion or extinction of one or both species) or that both series are stationary, which corresponds to the species coexisting with population levels fluctuating about an equilibrium level. In the stationary case, the AR(2) component offers the possibility that the autocorrelation function will either decay exponentially after lag 1 or that it will oscillate (see Section 5.1.1). The latter possibility corresponds to quasicyclical behaviour in the population time series, which is precisely what is observed for some species. Perhaps the best-known example of this involves the Canadian lynx, for which the snowshoe hare is a staple food. The populations of both lynx and hare show clear cycles, with a steady period of around 10 years. The lynx series is a classic example that has attracted a lot of attention in the time series literature (see, for example, Priestley, 1981, Section 5.5). The development above provides one possible explanation for the cyclical behaviour, in terms of the underlying population dynamics.

This explanation is not universally accepted, however; see Krebs (2001, page 226) and references therein.   ■

In the preceding example, there were potentially two variables of interest (predator numbers and prey numbers) so that a bivariate model was required. More generally, multivariate models are required whenever one is interested in studying two or more variables simultaneously. For an overview of modelling approaches for multivariate time series, see Section 6.2.

5.2   Trend removal via differencing

In the models considered so far in this chapter, the values of any autoregressive parameters are constrained by the requirement for (trend–)stationarity. For example, an AR(1) process is stationary only if $|\phi| < 1$. If we set $\phi = 1$ in the AR(1) model equation, we obtain a *random walk*:

$$Y_t = Y_{t-1} + \delta_t. \tag{5.23}$$

Figure 5.5(a) shows simulations, both of this process and of an AR(1) with $\phi = 0.95$. Notice that the stationary process fluctuates around its mean value of zero, whereas the

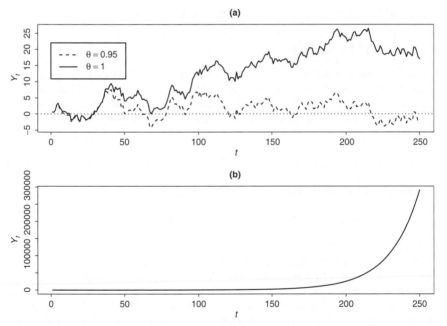

Figure 5.5   Simulated time series, each of length $T = 250$, from models of the form $Y_t = \phi Y_{t-1} + \delta_t$, with $\delta_t \sim N(0, 1)$. For each simulation, $Y_0 = 0$ and the same white noise sequence $\delta_1, \ldots, \delta_{50}$ is used. Panel (a) shows series with $\phi = 0.95$ (corresponding to a stationary series with mean zero) and $\phi = 1$ (a random walk). Panel (b) shows a series with $\phi = 1.05$. Notice the different vertical scale here.

random walk drifts away. To understand the latter behaviour, suppose that at time $t = 0$ the random walk is at y_0. Conditional on this value, we have $Y_1 = y_0 + \delta_1$, $Y_2 = Y_1 + \delta_2 = y_0 + \delta_1 + \delta_2$ and, in general, $Y_t = y_0 + \sum_{j=1}^{t} \delta_j$. The expectation of this quantity is y_0 and its variance is $t\sigma_\delta^2$. The linear increase of the variance with t suggests that, over time, the series (Y_t) will tend to drift further and further from its starting point of y_0. Formally, for any fixed M the probability that Y_t lies in the interval $(y_0 - M, y_0 + M)$ tends to zero as $t \to \infty$.

It is useful to rearrange (5.23) as

$$Y_t - Y_{t-1} = \delta_t,$$

which shows that for a random walk the differences between successive observations form a white noise process and are therefore stationary. This simple observation forms the basis for an alternative class of nonstationary processes in which the differences between successive observations are stationary. For obvious reasons, these processes are sometimes described as *difference–stationary*.

Using the backshift operator, the difference between successive observations can be written as $(1 - B)Y_t$. If we take the resulting time series and calculate differences again, we obtain the *second differences*:

$$(1 - B)[(1 - B)Y_t] = (1 - B)^2 Y_t = (1 - 2B + B^2)Y_t = Y_t - 2Y_{t-1} + Y_{t-2}.$$

In general, the dth *differences* can be calculated as $(1 - B)^d Y_t$.

5.2.1 ARIMA models

An important class of difference–stationary processes is the so-called *autoregressive integrated moving average* (ARIMA) class. A process (Y_t) is said to be ARIMA of order (p, d, q) if its dth difference is a stationary and invertible ARMA(p, q) process. The model equation for such a process can be deduced by substituting $(1 - B)^d Y_t$ for Y_t in (5.21):

$$(1 - \phi_1 B - \phi_2 B^2 - \cdots - \phi_p B^p)(1 - B)^d Y_t = (1 + \theta_1 B + \cdots + \theta_q B^q)\delta_t. \tag{5.24}$$

For example, for an ARIMA(1,1,0) process we have

$$(1 - \phi_1 B)(1 - B)Y_t = \delta_t \Rightarrow \left[1 - (1 + \phi_1)B + \phi_1 B^2\right]Y_t = \delta_t ,$$

so that $Y_t = (1 + \phi_1)Y_{t-1} - \phi_1 Y_{t-2} + \delta_t$.

We give here a very brief overview of the main features of ARIMA models. More details can be found in Brockwell and Davis (2002, Chapter 6), for example. Model identification is essentially the same as for ARMA models; the only change is the need to identify an appropriate degree of differencing, d. One way to do this is via visual inspection of time series plots of successive differences since, as in Figure 5.5(a), the appearance of series that require differencing is often very distinct from that of stationary processes. Very slow decay in the sample ACF is another common symptom of nonstationarity, as illustrated in an example below. In general, when using ARIMA models the aim is to find the smallest value of d for which the differences can plausibly be regarded as stationary, since *overdifferencing* can have undesirable consequences. For example, in the random walk model (5.23) the first differences form a white noise sequence, but the second differences are $(1 - B)^2 Y_t = \delta - \delta_{t-1}$. This is a noninvertible MA(1) process, which is not amenable to analysis (recall the discussions in Sections 5.1.5 and 5.1.7). Notice that in this case the variances of the first and second differences are σ_δ^2 and $2\sigma_\delta^2$ respectively: the overdifferenced series has a higher variance. This is a symptom of overdifferencing in general.

Once an appropriate degree of differencing has been identified, a stationary ARMA model can be fitted to the differences using the methods already described. Notice, however, that when calculating the first differences we cannot compute $Y_1 - Y_0$ since Y_0 is unobserved. Similarly, in general the dth differences can be computed only from time $d + 1$ onwards. Therefore, if an ARIMA(p, d, q) model is fitted to a time series of length T, the 'effective sample size' to be used in, for example, AIC calculations is $T - d$; residuals can also be calculated only from time $d + 1$ onwards. An unfortunate consequence of this is that models with different values of d cannot be compared using formal techniques such as AIC, since they are not fitted to the same data.

If an ARIMA model has been fitted with a view to forecasting future values of a process, perhaps the simplest way to proceed conceptually is to apply the theory of Section 5.1.7 to forecast the values of future differences (which follow a stationary ARMA process) and then to deduce the corresponding values for the original series. For example, if $d = 1$ and we observe data up to time T, we can forecast the first difference at time $T + 1$ as $\hat{D}_{T+1}$, say. However, we must have $\hat{D}_{T+1} = \hat{Y}_{T+1} - Y_T$, so that $\hat{Y}_{T+1} = Y_T + \hat{D}_{T+1}$. Subsequent forecasts can be calculated in a similar way, and forecast error variances calculated as described in Section 6.4 of Brockwell and Davis (2002). An alternative is to use state space methods, as implemented in the predict.Arima routine in R. A full treatment is beyond the scope of this book; however, it is worth noting an important difference between the long-range forecast behaviour of ARIMA models with $d > 0$ and that of stationary ARMA processes. In the stationary case, the long-range forecasts themselves tend to the overall mean of the process and the forecast error variance tends to a finite limit (see Section 5.1.7 before Example 5.3). For an ARIMA process, however, the limiting behaviour of the forecasts themselves is model-dependent and, perhaps more importantly, the forecast error variance grows without bound as the forecast lead time increases. This is because, as with the random walk, any realisation of a difference–stationary process can end up arbitrarily far from its starting point after enough time has elapsed; in practice, it means that prediction intervals at long lead times are often extremely wide. We return to this point in Section 5.3.

At first sight, the possibilities offered by differencing may seem limited – for example, there is no obvious way to deal with an AR(1) process with $\phi = -1$ or with $|\phi| > 1$. However, the types of nonstationarity that cannot be handled by differencing are either unlikely to be encountered in practical problems or can be handled using small modifications of the differencing idea (Box, Jenkins and Reinsel, 1994, Section 4.1 and Chapter 9). For example, an AR(1) process with $\phi > 1$ will always look like that in Figure 5.5(b), with the random fluctuations effectively swamped by exponential growth. This type of behaviour is rare in environmental applications and, where it does occur, it would make more sense to model it using a trend–stationary model in which the trend is an exponential function of time.

Another argument for the use of differencing is that explicit expressions for differences can all be written in the form $\sum_j w_j Y_{t-j}$, where the 'weights' $\{w_j\}$ sum to zero. From the discussion in Section 2.3, any such sequence of weights can be regarded as a high-pass filter, whence differencing can be interpreted as a trend removal device. In particular, it can be shown (Brockwell and Davis, 2002, Section 6.1) that if the original time series contains a polynomial trend of degree d, then this trend can be removed by differencing d times. To illustrate this, consider taking differences in the linear trend model

$$Y_t = \beta_0 + \beta_1 t + \varepsilon_t.$$

The first differences are given by

$$Y_t - Y_{t-1} = \beta_0 + \beta_1 t + \varepsilon_t - \left[\beta_0 + \beta_1(t - 1) + \varepsilon_{t-1}\right] = \beta_1 + \varepsilon_t - \varepsilon_{t-1}, \qquad (5.25)$$

which is equal to a constant plus the first differences of the (ε_t) series. Therefore, if (ε_t) itself is an ARIMA($p, 1, q$) process for some values of p and q, $Y_t - Y_{t-1}$ is simply an ARMA(p, q) process with nonzero mean β_1. Analogous results hold for higher-order polynomials. In general, if the dth difference of (Y_t) is an ARMA(p, q) process with nonzero mean, then (Y_t) itself can be regarded as an ARIMA(p, d, q) process superimposed on a polynomial trend of degree d. Notice, however, that the use of differencing to remove polynomial trends is only appropriate when the errors (ε_t) themselves form an ARIMA process of the appropriate degree. In (5.25), for example, if (ε_t) is already stationary then $(\varepsilon_t - \varepsilon_{t-1})$ is overdifferenced. Of course, in this case the process (Y_t) would be trend–stationary, and hence could be analysed using the methods described in Section 5.1.

Example 5.7
To illustrate the use of ARIMA models, we pay a final visit to the haddock biomass series of Section 1.3.2. In Section 5.1, we found a convincing trend–stationary model for this series. We now consider whether it could be modelled as an ARIMA process instead. For illustrative purposes, Figure 5.6(a) shows the sample autocorrelation function of the log biomass series. The correlations decay to zero rather more slowly than for the stationary cases in Figure 5.2: as noted above, this is a common symptom of nonstationarity (in fact, for most nonstationary series the decay is much slower than in this example). Of course, in this particular case the nonstationarity is apparent from the time series plot in Figure 1.2(a), so the sample ACF tells us nothing new. Sometimes, however, the ACF can provide important additional information regarding stationarity.

Figure 5.6(b) is a time series plot of the first differences of the log biomass series. In Section 1.3.2, we noted that these differences are actually the logarithms of the proportional increase in biomass each year. They look fairly stationary, apart from a couple of large values at the beginning of the series. Their mean is -0.0086; this is surprisingly small, since we have until now almost taken it for granted that there is a linear trend in this series – in this case, from (5.25) the differences should have nonzero mean. To understand what is happening, notice that the mean difference is

$$\frac{1}{T-1} \sum_{t=2}^{T} (y_t - y_{t-1}) = \frac{y_T - y_1}{T-1},$$

and therefore depends on just two of the undifferenced observations. In this particular series, Figure 1.2(a) shows that the first value y_1 is unusually low (this was also apparent in the residual analysis from the initial regression model, discussed in Section 3.1.1). We will bear this in mind during the subsequent analysis.

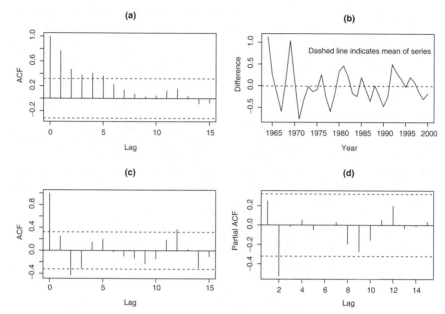

Figure 5.6   ARIMA model identification for the logarithms of the haddock biomass series. (a) ACF of the logarithms, (b) time series of the differences, (c) ACF of the differences, (d) PACF of the differences.

Having concluded that the differences are stationary, Figures 5.6(c) and (d) show their sample ACF and PACF. The ACF shows a clear oscillating pattern, suggesting an AR(2) model. The PACF confirms this, with a single large coefficient at lag 2. One might be tempted to conclude that, since the PACF at lag 1 does not differ significantly from zero, we should try to fit an AR(2) model with a zero coefficient of Y_{t-1}. In fact, such a conclusion is not justified since, for an AR(2) process, the theoretical autocorrelation at lag 1 is $\rho(1) = \phi_1/(1 - \phi_2)$ (Kendall and Ord, 1990, page 59). If ϕ_2 is negative, therefore, ρ_1 could be rather smaller than ϕ_1 and hence could appear insignificant even if ϕ_1 is nonzero.

If the first differences appear to follow an AR(2) process, the original log biomass series must follow an ARIMA(2,1,0). Despite the fact that the mean of the differences is close to zero, we may also wish to account for the apparent linear trend. In R, there are two possible ways to do this using the arima command. The first is to fit an ARIMA(2,1,0) to the original data with a covariate representing the trend – in this case, the result will correspond to that obtained by first differencing both the data and the covariate values and then fitting a stationary AR(2) to the result. A peculiar feature is that residuals will be computed for all observations – the first of these is essentially meaningless since, as noted above, the difference $y_1 - y_0$ cannot be computed (the calculations used by arima rely on a Bayesian argument, which will not be discussed here).

The second option for fitting the required ARIMA(2,1,0) model is to calculate the differences explicitly and then use arima to fit an AR(2) model with nonzero mean to these. The results will be identical to those from the first approach, except that in this case a residual will not be calculated for the first observation since it is lost in the differencing prior to calling arima. This option is arguably more transparent; however, if we subsequently wish to use the fitted model for prediction then the first approach should be used, since the computed predictions and standard errors will then relate directly to log biomass rather than to its differences.

Whichever option is used, the results indicate that the slope of the trend line (or equivalently the mean of the differenced series) is not significantly different from zero: the estimate is -0.117 with a standard error of 0.0419. In case this is influenced unduly by the first observation (see above), it may be worth refitting the model with this observation excluded. If this is done, the estimate changes to -0.0346, with a standard error of 0.0356. This estimate is much closer to that from the trend–stationary model in Table 5.1; however, it is still far from being statistically significant at any conventional level. We therefore drop this term from the model and refit. The results are shown in Table 5.2. Both remaining coefficients are highly significant, confirming that they are both needed in the model. Standard diagnostics, of the type discussed in Section 5.1.6, indicate that this model provides an excellent fit to the data.

Table 5.2   Edited R output for the ARIMA(2,1,0) model fitted to haddock biomass data.

```
Coefficients:
          ar1       ar2
       0.4926   -0.6597
s.e.   0.1451    0.1366

sigma^2 estimated as 0.08546
log likelihood = -7.61,   aic = 21.23
```

It is of interest to compare this model with the trend–stationary alternative considered in Table 5.1. As discussed in Section 1.3.2, the implied mechanisms behind the two models are rather different. The trend–stationary model asserts that, over the period of record, log biomass has followed a linear downward trend; this could reasonably be attributed to factors such as fishing activity. In contrast, the difference–stationary ARIMA model asserts that variation in biomass is essentially unsystematic. Setting aside for a moment the dreadful out-of-sample predictive performance of the trend–stationary model, it would be useful if, in general, we could determine which model fits better (and hence which of the two underlying mechanisms is better supported by the data). It is tempting to try and compare the models by using criteria such as AIC: the values in Tables 5.1 and 5.2 are 26.92 and 21.23 respectively, and appear to favour the ARIMA model. However, the AIC for the trend–stationary model is obtained from $T = 38$ observations whereas the second is from $T - 1 = 37$ differences – therefore, as discussed previously, we cannot use AIC to compare the models. We therefore confine ourselves to an informal comparison. The error variances for the two models are estimated as 0.08824 and 0.08546 respectively. Although these are very similar, the ARIMA model achieves a slightly smaller value with fewer parameters. Moreover, there seems to be slightly less autocorrelation in the ARIMA model residuals (not shown here) than in those for the trend–stationary model shown in Figure 5.3.

An alternative means of comparing the models is to examine their out-of-sample predictive performance. The predictions for the trend–stationary model have already been discussed in Section 5.1.7. Prediction intervals for the ARIMA model, for the period

2001–2010, have been computed in the same way and are shown in Figure 5.4(b). The first point to note here is that the intervals widen extremely rapidly; this is a general feature of ARIMA models, as discussed above. Nonetheless, the observations for 2001–2004 still fall outside the intervals from this model: its performance, although undeniably better than that of the trend–stationary model, is still somewhat unsatisfactory.

In summary, the ARIMA model appears to provide a slightly better fit to the data used for model fitting and to provide improved (but still inadequate) predictive performance. On the basis of the data alone, therefore, one should probably conclude that this model is preferable. However, more is known about North Sea fish stocks than simply the data used here. ICES (2001, page 16) suggests that the long-term decline may be due to a combination of high fishing effort and unfavourable climate conditions rather than to unsystematic year-on-year variation; in this case a trend–stationary model may be more appropriate, but with some index of fishing effort or climatic suitability substituted for the linear trend. Furthermore, as noted in Section 5.1, the 2000–2001 increase in biomass was primarily associated with an extremely good breeding season in 1999. To capture this kind of behaviour it would be necessary to account for the age structure of a population. Deterministic age-structured models are commonly used in fisheries management (e.g. Haddon, 2001, Section 2.8); stochastic versions can be handled using state space methods such as those discussed in Section 5.5 (see, for example, Gudmundsson, 2004). However, more detailed data would be required to implement either of these suggestions.   ∎

At one time, ARMA and ARIMA (AR(I)MA) models were quite popular in many areas of environmental science. For a small sample of applications see Edwards and Coull (1987), Milionis and Davies (1994), Mishra and Desai (2005), Robeson and Steyn (1990), Toth, Brath and Montanari (2000) and Zuur and Pierce (2004), and references therein. Their appeal stems largely from the availability of well-understood methods for identifying, checking and comparing models, as well as computationally efficient methods of fitting them. However, as illustrated in the haddock stocks example above, they can struggle to represent the complexity of some environmental time series. In many application areas, therefore, to deal with 'stochastic' trends the current move is towards more sophisticated and computationally intensive methods that aim to represent more directly the mechanisms underlying the series of interest, as well as accounting for features such as measurement error (which is likely to affect the haddock stocks example) and the potential availability of different sources of information. Such features are most easily handled within the framework of state space models, which are discussed below. However, it can be helpful to bear in mind the properties of AR(I)MA models when working with more sophisticated techniques that are often closely related. Furthermore, differencing is a simple trend removal device that can be extremely helpful if interest is not primarily in the trend but perhaps in the relationship between two series that both appear nonstationary.

5.2.2   Spurious regressions

To illustrate the last point above, we conclude this section with a cautionary example. Figure 5.7(a) shows two simulated random walks, each of length 250. The series have been generated independently; (Y_t) has already been seen in Figure 5.5(a). Although the series are unrelated, a casual inspection might suggest otherwise because of the apparent shared trend. This is typical of difference–stationary processes, essentially because their realisations are unbounded (see the previous subsection). If two such processes are observed for long enough, they will both change a lot over the period of observation – either in the

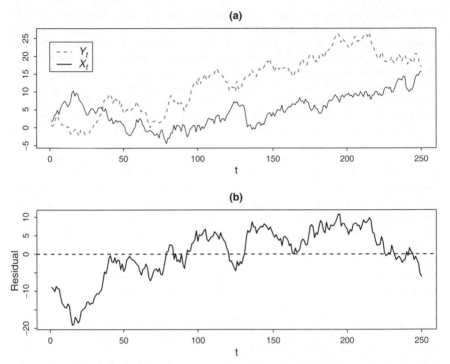

Figure 5.7   Illustration of the phenomenon of spurious regressions. (a) Simulated realisations of two uncorrelated random walks (X_t) and (Y_t). (b) Residuals from the regression of Y_t upon X_t.

same direction as in Figure 5.7(a) or in opposite directions. This gives the appearance of a strong association between them.

Consider now what happens if we regress (Y_t) upon (X_t) in Figure 5.7. Fitting the model $Y_t = \beta_0 + \beta_1 X_t + \varepsilon_t$ yields an estimate $\hat{\beta}_1 = 0.965$, with the standard error calculated as 0.101. The conventional t statistic for testing the null hypothesis $H_0 : \beta_1 = 0$ is 9.6 on 248 degrees of freedom, so that the null hypothesis is rejected overwhelmingly. This test result is correct, since the null hypothesis asserts that the series (Y_t) was generated as $Y_t = \varepsilon_t$ (i.e. that it is white noise). It is, however, wrong to conclude from this that there is a genuine relationship between (X_t) and (Y_t): we have learned that the null hypothesis is false, not that the alternative is true (see also the discussion of test procedures in Section 2.4). A careful analyst would avoid the potential pitfall via the use of appropriate diagnostics. For example, Figure 5.7(b) shows that the residuals from the regression bear no resemblance to a white noise sequence (in fact, their behaviour is similar to that of a random walk, which is unsurprising since any linear combination of independent random walks is itself just another random walk): this shows clearly that the assumptions underlying the analysis are violated.

This example illustrates a phenomenon known in the econometrics literature as 'spurious regressions' (Davidson and Mackinnon, 2004, Section 14.2). It shows that apparent correlation can arise from random mechanisms that are completely unconnected. The bottom line is that particular care is required when investigating relationships between difference–stationary series. The danger is not only of misinterpreting the results of test procedures in the absence of a genuine relationship: the theory underlying the calculation of standard errors and F statistics can also break down if, for example, a lagged response model (see Section 3.3.3 before Example 3.8) is fitted to a difference–stationary process. This is because, in this setting, such models can immediately become stationary if the coefficient(s) of the lagged variable(s) move away from their true values (for example if the coefficient of Y_{t-1} decreases from 1 in a random walk): this causes abrupt changes in the behaviour of the information matrix (see Section 3.1.6), which invalidates the usual asymptotic theory. Fortunately, the usual theory does hold in situations where a difference–stationary series (X_t) drives a response (Y_t) via simple relationships such as $Y_t = \beta_0 + \beta_1 X_t + \varepsilon_t$, where (ε_t) is a stationary process: here, (Y_t) is itself stationary if β_1 is zero and difference–stationary otherwise – but the information matrix depends solely on (X_t) and not on the regression parameters, so the previous problem does not arise. The theory holds more generally where some linear combination of the series under consideration is stationary: in this case, the series are said to be *cointegrated* (Davidson and Mackinnon, 2004, Section 14.5).[1] Overall, then, it can be argued that the spurious regressions phenomenon is often caused by misinterpreting mis-specified models and can thus be avoided by careful residual analyses. Alternatively, one could circumvent the problem entirely by differencing the series prior to analysis. For the series in Figure 5.7, for example, a regression of $(1 - B)Y_t$ upon $(1 - B)X_t$ correctly accepts the null hypothesis of no association (the p-value for this particular test is 0.774). Of course, differencing is useful only if relationships between differenced series (representing rates of change or growth) are of subject-matter interest in themselves; however, it is always worth bearing in mind as an option.

5.3   Long memory models

Although we have treated stationary and nonstationary processes separately in this chapter, in practice the distinction between them may not always be clear-cut. Figure 5.8(a) illustrates this: it shows annual values of the East Atlantic Jet (EAJ) index from 1958 to 1998, obtained from the Climate Prediction Center of the US National Oceanographic and Atmospheric Administration.[2] The EAJ is an index of large-scale atmospheric variability that has been related to various aspects of local climate, including summer melting of glaciers in Svalbard (Washington *et al.*, 2000) and spring precipitation in the Iberian peninsula (Martín *et al.*, 2004).

The EAJ series exhibits trend-like behaviour, with an extended period of low values from the mid 1960s to the mid 1980s and then a period of high values afterwards. The sample ACF, shown in Figure 5.8(b), shows a relatively long series of positive (although small) coefficients, which is typical of nonstationary processes. This may lead us to suspect

---

[1] There is a substantial body of rather technical literature suggesting that the usual asymptotic theory fails in regression models with difference–stationary covariates that are not cointegrated: see Banerjee *et al.* (1993, Chapter 6) for a summary. However, the crux of the argument seems to be that the normalised information tends to a random matrix rather than a constant, and it is not always obvious that this is relevant in applications. For example, this can be shown to occur when regressing on a random walk (X_t) as in the example above, but the usual arguments can be applied if one regards the inference as conditional on the observed values of (X_t). For a formal discussion of this in a wider context, see Sweeting (1992).

[2] http://www.cpc.noaa.gov.

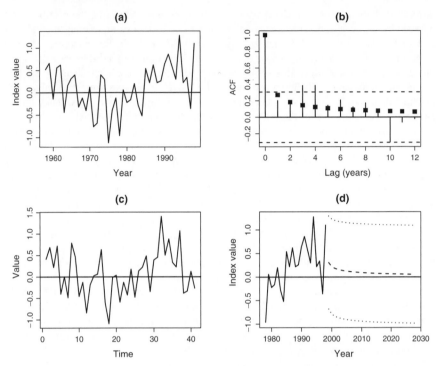

Figure 5.8   (a) Annual time series of the East Atlantic Jet index, 1958–1998. (b) Sample ACF of the East Atlantic Jet series, with a theoretical ACF for the fitted ARIMA(0, d, 0) model superimposed. (c) Specimen realisation of 41 values from the fitted model. (d) Forecasts and approximate 95 % prediction intervals from the fitted model.

nonstationarity in the series and hence to model it using ARIMA models. However, we have seen that realisations from such models will eventually move arbitrarily far from their starting point (equivalently, their long-run forecast error variances are unbounded). This is unrealistic in the present context since the behaviour of the climate system, of which the EAJ represents a part, is constrained by physical and chemical laws. The EAJ index therefore cannot take indefinitely large values, and a nonstationary ARIMA model is inappropriate.

Many climatic time series exhibit features similar to the EAJ. This has led to the concept of 'interdecadal variability', which is characterised, implicitly or explicitly, by (quasi-)periodic structures at low frequencies (see Moron, Vautard and Ghil, 1998 for a review). A first glance at Figure 5.8(a) might suggest, for example, that there is an underlying cycle of around 40 years. The existence of exact periodicities in naturally occurring annual series seems unlikely, however, and in any case it would be unwise to draw such conclusions in the absence of a much longer data series.

Stationarity is not a property of data, but rather of a stochastic model (see the definition in Section 2.1.3). It is therefore not appropriate to ask whether a particular series is stationary: rather, one should ask whether it is described better by a stationary or a nonstationary model. We have encountered two broad classes of nonstationary process: trend–stationary and difference–stationary. Neither seems appropriate for the EAJ series: it is difficult to identify a deterministic trend with any confidence on the basis of the available data, and difference–stationary models have an unrealistic long-run behaviour. This motivates the search for stationary models that exhibit the kind of behaviour seen in Figure 5.8(a). Perhaps the most satisfying such model uses the concept of 'fractional differencing', introduced by Granger and Joyeux (1980) and Hosking (1981). The idea is to extend the definition of the ARIMA(p, d, q) process in Equation (5.24) to noninteger values of the differencing parameter d: since the model is stationary if $d = 0$ and nonstationary with unbounded realisations if $d = 1$, then perhaps intermediate values of d will lead to a stationary process with trend-like features. Initially this seems unhelpful, since it is not clear that such 'fractional differences' have any meaning or that they could be calculated in practice. The key is to use the backshift operator to write the dth difference of a series (Y_t) as $(1 - B)^d Y_t$ and then to use a power series expansion:

$$(1 - B)^d = \sum_{j=0}^{\infty} \binom{d}{j} (-B)^j = \sum_{j=0}^{\infty} \frac{d!}{j!(d-j)!} (-B)^j$$

$$= B^0 + \sum_{j=1}^{\infty} \frac{d(d-1) \cdots (d-j+1)}{j!} (-B)^j = \sum_{j=0}^{\infty} a_j B^j, \text{ say.}$$

With this convention, the dth difference of the series Y_t can be written as $(1 - B)^d Y_t = \sum_{j=0}^{\infty} a_j Y_{t-j}$. If d is an integer, the only nonzero terms in this sum are those for $j = 0, \ldots, d - 1$ and we recover the usual differences. Otherwise, however, the sum involves an infinite number of past values of the process.

Having defined the fractional differencing operator, we can extend the class of ARIMA processes to allow noninteger values of the differencing parameter, as required. When d is not an integer, we refer to the resulting process as *fractionally integrated*. Hosking (1981) shows that for d in the range $(-1/2, 1/2)$ the resulting process is stationary and invertible, and derives a variety of properties including the autocovariance, autocorrelation and partial autocorrelation functions. The autocovariance function provides all of the information necessary to implement the innovations algorithm (see Section 5.1.5), and hence to carry out Gaussian likelihood based estimation and prediction for this class of models. Unfortunately, the computational burden is substantially greater than for the ARMA models considered previously; in practice, therefore, the likelihood is often approximated. In R, approximate likelihood based fitting of these models is available in libraries `fracdiff` (Fraley *et al.*, 2006) and `longmemo` (Beran and Maechler, 2009).

A key feature of fractionally integrated processes is that when d lies in the interval $(0, 1/2)$, at large lags the autocorrelation $\rho(k)$ is approximately proportional to k^{2d-1}: the autocorrelations therefore decay with lag according to a power law. This decay rate is much slower than that for stationary ARMA models, where the autocorrelations decay exponentially. A consequence of this is that in a fractionally integrated process with $d \in (0, 1/2)$, non-negligible correlation can persist between observations that are widely separated in time. In practice, this means that realisations of such processes can remain above or below their underlying mean level for extended periods. For practical examples of this in relation to climatic time series, see Percival, Overland and Mofjeld (2001).

Example 5.8

As just described, the characteristics of fractionally integrated processes are very similar to those of the EAJ series in Figure 5.8(a). By way of illustration, an ARIMA($0, d, 0$)

model has been fitted to the series using the `fracdiff` library in R: this implements a fast approximate maximum likelihood method due to Haslett and Raftery (1989). The differencing parameter d is estimated as $\hat{d} = 0.213$, with a reported standard error of 3×10^{-7} (although this seems unrealistically small, suggesting a problem with the standard error calculation).[3] The theoretical ACF of the fitted model is superimposed on the sample ACF in Figure 5.8(b), and is qualitatively very similar to the sample version. A time series plot and sample ACF of the standardised innovations (not shown) suggest that the model accounts for most of the apparent structure in Figure 5.8(a). Figure 5.8(c) shows a sequence of 41 values simulated from the fitted model: this demonstrates the ability of the fitted model to generate 'trend-like' features similar to those in the original data. Finally, Figure 5.8(d) shows forecasts from the fitted model for the period 1999–2028, along with approximate 95 % prediction intervals (these are based on the innovations algorithm, as discussed in Section 5.1.7). Notice that the forecasts decay very slowly to the mean of the process under the fitted model: this reflects the persistence of the correlation. Notice also that because the fitted model is stationary, the prediction intervals remain bounded and are hence physically plausible.   ∎

Because of the persistence of correlations in fractionally integrated processes, they are often described as having *long memory*. A long memory process may be defined formally (McLeod and Hipel, 1978) as a stationary process for which the autocorrelations are not absolutely summable:

$$\sum_{k=0}^{\infty} |\rho(k)| = \infty.$$

For this condition to hold, ultimately the autocorrelations must decay at a rate of $1/k$ or slower, which is the case for fractionally integrated processes with $d > 0$.

Long memory processes are often characterised via their Fourier representations (see Section 2.3.1). To appreciate this characterisation it is necessary to consider the variances of the normalised sample Fourier coefficients (e.g. $\sqrt{T}A(f)$ and $\sqrt{T}B(f)$ in the notation of Section 3.4) of a realisation of the process. For many stationary processes, as the sample size increases two things happen. Firstly, the discrete grid of Fourier frequencies becomes successively finer until it can be approximated by a continuous frequency range and, secondly, the variance of the normalised coefficient at frequency f tends to a frequency-dependent limit, $h(f)$, say (this exposition is deliberately heuristic: for a more rigorous account, see Priestley, 1981, Chapter 4). This limit is called the *spectral density* of the process. At frequencies where the spectral density is large, the Fourier coefficients of any process realisation will also tend to be large; hence the Fourier representation of that realisation will contain substantial contributions from sinusoids at those frequencies. At frequencies close to zero, the spectral density of any long-memory process has the form $h(f) \propto f^{-\beta}$ for some $\beta \in (0, 1)$ (Beran, 1994, Section 2.1). Hence any realisation of a long-memory process will be dominated by low-frequency variability and, in particular, may exhibit apparent long-term cycles. This formalises more precisely the concept of interdecadal variability discussed at the start of this section.

Since the study of Nile river flows by Hurst (1951) in which the phenomenon of long-memory was first observed, the development of long-memory models has been driven

---

[3] An exact maximum likelihood fit of a zero-mean ARIMA(0, d, 0) model, using the software accompanying this book, yields the estimate $\tilde{d} = 0.240$ with a more plausible standard error of 0.113. The exact fit is much slower, however.

largely by applications in environmental science, notably hydrology and climatology. The presence of long memory in many environmental time series is now generally recognised, although initially it may be difficult to envisage mechanisms that could give rise to such structure: for example, it is rather unlikely that any natural process is really generated as an infinite sum of past values, as suggested by the formula $(1 - B)^d Y_t = \sum_{j=0}^{\infty} a_j Y_{t-j}$. In fact, several plausible mechanisms exist. One is the aggregation of short-memory processes: Granger (1980) showed that a sum of AR(1) processes can have long memory if the parameters of the individual processes are sampled independently from appropriate probability distributions. In a complex system such as the climate it is easy to envisage large-scale structure as arising from a superposition of smaller-scale processes, many of which may have quite different dynamics: this could therefore be an explanation for the apparent long memory in the EAJ series considered earlier. Beran (1994, Section 1.3) reviews several other mechanisms that can give rise to long memory.

Apart from fractionally integrated processes, most models for long-memory processes are specified directly via the autocorrelation or spectral density function; see Beran (1994, Chapter 2) for a review. A variety of heuristic methods are available for parameter estimation in rather simple models. However, Gaussian likelihood based inference is more flexible and usually more efficient. As with fractionally integrated ARIMA models, exact likelihood calculations are computationally expensive; most of the available methods are therefore based on approximations. Some theoretical progress has been made in more complex settings, for example time series regression with long-memory errors (Beran, 1994, Chapter 9). An interesting, but perhaps unsurprising, result is that the presence of long memory substantially reduces the precision with which polynomial trends can be identified. Indeed, this is one case in which the standard least squares estimators are not optimal even in large samples (recall the discussion in Section 3.1.1): it is worth the extra effort of likelihood based or GLS estimation in such settings. Applications to the analysis of global temperatures have been considered by Bloomfield (1992) and Smith and Chen (1996): both papers showed that even allowing for the potential presence of long memory in the climate system, there was strong evidence for an increasing trend in global temperatures over the twentieth century. More recently, wavelet methods (see Section 4.3.2) have been advocated as an alternative means of separating long-memory errors from deterministic trends that can be approximated reasonably by polynomials (Craigmile, Guttorp and Percival, 2004, 2005): the approach works in essence because information on such trends is concentrated in a relatively small number of coefficients in the wavelet decomposition. Despite these advances, however, software for the routine application of long-memory models is generally less well developed than for most of the other material covered in Part I of this book.

5.4 Models for irregularly spaced series

The development of plausible stochastic models for irregularly spaced series is less straightforward than in the regular case. One reason for this is that the dependence between successive observations typically varies with their temporal separation, and it is not immediately obvious how to write down a simple and meaningful model that captures this feature. Broadly speaking, the available models fall into two main classes. The first is appropriate when the irregularity is an intrinsic feature of the process under investigation as when studying, for example, the intensities of tropical storms or earthquakes.

In this case it is often natural to consider models based on marked point processes (see Section 6.3 for a brief introduction).

The second class of models is more appropriate when the irregularity is not a feature of the process itself, but rather of the sampling regime: there is an underlying stochastic process $Y(\cdot)$ that could in principle be observed at any time, but for which observations are available only at times $t_1, t_2, \ldots, t_N$. Accordingly we now discuss models for continuous-time processes. A more detailed introductory treatment can be found in Ross (2003, Chapter 10).

In all of the models considered so far for regularly spaced series, a fundamental component has been the white noise sequence (δ_t): without this, the series would be completely predictable and hence uninteresting. If we attempt to generalise the concept of white noise to a process $\delta(\cdot)$ in continuous time, we must allow two values of $\delta(\cdot)$ separated by an arbitrarily small time interval to be completely uncorrelated. This has some unexpected consequences. For example, it can be shown (Priestley, 1981, Section 3.7) that if such a process has finite variance then, with probability 1, any weighted time average of the form $\int_a^b w(t)\delta(t)\mathrm{d}t$ is equal to zero: again, this is uninteresting. The only escape is to allow the process variance to be infinite. Mathematically, this enables the development of a satisfactory theory, but it is difficult to visualise the appearance of realisations from such a process.

In contrast to white noise, the random walk process can be extended to a continuous-time setting without difficulty. From the definition at (5.23), it is easy to show that for any nonoverlapping time intervals $(t_1, t_1 + h_1)$ and $(t_2, t_2 + h_2)$, the *increments* $Y_{t_1+h_1} - Y_{t_1}$ and $Y_{t_2+h_2} - Y_{t_2}$ are zero-mean, uncorrelated random variables with variances $h_1\sigma_\delta^2$ and $h_2\sigma_\delta^2$ respectively. If in addition the (δ_t) are Gaussian, then the increments are not just uncorrelated but independent. The extension to continuous time is now straightforward. The analogue of a Gaussian random walk is a process $W(\cdot)$ such that for $h > 0$ and all t, $W(t + h) - W(t) \sim N(0, h\sigma^2)$ and the increments in nonoverlapping time intervals are independent. Such a process is called a *Wiener process* or *Brownian motion*. Like the random walk, it is nonstationary. Its structure is also easy to visualise: given an initial value $Y(t_0) = y(t_0)$, values at any discrete set of time points $t_0 < t_1 < \cdots < t_N$ can be generated by cumulating increments that are sampled from the appropriate normal distributions.

Brownian motion forms the basis of many models for continuous-time processes. However, many of the model equations require mathematical notation that is beyond the scope of this book. We therefore give just a brief summary.

Probably the most widespread model for continuous-time stationary processes is the analogue of the AR(1), referred to variously as the CAR(1) or *Ornstein–Uhlenbeck* process. The autocorrelation function of this process is, for all lags $h \in \mathbb{R}$,

$$\rho(h) = \exp(-\alpha|h|) \tag{5.26}$$

for some parameter $\alpha > 0$. There is a close connection with the ordinary AR(1): to see this, consider what would happen if we observed a CAR(1) process every Δ time units. The lag 1 autocorrelation of the resulting regularly spaced series would be $\rho(\Delta) = \exp(-\alpha\Delta) = \phi$, say; at lag 2 we would obtain $\rho(2\Delta) = \exp(-2\alpha\Delta) = \phi^2$ and, in general, at lag k we would obtain $\rho(k\Delta) = \phi^{|k|}$. However, this has exactly the same form as the ACF of the AR(1) process given in Section 5.1.1. It follows that sampling a CAR(1) process at regular time intervals yields an ordinary AR(1) process. Notice, however, that since $\phi = \exp(-\alpha\Delta)$ here, we must have $0 < \phi < 1$: the process cannot produce negative autocorrelations.

Continuous-time analogues of more general ARMA models are also available (Priestley, 1981, Section 3.7; Brockwell and Davis, 2002, Section 10.4). As in the AR(1) case, their properties are broadly similar to those of the corresponding discrete-time versions: for example, the continuous-time AR(2) process can generate quasi-periodic behaviour. There are some surprises, however: for example, sampling a continuous ARMA(2,1) process at regular intervals can yield a discrete AR(1) with a negative coefficient – and sampling a continuous AR(p) process yields a discrete ARMA($p, p - 1$) (Harvey, 1989, Section 9.1). Long-memory models for continuous-time processes include *fractional Brownian motion* (Beran, 1994, Section 2.4), which generalises the idea of fractional differencing discussed in Section 5.3; continuous fractional ARIMA models have been developed by Tsai and Chan (2005). More generally, in the statistical literature there is currently considerable interest in models defined by *stochastic differential equations* (SDEs); Cox and Miller (1965, Chapter 5) remains an excellent foundational reference in this area. SDEs are derived essentially by considering the limiting behaviour of sampled versions of a process, as the sampling interval tends to zero. They are widely used in areas such as mathematical finance, where they are able to capture many of the features of erratic, nonstationary, high-frequency financial time series. Their use in environmental applications is currently limited, although they are starting to be used to represent uncertainty in deterministic physics based models: this provides an opportunity to incorporate the known dynamics of the underlying processes formally into an empirical analysis. For example, Tomassini *et al.* (2009) used a CAR(1) process representation to model variations in the parameterisation of a simple physics based model of the climate system: the goal here was to improve the model performance, with a particular focus on reproducing observed changes in global temperatures and ocean heat content over the nineteenth and twentieth centuries.

An important use of continuous-time models is to describe the error structure in regressions involving irregularly spaced time series. Under the assumption that the error process is stationary, knowledge of its autocorrelation structure enables the fitting of models by maximum likelihood or generalised least squares (see Section 3.3.3). In R, for example, the `nlme` library (Pinheiro *et al.*, 2008) offers the possibility of fitting regression models with a CAR(1) error structure; other continuous-time structures can be defined by the user if required. Identification of an appropriate structure can be achieved by inspection of the variogram (see Section 2.1.4) of residuals from an initial least squares fit. Equation (2.3) gives the relationship between the variogram and autocorrelation function of a stationary process: for example, in the CAR(1) case the residual variogram is given by $v(h) = \sigma^2 \left[1 - \exp\left(-\alpha|h|\right)\right]$, where σ^2 is the residual variance.

5.5   State space and structural models

5.5.1   Simple structural time series models

The models considered so far in this chapter have been descriptive, in the sense that they provide a means of summarising the behaviour of a series without necessarily considering its underlying structure. In this section we consider an alternative class of models that attempt to represent the underlying structure explicitly. The simplest nontrivial example of such a *structural time series model* is given by a pair of equations:

$$Y_t = \mu_t + \varepsilon_t, \tag{5.27}$$

$$\mu_t = \mu_{t-1} + h_t, \tag{5.28}$$

where (ε_t) and (h_t) are uncorrelated white noise sequences with variances σ_ε^2 and σ_h^2 respectively. The first equation here is the fundamental trend equation (3.1), according to which the observations can be regarded as a contaminated version of the underlying trend. The second equation asserts that the trend (μ_t) itself evolves stochastically as a random walk. If σ_h^2 is small, successive values of the trend will tend to be similar – although over longer periods of time, larger changes will occur. For this reason, the model defined by (5.27) and (5.28) is called the *local level model*.

An alternative to the local level model is the *local linear trend* model, which asserts that the trend at any time point is approximately linear. In place of (5.28), the local linear trend model substitutes a pair of equations:

$$\mu_t = \mu_{t-1} + \beta_{t-1} + h_t, \tag{5.29}$$

$$\beta_t = \beta_{t-1} + z_t, \tag{5.30}$$

where (h_t) and (z_t) are white noise sequences with variances σ_h^2 and σ_z^2 respectively, that are uncorrelated with each other and with (ε_t). The motivation for (5.29) and (5.30) is that if the trend was exactly linear then, for all t, we would have $\mu_t = \mu_{t-1} + \beta$, say, where β is the trend slope (i.e. change per unit time): (5.30) allows the slope to vary over time, which may be more realistic.

Structural models such as these are appealing since the model components are directly interpretable and often immediately relevant to the aim of an analysis. For this reason, the models are widely used in a variety of application areas. Environmental examples include an analysis of Nile river flows using a local level model with an added changepoint (Pole, West and Harrison, 1994, Section 7.1), an analysis of polychlorinated biphenyl (PCB) concentrations in salmonids using a local linear trend model (Conrad Lamon, Carpenter and Stow, 1998) and analysis of sea surface temperatures using a more complicated model incorporating seasonal effects (Young *et al.*, 1991). However, despite the immediate appeal of the models, some care is required since their properties may not always be appropriate for the problem at hand. To illustrate this for the local level model, consider that in (5.28) we can write $(1 - B)\mu_t = h_t$, where B is the backshift operator. Now, if we apply the operator $(1 - B)$ to both sides of (5.27), we obtain

$$(1 - B)Y_t = (1 - B)\mu_t + (1 - B)\varepsilon_t = h_t + \varepsilon_t - \varepsilon_{t-1}.$$

As a sum of white noise terms, the right-hand side here is clearly a stationary process, and it is easily verified that its autocorrelation function is

$$\rho(k) = \begin{cases} 1, & k = 0, \\ -\sigma_\varepsilon^2 / \left(\sigma_h^2 + 2\sigma_\varepsilon^2\right), & |k| = 1, \\ 0, & \text{otherwise.} \end{cases}$$

This is the autocorrelation function of an MA(1) process (see Section 5.1.2): we can therefore conclude that, for the local level model, the first differences of the observations

(Y_t) are indistinguishable from an MA(1) and hence that the observations themselves can be regarded as ARIMA(0,1,1). This argument is formalised in Harvey (1989, Section 2.5.3): the equivalent ARIMA specification is called the *reduced form* of the structural model. A consequence of this equivalence is that, in the long run, any realisation from a local level model will move arbitrarily far from its starting point (recall the discussion in Section 5.2.1): if this is deemed to be unrealistic, then the model should be applied with caution. Similar comments apply to the local linear trend model, for which the reduced form is ARIMA(0,2,2). Thus, for example, Zuur, Tuck and Bailey (2003) advocate that in the context of fisheries research, forecasts from such models should be restricted to relatively short lead times to avoid unrealistically large error variances. This is not true of all structural models, of course: in broad terms, the reason for the behaviour of the local level and local linear trend models is that they both have random walks embedded within them, and there are plenty of alternative models that do not share this feature. However, the discussion above illustrates the need to think carefully about the properties of such models before applying them indiscriminately.

5.5.2   The state space representation

Both of the models in the previous subsection can be written, using matrices, as a pair of equations of the form

$$\mathbf{Y}_t = \mathbf{Z}_t \boldsymbol{\alpha}_t + \boldsymbol{\varepsilon}_t, \tag{5.31}$$

$$\boldsymbol{\alpha}_t = \mathbf{M}_t \boldsymbol{\alpha}_{t-1} + \boldsymbol{\eta}_t, \tag{5.32}$$

where:

- $\mathbf{Y}_t$ is a $p \times 1$ vector of observations at time t: for both the local level and local linear trend models, there is only one observation at each time point so that $p = 1$.

- $\boldsymbol{\alpha}_t$ is an $m \times 1$ *state vector* at time t, which is typically unobserved. For the local level model, $m = 1$ and the state vector is $\alpha_t = \mu_t$. For the local linear trend model, $m = 2$ and the state vector is

$$\boldsymbol{\alpha}_t = \begin{pmatrix} \mu_t \\ \beta_t \end{pmatrix}.$$

- $\mathbf{Z}_t$ is a $p \times m$ matrix: for the local level model where $p = m = 1$, this is just a scalar taking the value 1 for all t, and for the local linear trend model where $p = 1$ and $m = 2$ it is given by $\begin{pmatrix} 1 & 0 \end{pmatrix}$.

- ($\boldsymbol{\varepsilon}_t$) is an uncorrelated sequence of zero-mean random vectors such that the $p \times p$ covariance matrix of $\boldsymbol{\varepsilon}_t$ is $\mathbf{H}_t$: since $p = 1$ in both of the models above, $\mathbf{H}_t = \sigma_\varepsilon^2$ is just a scalar.

- $\mathbf{M}_t$ is an $m \times m$ matrix: for the local level model this is equal to 1 for all t and for the local linear trend model it is given by

$$\mathbf{M}_t = \begin{pmatrix} 1 & 1 \\ 0 & 1 \end{pmatrix}.$$

- (η_t) is an uncorrelated sequence of zero-mean random vectors that is also uncorrelated with (ε_t), and is such that the $m \times m$ covariance matrix of η_t is Q_t. For the local level model we have $Q_t = \sigma_h^2$ and for the local linear trend model we have

$$Q_t = \begin{pmatrix} \sigma_h^2 & 0 \\ 0 & \sigma_z^2 \end{pmatrix}.$$

The rather general formulation given by (5.31) and (5.32) is called the *state space* form of a time series model. Equation (5.31) is called the *measurement equation*, and describes the relationship between the observations (Y_t) and the underlying state of the system as represented by the sequence (α_t). Equation (5.32) is called the *transition equation*, and describes the time evolution of the state vector itself.[4] By allowing Y_t to be a vector, the framework is capable of handling multivariate time series: in environmental applications these may arise in a variety of contexts including space–time data (in which observations are made on a variable of interest at several different locations), multispecies data and situations in which various types of indirect measurements are made on some underlying quantity of interest.

The transition equation (5.32) may at first sight appear restrictive, since the state vector α_t is only allowed to depend on a single previous time point. However, this apparent restriction is illusory since dependence on any number of lags can be represented by appropriate augmentation of the state vector. This is illustrated in the following example.

Example 5.9

Suppose that we are interested in observing a stationary zero-mean AR(2) process (X_t), but that the values of (X_t) cannot be measured exactly: instead, at each time t, we observe

$$Y_t = X_t + \varepsilon_t, \tag{5.33}$$

where (ε_t) is a white noise sequence. In this case, we could think of X_t as the unobservable state of the system at time t. As an AR(2) process, (X_t) satisfies

$$X_t = \phi_1 X_{t-1} + \phi_2 X_{t-2} + \delta_t,$$

which is not in the form (5.32) of a state space transition equation. However, if we write

$$\alpha_t = \begin{pmatrix} X_t \\ X_{t-1} \end{pmatrix}, \tag{5.34}$$

then it is easily verified that the sequence (α_t) satisfies

$$\alpha_t = \begin{pmatrix} \phi_1 & \phi_2 \\ 1 & 0 \end{pmatrix} \alpha_{t-1} + \begin{pmatrix} \delta_t \\ 0 \end{pmatrix},$$

which is in the required form. The representation is completed via the measurement equation

$$Y_t = \begin{pmatrix} 1 & 0 \end{pmatrix} \alpha_t + \varepsilon_t.$$

---

[4] Some authors use a slightly different, but equivalent, formulation. For example, Durbin and Koopman (2001) use different time subscripts on the state vectors in the transition equation (their equation 4.1). Effectively, their η_t is our η_{t+1} and their T_t is our M_{t+1}.

It is instructive to obtain the reduced form of the model for the observations (Y_t) in this example. As with the local level model, this can be done using the backshift operator: first note that $(1 - \phi_1 B - \phi_2 B^2)X_t = \delta_t$. Applying the operator $(1 - \phi_1 B - \phi_2 B^2)$ to both sides of (5.33) now yields

$$(1 - \phi_1 B - \phi_2 B^2)Y_t = (1 - \phi_1 B - \phi_2 B^2)X_t + (1 - \phi_1 B - \phi_2 B^2)\varepsilon_t,$$

so that $Y_t = \phi_1 Y_{t-1} + \phi_2 Y_{t-2} + \delta_t + \varepsilon_t - \phi_1 \varepsilon_{t-1} - \phi_2 \varepsilon_{t-2}.$

This expression contains two autoregressive terms, along with contributions from white noise sequences at lags 1 and 2. It may be inferred, therefore, that (Y_t) is stochastically equivalent to an ARMA(2,2) process. A rigorous demonstration of this is beyond the scope of this book – the argument is essentially the same as that required for the predator–prey model in Example 5.6 and for the local level model above. Thus we have learned that if we observe an AR(2) process contaminated with white noise (due, for example, to measurement error), the observations can be regarded as ARMA(2,2). Similar methods can be used to find the properties of more general 'contaminated ARMA' processes. ■

As this example suggests, a wide range of dependence structures can be handled in the state space framework. Indeed, it can be shown that any model in the AR(I)MA family can be represented in this way, so that the class of state space models is at least as wide as the class of AR(I)MA models. The state space representation of AR(I)MA models is not unique: see Brockwell and Davis (2002, Section 8.2) and Gardner, Harvey and Phillips (1980) for some alternative formulations.

The local level and local linear trend models can be regarded as 'dynamical', in the sense that the underlying parameters (for example the mean level and the trend slope) are allowed to vary over time. The state space form allows many other common structures to be handled in a similar way: for example, dynamic representations of seasonality and regression relationships are available (Durbin and Koopman, 2001, Sections 3.2 and 3.6). Dynamic regressions can be regarded as an alternative means of specifying varying coefficient models (Section 4.3.3) to allow the strength of a regression relationship to change over time: this is often a natural way of representing the fact that relationships within a system may respond to changes in the wider environment (Pole, West and Harrison, 1994, Section 1.4; Young, 1998). It is achieved by allowing the coefficient vector $\boldsymbol{\beta}$ in a model such as (3.10) to vary so that the regression equation at time t becomes

$$Y_t = \mathbf{x}_t \boldsymbol{\beta}_t + \varepsilon_t.$$

This can be accommodated within the state space framework by setting $\boldsymbol{\alpha}_t = \boldsymbol{\beta}_t$ and $\mathbf{Z}_t = \mathbf{x}_t$ in (5.31), and specifying a model for the time evolution of the regression coefficients $(\boldsymbol{\beta}_t)$. Notice that in this case the sequence of matrices $(\mathbf{Z}_t)$ is itself time-varying, by contrast with the state space representation of the local level and local linear trend models. There are also situations in which the sequence $(\mathbf{M}_t)$ in the transition equation (5.32), and the covariance matrices $(\mathbf{H}_t)$ and $(\mathbf{Q}_t)$, may be time-varying. A notable example is in the state space representation of continuous-time ARMA processes observed at irregular intervals (Harvey, 1989, Section 9.1). For environmental applications of time-varying regression, see van der Wal and Janssen (2000) and Conrad Lamon, Carpenter and Stow (2000). The latter authors fit an extension of the local linear trend model to PCB concentration data from eight species of fish and one species of bird: their model represents an underlying trend that is common to all species, with individual species-specific departures from this underlying trend that themselves are allowed to change over time. *Dynamic factor analysis* is a related approach that uses a state space representation to identify shared components (such as common trends) in multivariate series; this is discussed in Section 6.2 and explored further in Chapter 9.

As well as having the flexibility to represent a wide variety of statistical model structures, the state space form may be used to represent the known dynamics of a system, particularly when these dynamics are highly structured. The following example, which is a slightly adapted and simplified version of a model from Besbeas *et al.* (2002) for bird census data, illustrates this.

Example 5.10
Suppose we want to estimate population numbers of some species of bird for which both males and females reach sexual maturity one year after hatching. Let $N_t^{(J)}$, $N_t^{(F)}$ and $N_t^{(M)}$ denote respectively the total number of juvenile, adult female and adult male birds in year t. Also, let μ denote the average number of young per female, and let π denote the proportion of females in the population. Finally, let λ_J and λ_A denote the annual survival probabilities of juvenile and adult birds respectively, which are assumed to be the same for both males and females. Then, assuming no migration into or out of the population, the number of adult females in year t is equal to the number of adult survivors from the previous year, plus the number of juvenile females surviving to sexual maturity. On average, we therefore have

$$N_t^{(F)} = \lambda_A N_{t-1}^{(F)} + \lambda_J \pi N_{t-1}^{(J)}. \tag{5.35}$$

For adult males we obtain a similar expression:

$$N_t^{(M)} = \lambda_A N_{t-1}^{(M)} + \lambda_J (1 - \pi) N_{t-1}^{(J)}. \tag{5.36}$$

For juveniles, we know that $N_t^{(J)} = \mu N_t^{(F)}$ on average, and hence we expect

$$N_t^{(J)} = \mu \lambda_A N_{t-1}^{(F)} + \mu \lambda_J \pi N_{t-1}^{(J)}. \tag{5.37}$$

Now define the 'state' of the population in year t to comprise the numbers of adult females, adult males and juveniles. This can be represented via a state vector

$$\alpha_t = \left(\begin{array}{ccc} N_t^{(F)} & N_t^{(M)} & N_t^{(J)} \end{array} \right)',$$

and its time evolution is given by the relationship

$$\alpha_t = \left(\begin{array}{c} N_t^{(F)} \\ N_t^{(M)} \\ N_t^{(J)} \end{array} \right) = \left(\begin{array}{ccc} \lambda_A & 0 & \lambda_J \pi \\ 0 & \lambda_A & \lambda_J (1 - \pi) \\ \mu \lambda_A & 0 & \mu \lambda_J \pi \end{array} \right) \alpha_{t-1} + \left(\begin{array}{c} \eta_t^{(F)} \\ \eta_t^{(M)} \\ \eta_t^{(J)} \end{array} \right). \tag{5.38}$$

The final term here is a random vector accounting for deviations from the average relationships given by (5.35) to (5.37). In general, one might expect these deviations to be correlated due, for example, to external factors (such as weather conditions) that simultaneously affect the survival of all birds.

Equation (5.38) is a transition equation in the same form as (5.32). To complete the state space representation of the system, we must derive a measurement equation that reflects what is actually observed. Typically, bird numbers are estimated on the basis of surveys or censuses that may gather more or less detailed information. For instance, a survey based on the number of nest sites can be regarded as estimating the number of breeding (i.e. adult) females: in this case, we could denote by Y_t the estimated number of adult females in year t and write

$$Y_t = \begin{pmatrix} 1 & 0 & 0 \end{pmatrix} \boldsymbol{\alpha}_t + \varepsilon_t,$$

where ε_t accounts for the survey error. If, on the other hand, the survey estimates the total number of adult birds each year $(N_t^{(F)} + N_t^{(M)})$ without differentiating between males and females, we could write

$$Y_t = \begin{pmatrix} 1 & 1 & 0 \end{pmatrix} \boldsymbol{\alpha}_t + \varepsilon_t.$$

A third possibility is that a survey estimates numbers of adult females and males separately. In this case, the observations $(\mathbf{Y}_t)$ are bivariate and, in an obvious notation, the survey estimates can be linked to the underlying population numbers as

$$\mathbf{Y}_t = \begin{pmatrix} Y_t^{(F)} \\ Y_t^{(M)} \end{pmatrix} = \begin{pmatrix} 1 & 0 & 0 \\ 0 & 1 & 0 \end{pmatrix} \boldsymbol{\alpha}_t + \begin{pmatrix} \varepsilon_t^{(F)} \\ \varepsilon_t^{(M)} \end{pmatrix}. \qquad \blacksquare$$

The above example illustrates the ability of the state space form both to represent known process dynamics and to link the process of interest with what is actually observed. This latter feature makes the state space framework particularly useful whenever quantities of interest cannot be measured directly. This has been recognised in the field of ecological population dynamics in particular: with appropriate modifications, models similar to that described above have been used in a variety of ecological contexts (see Section 5.6 for references).

5.5.3   The Kalman filter

Suppose now that we have a time series model in state space form and that the various quantities in the model equations (5.31) and (5.32) are known (estimation of these quantities is covered in the next subsection). Suppose also that observations $\mathbf{Y}_1 = \mathbf{y}_1, \ldots, \mathbf{Y}_T = \mathbf{y}_T$ are available. Depending on the application, one or more of the following may be of interest:

- *Prediction* of $\boldsymbol{\alpha}_{T+1}, \ldots, \boldsymbol{\alpha}_{T+\ell}$ or of $\mathbf{Y}_{T+1}, \ldots, \mathbf{Y}_{T+\ell}$.

- *Filtering*, which in this context means estimating the current state of the system $\boldsymbol{\alpha}_T$.

- *Smoothing*, which means using the data $\mathbf{y}_1, \ldots, \mathbf{y}_T$ to estimate the entire history $\boldsymbol{\alpha}_1, \ldots, \boldsymbol{\alpha}_T$ of the underlying state vector.

In the subsequent development we will use the notation $\hat{\boldsymbol{\alpha}}_{t_1 | t_2}$ to denote an estimate of the state of the system at time t_1, derived from all observations up to and including time t_2 (which may be greater than, less than or equal to t_1). We will denote by $\boldsymbol{P}_{t_1 | t_2}$ the covariance matrix of the associated estimation error $\boldsymbol{\alpha}_{t_1} - \hat{\boldsymbol{\alpha}}_{t_1 | t_2}$. Similarly, for $t_1 > t_2$, $\hat{\mathbf{Y}}_{t_1 | t_2}$ will denote a prediction of $\mathbf{Y}_{t_1}$ made at time t_2, with associated error covariance matrix $\boldsymbol{F}_{t_1 | t_2}$.

We consider the prediction problem first. As when forecasting with ARMA models (see Section 5.1.7), it seems reasonable to base predictions on the conditional expectations $\left\{ \mathrm{E}\left(\boldsymbol{\alpha}_{T+j} | \mathbf{Y}_1, \ldots, \mathbf{Y}_T\right) : j = 1, \ldots, \ell \right\}$. Suppose that at time T we have computed $\hat{\boldsymbol{\alpha}}_{T|T} = \mathrm{E}\left(\boldsymbol{\alpha}_T | \mathbf{Y}_1, \ldots, \mathbf{Y}_T\right)$ (this computation is discussed below). According to the notation just defined, the covariance matrix of the associated error $\boldsymbol{\alpha}_T - \hat{\boldsymbol{\alpha}}_{T|T}$ is $\boldsymbol{P}_{T|T}$. Then, from the transition equation (5.32), the conditional expectations of $\boldsymbol{\alpha}_{T+1}, \ldots, \boldsymbol{\alpha}_{T+\ell}$ are

$$\hat{\boldsymbol{\alpha}}_{T+1|T} = \boldsymbol{M}_{T+1} \hat{\boldsymbol{\alpha}}_{T|T},$$

$$\hat{\boldsymbol{\alpha}}_{T+2|T} = \boldsymbol{M}_{T+2} \hat{\boldsymbol{\alpha}}_{T+1|T} = \boldsymbol{M}_{T+2} \boldsymbol{M}_{T+1} \hat{\boldsymbol{\alpha}}_{T|T} \tag{5.39}$$

and so on. The one-step-ahead prediction error is thus

$$\boldsymbol{\alpha}_{T+1} - \hat{\boldsymbol{\alpha}}_{T+1|T} = \boldsymbol{M}_{T+1} \boldsymbol{\alpha}_T + \boldsymbol{\eta}_{T+1} - \boldsymbol{M}_{T+1} \hat{\boldsymbol{\alpha}}_{T|T}$$

$$= \boldsymbol{M}_{T+1} \left(\boldsymbol{\alpha}_T - \hat{\boldsymbol{\alpha}}_{T|T}\right) + \boldsymbol{\eta}_{T+1}$$

and the covariance matrix of this quantity is

$$\boldsymbol{P}_{T+1|T} = \boldsymbol{M}_{T+1} \mathrm{E}\left\{\left(\boldsymbol{\alpha}_T - \hat{\boldsymbol{\alpha}}_{T|T}\right)\left(\boldsymbol{\alpha}_T - \hat{\boldsymbol{\alpha}}_{T|T}\right)'\right\} \boldsymbol{M}'_{T+1} + \boldsymbol{Q}_{T+1}$$

$$= \boldsymbol{M}_{T+1} \boldsymbol{P}_{T|T} \boldsymbol{M}'_{T+1} + \boldsymbol{Q}_{T+1} \tag{5.40}$$

(this follows from the transition equation (5.32), coupled with the fact that $\boldsymbol{\eta}_{T+1}$ is uncorrelated with anything prior to time $T + 1$). The covariance matrix of subsequent prediction errors can be computed in a similar way.

From the predictions $\hat{\boldsymbol{\alpha}}_{T+1|T}, \ldots, \hat{\boldsymbol{\alpha}}_{T+\ell|T}$, the measurement equation (5.31) allows predictions of the associated observation vectors $\mathbf{Y}_{T+1}, \ldots, \mathbf{Y}_{T+\ell}$ to be computed as $\hat{\mathbf{Y}}_{T+j|T} = \boldsymbol{Z}_{T+j} \hat{\boldsymbol{\alpha}}_{T+j|T}$. The error associated with $\hat{\mathbf{Y}}_{T+j|T}$ is

$$\mathbf{Y}_{T+j|T} - \hat{\mathbf{Y}}_{T+j|T} = \boldsymbol{Z}_{T+j} \boldsymbol{\alpha}_{T+j} + \boldsymbol{\varepsilon}_{T+j} - \boldsymbol{Z}_{T+j} \hat{\boldsymbol{\alpha}}_{T+j|T}$$

$$= \boldsymbol{Z}_{T+j} \left(\boldsymbol{\alpha}_{T+j} - \hat{\boldsymbol{\alpha}}_{T+j|T}\right) + \boldsymbol{\varepsilon}_{T+j},$$

which has covariance matrix

$$\boldsymbol{F}_{T+j|T} = \boldsymbol{Z}_{T+j} \boldsymbol{P}_{T+j|T} \boldsymbol{Z}'_{T+j} + \boldsymbol{H}_{T+j}.$$

We now turn to the problem of filtering (i.e. estimating $\boldsymbol{\alpha}_T$ on the basis of observations $\mathbf{y}_1, \ldots, \mathbf{y}_T$). Suppose first that we have calculated $\hat{\boldsymbol{\alpha}}_{T-1|T-1} = \mathrm{E}\left(\boldsymbol{\alpha}_{T-1} | \mathbf{Y}_1, \ldots, \mathbf{Y}_{T-1}\right)$ and the associated error covariance matrix $\boldsymbol{P}_{T-1|T-1}$. From these, we can calculate one-step-ahead predictions $\hat{\boldsymbol{\alpha}}_{T|T-1}$ and $\hat{\mathbf{Y}}_{T|T-1}$ and associated error covariance matrices as described above. In principle, we can also obtain the conditional expectation and covariance matrix of the combined state and observation vectors at time T: for example, the conditional expectation is

$$\mathrm{E}\left[\left(\begin{array}{c} \boldsymbol{\alpha}_T \\ \mathbf{Y}_T \end{array}\right) \middle| \mathbf{Y}_1, \ldots, \mathbf{Y}_{T-1}\right] = \left(\begin{array}{c} \hat{\boldsymbol{\alpha}}_{T|T-1} \\ \hat{\mathbf{Y}}_{T|T-1} \end{array}\right).$$

For the moment, suppose also that the quantities $(\boldsymbol{\varepsilon}_t)$ and $(\boldsymbol{\eta}_t)$ in (5.31) and (5.32) have multivariate normal distributions. In this case, the conditional joint distribution of $\left(\begin{array}{cc} \boldsymbol{\alpha}'_T & \mathbf{Y}'_T \end{array}\right)'$ is itself multivariate normal, and is fully specified by its expectation and covariance matrix. To calculate $\hat{\boldsymbol{\alpha}}_{T|T}$, all that is required is to condition additionally on $\mathbf{Y}_T$ in this joint distribution. This can be done using standard results for the multivariate normal distribution (see, for example, Durbin and Koopman, 2001, Section 2.13). We find (Harvey, 1989, Section 3.2)

$$\hat{\boldsymbol{\alpha}}_{T|T} = \hat{\boldsymbol{\alpha}}_{T|T-1} + \boldsymbol{P}_{T|T-1} \boldsymbol{Z}_T' \boldsymbol{F}_{T|T-1}^{-1} \left(\mathbf{y}_T - \hat{\mathbf{Y}}_{T|T-1} \right), \tag{5.41}$$

with associated error covariance matrix

$$\boldsymbol{P}_{T|T} = \boldsymbol{P}_{T|T-1} - \boldsymbol{P}_{T|T-1} \boldsymbol{Z}_T' \boldsymbol{F}_{T|T-1}^{-1} \boldsymbol{Z}_T \boldsymbol{P}_{T|T-1}. \tag{5.42}$$

In fact, these results do not depend critically on the multivariate normal assumption: if this assumption does not hold then the estimated state vectors are no longer conditional expectations, but instead are *best linear unbiased predictors* (BLUPs) in the sense that they have the smallest possible mean squared error among all linear functions of the observations – see Harvey (1989, Section 3.2.3), Brockwell and Davis (2002, Section 8.4) and Durbin and Koopman (2001, Section 4.2).

Equations (5.39), (5.40), (5.41) and (5.42) together make up the *Kalman filter*. This was originally developed for efficient online updating in response to new data: the idea is to start with an initial assessment of the system state and associated error covariance matrix, say $\hat{\boldsymbol{\alpha}}_{0|0}$ and $\boldsymbol{P}_{0|0}$, use these to calculate $\hat{\boldsymbol{\alpha}}_{1|0}$ and $\boldsymbol{P}_{1|0}$ and then, as successive observations become available, calculate the sequence

$$\left\{ \hat{\boldsymbol{\alpha}}_{1|1}, \boldsymbol{P}_{1|1}, \hat{\boldsymbol{\alpha}}_{2|1}, \boldsymbol{P}_{2|1} \right\}, \left\{ \hat{\boldsymbol{\alpha}}_{2|2}, \boldsymbol{P}_{2|2}, \hat{\boldsymbol{\alpha}}_{3|2}, \boldsymbol{P}_{3|2} \right\}, \ldots.$$

A particularly attractive feature of this algorithm is its computational tractability. In general, the number of calculations required to process a series of length T is roughly proportional to T, because the calculations required to incorporate each new observation depend only on that observation itself and on the latest existing prediction: it is not necessary to reprocess the entire data set each time. Notice also that a small modification of the algorithm will handle missing observations: if $\mathbf{Y}_t$ is not observed then there is no information from time t to update $\hat{\boldsymbol{\alpha}}_{t|t-1}$, so we have $\hat{\boldsymbol{\alpha}}_{t|t} = \hat{\boldsymbol{\alpha}}_{t|t-1}$ and $\boldsymbol{P}_{t|t} = \boldsymbol{P}_{t|t-1}$. Slightly more complicated techniques are required if some, but not all, observations are missing at a particular time point – see Harvey (1989, Section 3.4.7).

To implement the Kalman filter in practice, the only remaining problem is to choose the initial values $\hat{\boldsymbol{\alpha}}_{0|0}$ and $\boldsymbol{P}_{0|0}$. If the transition equation (5.32) defines a stationary process, these values can be taken from the overall properties of the process. For example, for the contaminated AR(2) process in Example 5.9, the state vector is given by (5.34). Since the elements of this state vector are consecutive values in a stationary AR(2) process with zero mean, in this case we could take

$$\hat{\boldsymbol{\alpha}}_{0|0} = \begin{pmatrix} 0 \\ 0 \end{pmatrix} \quad \text{and} \quad \boldsymbol{P}_{0|0} = \begin{pmatrix} \gamma(0) & \gamma(1) \\ \gamma(1) & \gamma(0) \end{pmatrix},$$

where $\gamma(\cdot)$ denotes the autocovariance function of the process. Gardner, Harvey and Phillips (1980) give a general algorithm for finding the value of $\boldsymbol{P}_{0|0}$ in stationary state space models; see also Harvey (1989, p. 121).

If the transition equation does not define a stationary process, initialisation is slightly more complicated. The simplest solution is to set any nonstationary elements of $\hat{\boldsymbol{\alpha}}_{0|0}$ to some arbitrary value and to set the corresponding submatrix of $\boldsymbol{P}_{0|0}$ to $\kappa \boldsymbol{I}$, where $\boldsymbol{I}$ is an identity matrix of appropriate dimension and κ is a large positive number. The allocation of large variances to the nonstationary components represents, in a crude sense, the fact that their values are highly uncertain. Although the arbitrary choice of initial values may seem unsatisfactory, the effect is typically confined to the first few time points, so that the approach may perform acceptably if T is large. A more rigorous solution extends the idea to obtain the *exact initial Kalman filter*: this allows κ to tend to infinity (thereby representing complete ignorance of the initial state of the nonstationary components) and explicitly identifies a set of time points $\{0, \ldots, t_{\text{INIT}}\}$ that are affected by the unknown initial conditions. In general, the details are nontrivial; see Durbin and Koopman (2001, Chapter 5). However, the following example illustrates the basic principle.

Example 5.11
Consider the local level model introduced at the beginning of Section 5.5.1. We have seen that in the state space representation of this model, the system matrices Z_t and M_t in (5.31) and (5.32) are both scalars that are equal to 1 and that the variances H_t and Q_t are equal to σ_ε^2 and σ_h^2 respectively. Starting with $\hat{\alpha}_{0|0}$ and $P_{0|0}$, the Kalman scheme for this model involves the recursive calculation of

$$\hat{\alpha}_{t|t-1} = \hat{\alpha}_{t-1|t-1} \quad \text{and} \quad P_{t|t-1} = P_{t-1|t-1} + \sigma_h^2;$$

$$\hat{Y}_{t|t-1} = \hat{\alpha}_{t|t-1} \quad \text{and} \quad F_{t|t-1} = P_{t|t-1} + \sigma_\varepsilon^2;$$

$$\hat{\alpha}_{t|t} = \hat{\alpha}_{t|t-1} + \frac{P_{t|t-1}\left(y_t - \hat{Y}_{t|t-1}\right)}{F_{t|t-1}} \quad \text{and} \quad P_{t|t} = P_{t|t-1}\left(1 - \frac{P_{t|t-1}}{F_{t|t-1}}\right).$$

If we initialise this scheme with $\hat{\alpha}_{0|0} = 0$ and $P_{0|0} = \kappa$, then we obtain

$$\hat{\alpha}_{1|0} = 0; \qquad P_{1|0} = \kappa + \sigma_h^2;$$

$$\hat{Y}_{1|0} = 0; \qquad F_{1|0} = \kappa + \sigma_h^2 + \sigma_\varepsilon^2;$$

$$\hat{\alpha}_{1|1} = \frac{\left(\kappa + \sigma_h^2\right) y_1}{\kappa + \sigma_h^2 + \sigma_\varepsilon^2} \quad \text{and} \quad P_{1|1} = \frac{\sigma_\varepsilon^2 \left(\kappa + \sigma_h^2\right)}{\kappa + \sigma_h^2 + \sigma_\varepsilon^2}.$$

If we now let κ tend to infinity, we see that $P_{1|0}$ and $F_{1|0}$ both tend to infinity themselves; however, $\hat{\alpha}_{1|1}$ tends to Y_1 and $P_{1|1}$ tends to σ_ε^2. Thus the Kalman filter for this particular model can be initialised exactly by setting $\hat{\alpha}_{1|1} = Y_1$ and $P_{1|1} = \sigma_\varepsilon^2$. This contrasts with the stationary situation, in which the filter can be initialised exactly at time $t = 0$.   ■

The filtering problem delivers an estimate of $\boldsymbol{\alpha}_t$ based on observations up to time t: as such, it is relevant for online assessment of the state of a system. However, subsequent observations may also contain relevant information, which provides the opportunity for online assessments to be amended retrospectively. This procedure is known as *smoothing*, for reasons that will become apparent below. Here we discuss *offline* or *fixed-interval* smoothing, in which the aim is to obtain a single set of estimates of $\boldsymbol{\alpha}_1, \ldots, \boldsymbol{\alpha}_T$ based on observations $\mathbf{Y}_1, \ldots, \mathbf{Y}_T$. In the context of a trend analysis, it provides a means of assessing the underlying changes in a system on the basis of all the available data.

In the notation used above for the filtering problem, the aim of fixed-interval smoothing is to find estimates $\{\hat{\boldsymbol{\alpha}}_{t|T} : t = 1, \ldots, T\}$ with associated error covariance matrices $\{\boldsymbol{P}_{t|T} : t = 1, \ldots, T\}$. Given observations to time T, the Kalman filter provides a set of estimates $\{\hat{\boldsymbol{\alpha}}_{t|t} : t = 1, \ldots, T\}$ and covariance matrices $\{\boldsymbol{P}_{t|t} : t = 1, \ldots, T\}$: for the final time point T, the filtered and smoothed quantities are the same. For time $T - 1$, however, we have an opportunity to update $\hat{\boldsymbol{\alpha}}_{T-1|T-1}$ using the relevant information from time T (which, due to the specific form of the state space model, is entirely contained in the Kalman filter output and does not depend further upon $\mathbf{Y}_T$). This new estimate can then be used to update $\hat{\boldsymbol{\alpha}}_{T-2|T-2}$, and we can proceed in turn to update all of the remaining estimates. For $t = T - 1, T - 2, \ldots, 1$ the required calculations are (Harvey, 1989, Section 3.6.2)

$$\hat{\boldsymbol{\alpha}}_{t|T} = \hat{\boldsymbol{\alpha}}_{t|t} + \boldsymbol{P}_{t|t}\boldsymbol{M}'_{t+1}\boldsymbol{P}^{-1}_{t+1|t}\left(\hat{\boldsymbol{\alpha}}_{t+1|T} - \boldsymbol{M}_{t+1}\hat{\boldsymbol{\alpha}}_{t|t}\right)$$

and $\qquad \boldsymbol{P}_{t|T} = \boldsymbol{P}_{t|t} + \boldsymbol{P}_{t|t}\boldsymbol{M}'_{t+1}\boldsymbol{P}^{-1}_{t+1|t}\left(\boldsymbol{P}_{t+1|T} - \boldsymbol{P}_{t+1|t}\right)\boldsymbol{P}^{-1}_{t+1|t}\boldsymbol{M}_{t+1}\boldsymbol{P}_{t|t}.$ (5.43)

The resulting estimates $\{\hat{\boldsymbol{\alpha}}_{t|T}\}$ can be regarded as conditional expectations $\{\mathrm{E}(\boldsymbol{\alpha}_t | \mathbf{Y}_1, \ldots, \mathbf{Y}_T)\}$ under the assumption of multivariate normality, or as BLUPs otherwise. Missing observations can also be estimated, using a slight modification of the algorithm (Durbin and Koopman, 2001, Section 4.8). Notice that the algorithm requires storage of the quantities $\{\boldsymbol{P}_{t+1|t}\}$ in addition to $\{\hat{\boldsymbol{\alpha}}_{t|t}\}$ and $\{\boldsymbol{P}_{t|t}\}$ during the initial application of the Kalman filter.

In principle, the formulae given above provide a means of filtering and smoothing in any state space model for which the parameters are known. Modern computer implementations, however, contain refinements designed to speed up the calculations and improve the numerical stability of the algorithms – see Durbin and Koopman (2001, Sections 4.3 and 6.4) for examples.

5.5.4   Parameter estimation

In practice, of course, state space models usually contain parameters that must be estimated from the data. As in Section 5.1.5, we consider the method of maximum likelihood under the assumption of Gaussianity, which here means that the $(\boldsymbol{\varepsilon}_t)$ and $(\boldsymbol{\eta}_t)$ in the models have multivariate normal distributions. Let $\boldsymbol{\theta}$ denote the vector of unknown parameters, let $\mathbf{y}$ denote the combined vector of all observations available to time T and let $\mathcal{D} \subseteq \{1, \ldots, T\}$ denote the set of all time indices for which at least one element of $\mathbf{Y}_t$ is observed. In addition, suppose for the moment that the initial state vector $\boldsymbol{\alpha}_0$ is known. Then, in exactly the same way as for univariate observations (see Equation (3.65)), the log-likelihood can be written as a sum of contributions from each time point:

$$\ell(\boldsymbol{\theta}; \mathbf{y}, \boldsymbol{\alpha}_0) = \sum_{t \in \mathcal{D}} \log f_t\left(\mathbf{y}_t \mid \{\mathbf{y}_j : j < t\}; \boldsymbol{\theta}, \boldsymbol{\alpha}_0\right), \qquad (5.44)$$

where $f_t(\cdot; \boldsymbol{\theta}, \boldsymbol{\alpha}_0)$ (which depends explicitly upon $\boldsymbol{\alpha}_0$ only at time $t = 1$) denotes the conditional density of the observations at time t. Now under the Gaussian state space formulation, this conditional density is multivariate normal with mean vector $\hat{\mathbf{Y}}_{t|t-1}$ and covariance matrix $\boldsymbol{F}_{t|t-1}$ (both of which potentially depend on elements of $\boldsymbol{\theta}$). For a given value of $\boldsymbol{\theta}$ the Kalman filter can be used to calculate these quantities at all observation times, and hence the Gaussian log-likelihood can be evaluated conditional on $\boldsymbol{\alpha}_0$.

It remains to deal with the fact that $\boldsymbol{\alpha}_0$ is unknown. If the transition equation (5.32) defines a stationary process, then the exact log-likelihood function can be obtained for any value of $\boldsymbol{\theta}$ by replacing the term $\log f_1(\mathbf{y}_1; \boldsymbol{\theta}, \boldsymbol{\alpha}_0)$ in (5.44) with $\log f_1(\mathbf{y}_1; \boldsymbol{\theta}, \hat{\boldsymbol{\alpha}}_0, \boldsymbol{P}_{0|0})$, where $\hat{\boldsymbol{\alpha}}_{0|0}$ and $\boldsymbol{P}_{0|0}$ are the stationary mean and covariance matrix of $(\boldsymbol{\alpha}_t)$ as in the previous subsection (without loss of generality, we are assuming that $1 \in \mathcal{D}$ here so that $t = 1$ is the time of the first available observation). If the state vector contains nonstationary elements, however, this approach is not possible. In this case, one possibility is to treat α_0 as part of the model parameter vector and to maximise the Gaussian log-likelihood with respect to both θ and α_0. However, if the dimension of α_0 is relatively large compared with that of θ then this can considerably increase the computational cost (Durbin and Koopman, 2001, Section 5.7.5). As an alternative, one might consider using the exact initial Kalman filter of the previous subsection to handle the first few terms in the log-likelihood; however, this leads to problems of a different kind. To illustrate the nature of these, consider the example of the local level model where we have seen that exact initialisation yields $\hat{Y}_{1|0} = 0$ and $F_{1|0} = \infty$. For this model, the Gaussian density of Y_1 is

$$f_1(y) = \frac{1}{\sqrt{2\pi F_{1|0}}} \exp\left[-\frac{\left(y - \hat{Y}_{1|0}\right)^2}{2F_{1|0}}\right]$$

and this tends to zero as $F_{1|0} \to \infty$. Its logarithm (which is the term corresponding to $t = 1$ in the log-likelihood) therefore tends to $-\infty$, and the log-likelihood itself tends to $-\infty$ for all parameter values. The implication that the data contain no information about the parameters clearly is not correct since, as we have seen, this model is stochastically equivalent to an ARIMA(0,1,1): in principle, therefore, the parameters could be estimated by fitting an MA(1) model to the first differences of the observations. One possible approach is to use a modified *diffuse log-likelihood* to handle the initialisation of nonstationary state space models (Durbin and Koopman, 2001, Chapter 7). An alternative is simply to omit any observations from the computed likelihood if they are affected by initialisation in this way (Harvey, 1989, page 127). In the context of the local level model this is consistent with estimation under the alternative ARIMA specification since, in this case, the initial difference at time $t = 1$ cannot be computed (see Section 5.2.1).

Having established that the log-likelihood can be computed for any value of θ, numerical methods can be used to maximise it. The expected information matrix (see Section 3.1.6) can then be estimated by evaluating the matrix of second derivatives at the maximum and used to construct approximate standard errors, tests and confidence intervals for model parameters.

The computational cost of evaluating the log-likelihood at a single value of θ is essentially the same as that required for the Kalman filter to process the available observations: we have seen that this is roughly proportional to the number of distinct observation times. This makes the approach suitable for embedding within a numerical optimisation routine that will typically require log-likelihood evaluations at many different values of θ. Notice also that missing data are easily handled in this framework, because the Kalman filter provides the necessary adjustments automatically, as described in the previous subsection. This is one argument for using state space methods to fit AR(I)MA models, as described in Section 5.1.5. In R, the `arima` command uses state space methods to fit AR(I)MA models. In addition, the `StructTS` command implements maximum likelihood fitting of simple structural time series models including the local level and local linear trend models; this can be used in conjunction with the `tsSmooth` command to carry out fixed-interval smoothing based on the fitted models. The `dlm` library (Petris, 2009) provides access to more general state space models.

Having fitted a state space model, standardised residuals can be computed for diagnostic purposes (see Section 5.1.6). For a univariate series, the standardised residual at time t is $(Y_t - \hat{Y}_{t|t-1})/\sqrt{F_{t|t-1}}$. For multivariate series, a vector of standardised residuals can be defined as

$$\mathbf{e}_t^* = \mathbf{F}_{t|t-1}^{-1/2} \left(\mathbf{Y}_t - \hat{\mathbf{Y}}_{t|t-1} \right),$$

where $\mathbf{F}_{t|t-1}^{-1/2}$ is a matrix square root of $\mathbf{F}_{t|t-1}^{-1}$ such that $\mathbf{F}_{t|t-1}^{-1/2} \left[\mathbf{F}_{t|t-1}^{-1/2} \right]' = \mathbf{F}_{t|t-1}^{-1}$. In practice a Choleski square root (Section 3.3.3) is often used. Under the fitted model, the components of $\mathbf{e}_t^*$ are uncorrelated random variables with zero mean and unit variance; further, $\mathbf{e}_t^*$ and $\mathbf{e}_s^*$ are uncorrelated for $t \neq s$.

Maximum likelihood is not the only method available for fitting state space models: Bayesian methods (Section 3.1.6) are also widely used in this context (Pole, West and Harrison, 1994; Durbin and Koopman, 2001, Chapter 8) and, in R, the dlm library has routines for Markov chain Monte Carlo analysis of state space models. The use of a Bayesian approach has the advantage that parameter uncertainty can be accounted for when making predictions, as described by Davison (2003, page 568) for example: this contrasts with the classical approach to prediction from ARMA models, in which the parameters are treated as known (see Section 5.1.7).

5.5.5   Connection with nonparametric smoothing

If the parameters in a state space model are known, the calculations in the Kalman filter and in fixed-interval smoothing all involve linear transformations. The smoothed estimates $\{\hat{\boldsymbol{\alpha}}_{t|T}\}$ defined by (5.43) are therefore linear functions of the observations $\mathbf{y}_1, \ldots, \mathbf{y}_t$, so that we can write $\hat{\boldsymbol{\alpha}}_{t|T} = \sum_{j=1}^{T} \mathbf{W}_{jt}\mathbf{y}_j$, say, where $\{\mathbf{W}_{jt} : j = 1, \ldots, T\}$ is a collection of $m \times p$ weighting matrices. In particular, if $p = 1$ then the $\{\mathbf{W}_{jt}\}$ are $m \times 1$ vectors and each element of $\hat{\boldsymbol{\alpha}}_{t|T}$ is a weighted average of the univariate observations. This suggests that there may be a connection with some of the nonparametric trend estimation techniques discussed in Chapter 4. We now explore this briefly. For more details, see the thorough review by Fahrmeir and Knorr-Held (2000).

Let $\boldsymbol{\alpha} = \left(\boldsymbol{\alpha}_0' \ldots \boldsymbol{\alpha}_T' \right)'$ be a vector containing all of the unobserved state variables from times 0 to T. We have seen that the Kalman smoother computes the conditional expectation $\hat{\boldsymbol{\alpha}} = \mathrm{E}(\boldsymbol{\alpha}|\mathbf{Y} = \mathbf{y})$, under the assumption that $\mathbf{Y}$ and $\boldsymbol{\alpha}$ have a joint multivariate normal distribution. In this case, the conditional density of $\boldsymbol{\alpha}$ is itself multivariate normal. Using Bayes' theorem, we can write this conditional density as

$$f(\boldsymbol{\alpha}|\mathbf{y}) = \frac{f(\mathbf{y}|\boldsymbol{\alpha}) \, f(\boldsymbol{\alpha})}{f(\mathbf{y})} \tag{5.45}$$

where $f(\cdot)$ here denotes a generic density. Using the fact that the mean and mode of a multivariate normal distribution are equal, $\hat{\boldsymbol{\alpha}}$ is the point at which (5.45), or equivalently its logarithm

$$\log f(\boldsymbol{\alpha}|\mathbf{y}) = \log f(\mathbf{y}|\boldsymbol{\alpha}) + \log f(\boldsymbol{\alpha}) - \log f(\mathbf{y}),$$

is maximised. The last term here does not depend on $\boldsymbol{\alpha}$, so $\hat{\boldsymbol{\alpha}}$ maximises

$$\log f(\boldsymbol{\alpha}|\mathbf{y}) = \log f(\mathbf{y}|\boldsymbol{\alpha}) + \log f(\boldsymbol{\alpha}).$$

Now the measurement equation (5.31) implies that, under the normality assumption, the univariate conditional distribution of Y_t given $\boldsymbol{\alpha}$ is normal with mean $\mathbf{Z}_t \boldsymbol{\alpha}_t \, (= \mu_t,$ say) and variance H_t. Furthermore, the observations are conditionally independent given $\boldsymbol{\alpha}$. We therefore have

$$\log f(\mathbf{y}|\boldsymbol{\alpha}) = -\frac{1}{2} \sum_{t=1}^{T} \left[\log (2\pi H_t) + \frac{(y_t - \mu_t)^2}{H_t} \right].$$

Similarly, the joint density of $\boldsymbol{\alpha}$ can be factorised as

$$f(\boldsymbol{\alpha}) = f(\boldsymbol{\alpha}_0) \prod_{t=1}^{T} f(\boldsymbol{\alpha}_t | \boldsymbol{\alpha}_{t-1})$$

and, under normality, all of the factors are multivariate normal densities. We therefore have (ignoring terms that do not involve elements of $\boldsymbol{\alpha}$)

$$\log f(\boldsymbol{\alpha}) = -\frac{1}{2} \left[\left(\boldsymbol{\alpha}_0 - \hat{\boldsymbol{\alpha}}_{0|0} \right)' \boldsymbol{P}_{0|0}^{-1} \left(\boldsymbol{\alpha}_0 - \hat{\boldsymbol{\alpha}}_{0|0} \right) \right.$$

$$\left. + \sum_{t=1}^{T} (\boldsymbol{\alpha}_t - \boldsymbol{M}_t \boldsymbol{\alpha}_{t-1})' \boldsymbol{Q}_t^{-1} (\boldsymbol{\alpha}_t - \boldsymbol{M}_t \boldsymbol{\alpha}_{t-1}) \right].$$

Combining these results, we see that maximising $f(\boldsymbol{\alpha}|\mathbf{Y})$ is equivalent to minimising

$$\left(\boldsymbol{\alpha}_0 - \hat{\boldsymbol{\alpha}}_{0|0} \right)' \boldsymbol{P}_{0|0}^{-1} \left(\boldsymbol{\alpha}_0 - \hat{\boldsymbol{\alpha}}_{0|0} \right)$$

$$+ \sum_{t=1}^{T} \left[\frac{(y_t - \mu_t)^2}{H_t} + (\boldsymbol{\alpha}_t - \boldsymbol{M}_t \boldsymbol{\alpha}_{t-1})' \boldsymbol{Q}_t^{-1} (\boldsymbol{\alpha}_t - \boldsymbol{M}_t \boldsymbol{\alpha}_{t-1}) \right], \qquad (5.46)$$

which can be regarded as a penalised sum of squares. We have encountered the minimisation of such expressions before, in the context of spline smoothers (Section 4.1.3). In fact, it can be shown that the cubic spline smoother is exactly equivalent to the Kalman smoother for a local linear trend model with $\sigma_h^2 = 0$ in (5.29). The smoothing parameter λ in the cubic spline formulation is equal to $\sigma_\varepsilon^2/\sigma_z^2$ in the local linear trend model. This viewpoint provides an alternative way of choosing a smoothing parameter for spline smoothing, since both σ_z^2 and σ_ε^2 can be estimated using the techniques discussed above. For further discussion of this, see Durbin and Koopman (2001, Sections 3.11 and 9.5).

Expression (5.46) can be used to derive smoothing weights for other simple state space models. For example, in the local level model of Section 5.5.1, the state vector $\boldsymbol{\alpha}_t$ is just the underlying trend μ_t. Let $\tilde{\boldsymbol{\mu}} = (\tilde{\mu}_0 \ \tilde{\mu}_1 \cdots \tilde{\mu}_T)'$ be the trend estimate from the Kalman smoother for this model. Then, with exact initialisation so that $\boldsymbol{P}_{0|0}$ is infinite (see Section 5.5.3 before Example 5.11), (5.46) is equal to

$$\sum_{t=1}^{T} \left[(y_t - \mu_t)^2 / \sigma_\varepsilon^2 + (\mu_t - \mu_{t-1})^2 / \sigma_h^2 \right].$$

Minimising this quantity with respect to $\boldsymbol{\mu}$ is equivalent to minimising

$$\sum_{t=1}^{T} \left[\lambda (y_t - \mu_t)^2 + (\mu_t - \mu_{t-1})^2 \right],$$

where now $\lambda = \sigma_\eta^2 / \sigma_\varepsilon^2$ is the *signal-to-noise ratio* for this model. Equating the derivatives of this expression to zero yields the following equations for $\tilde{\mu}$:

$$\tilde{\mu}_1 - \tilde{\mu}_0 = 0; \tag{5.47}$$

$$\lambda y_t - (2 + \lambda)\tilde{\mu}_t + \tilde{\mu}_{t-1} + \tilde{\mu}_{t+1} = 0 \qquad (t = 1, \ldots, T - 1); \tag{5.48}$$

$$\lambda y_T - (1 + \lambda)\tilde{\mu}_T + \tilde{\mu}_{T-1} = 0. \tag{5.49}$$

Substituting (5.47) into (5.48) for $t = 1$ gives

$$\lambda y_1 - (1 + \lambda)\tilde{\mu}_1 + \tilde{\mu}_2 = 0. \tag{5.50}$$

In matrix form, (5.48) to (5.50) can be written as $A\tilde{\mu} = \lambda \mathbf{y}$, say, where

$$A = \begin{pmatrix} 1+\lambda & -1 & 0 & 0 & \cdots & 0 \\ -1 & 2+\lambda & -1 & 0 & \cdots & 0 \\ 0 & -1 & 2+\lambda & -1 & \cdots & 0 \\ \vdots & \vdots & \ddots & \ddots & \ddots & \vdots \\ 0 & 0 & \cdots & -1 & 2+\lambda & -1 \\ 0 & 0 & \cdots & 0 & -1 & 1+\lambda \end{pmatrix}.$$

We therefore have $\tilde{\mu} = \lambda A^{-1}\mathbf{y}$, so that $\tilde{\mu}_t = \lambda \sum_{j=1}^{T} A^{-1}(t, j)y_j$, where $A^{-1}(i, j)$ is the (i, j)th element of A^{-1}. It can be shown[5] that

$$A^{-1}(i, j) = \left[\rho^{T-|i-j|} + \rho^{|i-j|-T} + \rho^{T+1-(i+j)} + \rho^{i+j-(T+1)}\right]$$

$$\left/ \left[\left(\rho^{-1} - \rho\right)\left(\rho^{-T} - \rho^{T}\right)\right], \right. \tag{5.51}$$

$$\text{where} \qquad \rho = \left[2 + \lambda - \sqrt{\lambda(\lambda + 4)}\right]/2. \tag{5.52}$$

---

[5] This is not straightforward: it is done by writing

$$A = A_0 + \text{diag}\left(1+\lambda-\rho^{-1} \ 0 \cdots 0 \ 1+\lambda-\rho^{-1}\right)$$

where

$$A_0 = \rho^{-1} \begin{pmatrix} 1 & -\rho & 0 & 0 & \cdots & 0 \\ -\rho & 1+\rho^2 & -\rho & 0 & \cdots & 0 \\ 0 & -\rho & 1+\rho^2 & -\rho & \cdots & 0 \\ \vdots & \vdots & \ddots & \ddots & \ddots & \vdots \\ 0 & 0 & \cdots & -\rho & 1+\rho^2 & -\rho \\ 0 & 0 & \cdots & 0 & -\rho & 1 \end{pmatrix}$$

and using the Sherman–Morrison formula (Press *et al.*, 1992, Section 2.7) along with the fact that the (i, j)th element of A_0^{-1} is $\rho^{1+|i-j|}/(1 - \rho^2)$. Equating $\rho^{-1}(1 + \rho^2)$ with $2 + \lambda$ leads to a quadratic equation for ρ, with two positive roots that are the reciprocals of each other; (5.52) defines the smaller of these, lying between 0 and 1. However, (5.51) clearly yields the same result regardless of which root is chosen.

From (5.51) it may be verified that, for each t, $\sum_{j=1}^{T} A^{-1}(t, j) = \lambda^{-1}$. Thus $\tilde{\mu}_t$ is a weighted average of the observations, with weights summing to unity. Notice also that (5.51) can be rewritten as

$$\left[\left(\rho^{-1} - \rho\right)\left(1 - \rho^{2T}\right)\right]^{-1} \left[\rho^{2T-|i-j|} + \rho^{|i-j|} + \rho^{2T+1-(i+j)} + \rho^{i+j-1}\right].$$

If T is large and $0 \ll i + j \ll T$, this is well approximated by $\rho^{|i-j|} / \left(\rho^{-1} - \rho\right)$ (since $|\rho| < 1$ from (5.52)). Thus, away from the ends of the series, the Kalman smoother for the local level model is effectively a Nadaraya–Watson estimator (Section 4.1.2) with a two-sided exponential kernel.

Example 5.12

To illustrate the use of state space models for nonparametric smoothing, consider once more the ozone concentration data of Section 1.3.4. As described there, it is of interest to characterise potential changes in both the mean and the seasonal cycle of these data, so a decomposition into 'trend', 'seasonal' and 'irregular' components would be useful. For regularly spaced series, a simple method for carrying out such a decomposition was described in Section 2.3.4; the STL procedure in Section 4.3.1 puts this on a more formal footing. However, the `stl` command in R requires regularly spaced series with no missing values. When there are large amounts of missing data (as with the ozone series), the close connection between nonparametric regression and fixed-interval smoothing provides an appealing alternative framework within which to think about time series decomposition. Continuous-time state space models (Harvey, 1989, Chapter 9) also provide the opportunity to handle 'genuinely' irregular series, although the implementation is slightly more complicated.

For application to daily ozone concentrations, Gaussian state space models have the potential drawback that they can generate negative values. Data transformation might be considered to avoid this problem, for example by taking logarithms as discussed in Section 3.3. However, the logarithms of the ozone data have a distribution (not shown) that is strongly negatively skewed. On the whole, then, since parameter estimation will be carried out using Gaussian likelihood methods, it seems more appropriate to model the untransformed data in the first instance. For the purposes of decomposing the series into interpretable components, this is likely to be perfectly adequate. Some caution is needed, however, if the fitted models are to be used to produce long-term predictions of future ozone concentrations, since there is no guarantee that these predictions will remain non-negative.

To decompose the series into its components, we start by writing down a plausible representation of the system. Let Y_t^* denote the actual mean ozone concentration on day t; then one might consider a model for (Y_t^*) of the form

$$Y_t^* = \mu_t + s_t + \eta_t, \tag{5.53}$$

where μ_t, s_t and η_t represent respectively the trend, seasonal and irregular components. However, in practice it is difficult to measure ozone concentrations accurately, so the observed ozone concentration Y_t can be written as

$$Y_t = Y_t^* + \varepsilon_t, \tag{5.54}$$

where ε_t is the measurement error. As usual, initially the measurement errors are assumed to form a white noise sequence with variance σ_ε^2.

Having written down the skeleton of a plausible structural model, we consider the individual components in more detail as follows:

1. *Trend component*. Since (μ_t) represents a trend it is natural to consider a non-stationary representation of it. Furthermore, for monitoring purposes it is helpful to use models that will support statements such as 'ozone concentrations are currently changing at an underlying rate of x units per year'. The local linear trend model (see Section 5.5.1) is particularly convenient in this respect, since the slope component (β_t) is directly interpretable as the 'underlying rate' of such statements.

   In the context of a trend–seasonal–irregular decomposition, it is also natural to require that the trend component varies smoothly in time. With a local linear trend model for (μ_t), this can be achieved by setting the variance σ_η^2 of the level innovations to zero (Durbin and Koopman, 2001, Section 3.2.1). The connection with spline smoothing, as discussed above, also makes this an appealing choice. Thus, to model the trend μ_t, we introduce an extra unobserved slope component β_t and write

$$\begin{pmatrix} \mu_t \\ \beta_t \end{pmatrix} = \begin{pmatrix} 1 & 1 \\ 0 & 1 \end{pmatrix} \begin{pmatrix} \mu_{t-1} \\ \beta_{t-1} \end{pmatrix} + \begin{pmatrix} 0 \\ z_t \end{pmatrix}, \tag{5.55}$$

   where (z_t) is a white noise sequence with variance σ_z^2.

2. *Seasonal component*. Figure 1.4 suggests that the seasonality in the ozone series is rather smooth and regular. If the seasonal cycle, represented by s_t in (5.53), were identical from one year to the next then, as in Section 3.2, it could be modelled using a cosine function: $s_t = \cos[2\pi\omega(t - t_0)]$, say, where $\omega = 1/365.25$.[6] It is not immediately obvious that this can be represented in the state space framework. Notice, however, that $s_t = \cos[2\pi\omega(t - t_0)]$ can be written as

$$\cos\{2\pi\omega[(t - 1) - t_0] + 2\pi\omega\}$$
$$= \cos\{2\pi\omega[(t - 1) - t_0]\}\cos 2\pi\omega - \sin\{2\pi\omega[(t - 1) - t_0]\}\sin 2\pi\omega$$
$$= s_{t-1}\cos 2\pi\omega - s_{t-1}^*\sin 2\pi\omega, \text{ say,}$$

   where $s_t^* = \sin[2\pi\omega(t - t_0)]$. A similar application of trigonometric identities yields

$$s_t^* = s_{t-1}\sin 2\pi\omega + s_{t-1}^*\cos 2\pi\omega.$$

   Thus, a fixed seasonal cycle (s_t) can be modelled within the state space framework by introducing an extra component (s_t^*) and writing

$$\begin{pmatrix} s_t \\ s_t^* \end{pmatrix} = \begin{pmatrix} \cos 2\pi\omega & -\sin 2\pi\omega \\ \sin 2\pi\omega & \cos 2\pi\omega \end{pmatrix} \begin{pmatrix} s_{t-1} \\ s_{t-1}^* \end{pmatrix} = \mathbf{\Omega} \begin{pmatrix} s_{t-1} \\ s_{t-1}^* \end{pmatrix}, \text{ say.} \tag{5.56}$$

   To allow the seasonal cycle to evolve over time, one possible approach is to add a random innovation term to the right-hand side of (5.56). However, the process so defined is nonstationary (Harvey, 1989, page 40): in the long run, therefore, it will move arbitrarily far from its starting values. This seems unnatural in the present context, where the seasonality in ozone is attributable primarily to the

---

[6] The cycle length of $\omega^{-1} = 365.25$ days accounts for leap years.

intensity of ultraviolet radiation (see Section 1.3.4) and is therefore unlikely to change much over time. A stationary process can be obtained by replacing $\boldsymbol{\Omega}$ with $\rho\boldsymbol{\Omega}$ before adding a random innovation, for some scalar ρ with $|\rho| < 1$. Unfortunately, this is not satisfactory either: being stationary, long-run forecasts from such a process converge to zero rather than to an average seasonal cycle. A solution is to write the seasonal component of the ozone series explicitly as a sum of two processes, the first being a deterministic cycle of the form (5.56) and the second a stationary process representing time-varying deviations from this deterministic cycle. Thus we write $s_t = u_t + v_t$, introduce two new processes (u_t^*), (v_t^*) in place of (s_t^*) in (5.56) and consider (u_t), (v_t), (u_t^*) and (v_t^*) to evolve according to the transition equations

$$\begin{pmatrix} u_t \\ u_t^* \end{pmatrix} = \begin{pmatrix} \cos 2\pi\omega & -\sin 2\pi\omega \\ \sin 2\pi\omega & \cos 2\pi\omega \end{pmatrix} \begin{pmatrix} u_{t-1} \\ u_{t-1}^* \end{pmatrix}$$

and

$$\begin{pmatrix} v_t \\ v_t^* \end{pmatrix} = \rho \begin{pmatrix} \cos 2\pi\omega & -\sin 2\pi\omega \\ \sin 2\pi\omega & \cos 2\pi\omega \end{pmatrix} \begin{pmatrix} v_{t-1} \\ v_{t-1}^* \end{pmatrix} + \begin{pmatrix} \tau_t \\ \tau_t^* \end{pmatrix}$$

for some $|\rho| < 1$. Here, (τ_t) and (τ_t^*) are further white noise sequences, each with variance σ_τ^2.

3. *Irregular component*. The variograms in Figure 2.4 suggest that ozone concentrations on successive days are autocorrelated. To keep things simple in the first instance, we therefore consider an AR(1) model for the irregular component (η_t) in (5.53):

$$\eta_t = \phi\eta_{t-1} + \delta_t,$$

where $|\phi| < 1$ and (δ_t) is a white noise sequence with variance σ_δ^2.

Putting all of these components together, we can define a state vector for the required decomposition as $\boldsymbol{\alpha}_t = \left(\mu_t \ \beta_t \ u_t \ u_t^* \ v_t \ v_t^* \ \eta_t \right)'$. In this case the observation equation (5.54) can be written as

$$Y_t = (1 \ 0 \ 1 \ 0 \ 1 \ 0 \ 1) \boldsymbol{\alpha}_t + \varepsilon_t$$

and the transition equation for $(\boldsymbol{\alpha}_t)$ is $\boldsymbol{\alpha}_t = \boldsymbol{M}\boldsymbol{\alpha}_{t-1} + \boldsymbol{\eta}_t$, with

$$\boldsymbol{M} = \begin{pmatrix} 1 & 1 & 0 & 0 & 0 & 0 & 0 \\ 0 & 1 & 0 & 0 & 0 & 0 & 0 \\ 0 & 0 & \cos 2\pi\omega & -\sin 2\pi\omega & 0 & 0 & 0 \\ 0 & 0 & \sin 2\pi\omega & \cos 2\pi\omega & 0 & 0 & 0 \\ 0 & 0 & 0 & 0 & \rho\cos 2\pi\omega & -\rho\sin 2\pi\omega & 0 \\ 0 & 0 & 0 & 0 & \rho\sin 2\pi\omega & \rho\cos 2\pi\omega & 0 \\ 0 & 0 & 0 & 0 & 0 & 0 & \phi \end{pmatrix}$$

and $\boldsymbol{\eta}_t = \left(0 \ z_t \ 0 \ 0 \ \tau_t \ \tau_t^* \ \delta_t \right)'$. The white noise terms z_t, τ_t, τ_t^* and δ_t are taken to be mutually uncorrelated, as is standard in this kind of setting (see, for example Harvey, 1989, page 171).

This state space model contains six unknown parameters: four white noise variances $\sigma_\varepsilon^2, \sigma_z^2, \sigma_\tau^2$ and σ_δ^2, and two further parameters ρ and ϕ. These can be estimated using maximum likelihood as described in Section 5.5.4. This has been done in R using the dlmMLE routine in the dlm library, which uses a numerical optimisation routine to maximise the log-likelihood. Optimisation was done on transformed parameters to ensure that the estimated variances were all non-negative and that the estimated values of ρ and ϕ were between zero and 1. Specifically, the log-likelihood was maximised with respect to $\log\sigma_\varepsilon^2$, $\log\sigma_z^2$, $\log\sigma_\tau^2$, $\log\sigma_\delta^2$, arctanhρ and arctanhϕ, where arctanh is the inverse hyperbolic tangent: $\text{arctanh}(x) = \log\left[(1 + x)/(1 - x)\right]$. Most numerical optimisation routines require an initial guess at the position of the optimum, and in 'difficult' problems the results can be sensitive to this choice, so it is helpful to select good starting values. Here, initial values for most of the parameters were obtained by fitting simpler models to obtain a rough idea of what kinds of values could be expected. The exception is the parameter ρ, which controls the persistence of departures from the average seasonal cycle. It can be shown (Harvey, 1989, page 57) that the autocorrelation function of the (v_t) component of our model is

$$\text{Corr}(v_t, v_{t-k}) = \rho^k \cos 2\pi\omega k.$$

At a lag of one year, therefore, the autocorrelation is $\rho^{365} \cos(2\pi \times 365/365.25)$, which is equal to ρ^{365} for all practical purposes. Since any departures from the average seasonal cycle will be of little interest unless they last for several months (otherwise they would belong in the 'irregular' component of the decomposition), we consider that (v_t) should exhibit substantial correlation at a 1-year lag. To obtain a 1-year autocorrelation of 0.5, for example, we would need to set $\rho^{365} = 0.5$ so that $\rho = 0.5^{1/365} = 0.998$. This suggests that ρ should be very close to 1, and therefore we set a high starting value of 0.999 in the optimisation routine.

Table 5.3   Parameter estimates for the state space model fitted to the daily ozone concentration series.

Parameter	σ_ε^2	σ_z^2	σ_τ^2	σ_δ^2	ρ	ϕ
Estimate	23.4	3.3×10^{-8}	0.659	41.6	0.982	0.572

The parameter estimates are shown in Table 5.3. Approximate standard errors for the transformed parameters (not shown) suggest that the estimates are fairly precise. The very small value of σ_z^2 may seem surprising. However, recall that this corresponds to the variance of innovations in the process (β_t) of (5.55); this process must vary slowly if it is to maintain the desired interpretation as a contribution to the trend component of the series. It may be helpful to consider the evolution of (β_t) over a period of a year: since it is modelled as a random walk, we have (see Section 5.2)

$$\beta_{t+365} - \beta_t = \sum_{j=1}^{365} z_t,$$

so that the variance of annual changes in (β_t) is $365\sigma_z^2$. From Table 5.3, this is estimated as $365 \times 3.3 \times 10^{-8} = 1.2 \times 10^{-5}$. The distribution of $\beta_{t+365} - \beta_t$ is therefore approximately normal with mean zero and variance $1.2 \times 10^{-5} = 0.003^2$; annual changes in the underlying slope are therefore likely to be within the range $\pm 2 \times 0.003 = (-0.006, 0.006)$. Although these values are still small, they are large enough to allow the slope to change gradually over time and hence to produce a smooth trend that adapts to changes in the data.

Table 5.3 also shows that the observation error variance σ_ε^2 and the innovation variance σ_δ^2 for the irregular component are both estimated to be quite large. In particular, the estimate of σ_ε^2 suggests that the daily ozone series contains a substantial component of variation that is independent from day to day; this may be attributed, at least in part, to measurement error. In fact, this can be seen in the variogram for deseasonalised data in Figure 2.4(b): a smooth curve drawn through the variogram and extrapolated back to lag zero would not yield a value of zero as might be expected. In geostatistics, this phenomenon is known as the *nugget effect* and is attributed both to measurement error and to fine-scale variation in the process under study (Webster and Oliver, 2001, Section 6.1). In the present context, the deseasonalised (and detrended) series can also be regarded as a contaminated AR(1) process: using the method illustrated in Section 5.5.2, it can be shown that this is stochastically equivalent to an ARMA(1,1) so that the autocorrelation function shows a step down to lag 1 and exponential decay thereafter (see Section 5.1.3). Since the variogram is a scaled mirror-image of the ACF, the step down in the ACF corresponds to the nugget effect in the variogram.

Figure 5.9 shows some diagnostics for the fitted model. Panel (a) is a time series plot of standardised residuals, which should behave as a white noise sequence with unit variance. There is no evidence of trend or seasonality here although there are some large outliers, both positive and negative. Panel (b) shows the variogram of these standardised residuals, which is essentially flat: there is no evidence of any temporal dependence

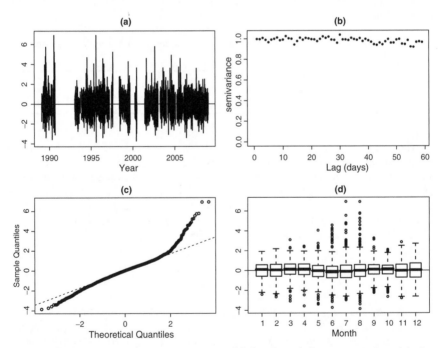

Figure 5.9   Diagnostics for state space model fitted to daily ozone series: (a) time series of standardised residuals, (b) variogram of standardised residuals, (c) normal quantile–quantile plot of standardised residuals, (d) boxplots of standardised residuals by month.

here. Panel (c) is a normal quantile–quantile plot, which shows that the tails of the residual distribution are far from Gaussian. This is connected with the outliers already noted; it should not affect the results of the estimation too much, although we should be aware that prediction intervals computed under the assumption of Gaussianity may be inaccurate (see Section 3.1.1). Finally, panel (d) is a boxplot of standardised residuals by month; this shows that although there is little seasonal structure in the main body of the residual distribution, the outliers occur mainly during the months of June, July and August. These may be associated with specific weather conditions affecting atmospheric ozone concentrations; to improve the model, therefore, a natural strategy would be to incorporate relevant covariate information such as daily temperatures. Nonetheless, the model seems to account well for the trend and for the seasonality in the main body of the data, so a decomposition based on this model will provide useful information.

The decomposition is shown in Figure 5.10. Here, the estimated trend component at each time point is $\hat{\mu}_t = (1\ 0\ 0\ 0\ 0\ 0\ 0)\,\hat{\boldsymbol{\alpha}}_{t|T}$ and the variance of the associated estimation error is

$$(1\ 0\ 0\ 0\ 0\ 0\ 0)\,\boldsymbol{P}_{t|T}\,(1\ 0\ 0\ 0\ 0\ 0\ 0)'.$$

This can be used to construct an approximate 95 % confidence interval for μ_t; the grey band in the second panel of Figure 5.10 shows the evolution of these confidence intervals through time. The smoothed values ($\hat{\boldsymbol{\alpha}}_{t|T}$) and $\boldsymbol{P}_{t|T}$ were obtained using the dlmSmooth routine in R. A similar approach is used to extract the estimated seasonal component $\hat{u}_t + \hat{v}_t$ and associated confidence intervals. Finally, the plotted irregular component is

$$y_t - \hat{\mu}_t - \hat{u}_t - \hat{v}_t.$$

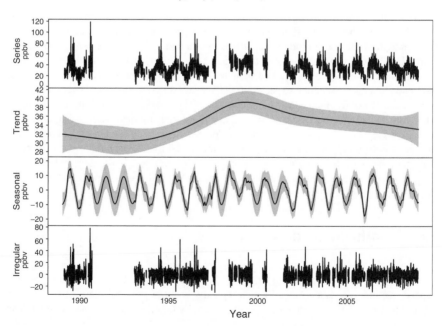

Figure 5.10   Trend–seasonal–irregular decomposition of the daily ozone series, obtained using fixed-interval smoothing with the fitted state space model. Grey bands for the trend and seasonal components indicate pointwise 95 % confidence intervals.

Note that this includes contributions from both the irregular component of the state vector (η_t) and the observation error (ε_t).

Figure 5.10 shows that the estimated trend is very smooth, rising to a maximum value in 1999 and decreasing since then in a roughly linear fashion. The estimate of the trend slope on 31 December 2008 is -0.542 ppbv per year, with an associated standard error of 1.67.[7] This suggests that underlying ozone concentrations should continue to decrease in the short term, although the standard error suggests that there is considerable uncertainty in any such projection. In the long run, of course, the nonstationary nature of the model means that it is effectively impossible to make meaningful predictions. However, the estimated trend in Figure 5.10 provides a useful summary of overall change over the 20-year record. The estimated seasonal component shows that departures from the average seasonal cycle tend to be short-lived. It would be interesting to see whether any of these departures can be related to known weather events that may have influenced the ozone concentrations; this in turn may suggest ways in which the model itself could be improved. Notice that where there are long periods of consecutive missing values (e.g. in the early 1990s), the estimated seasonal cycle reverts to a cosine function – this is the deterministic (u_t) contribution to the seasonal cycle, the stationary (v_t) contribution being estimated as zero in the absence of neighbouring data values.

Finally, the irregular component in Figure 5.10 shows some interesting structure with respect to short-lived extreme ozone events. There are several of these during the 20-year period, indicated by peaks in the irregular component. However, the magnitude of these events seems to have decreased steadily over time. To assess this formally, one could use extreme value methods (see Section 6.4).   ■

The equivalence between these state space models and nonparametric smoothing techniques also sheds some light on forecasting future values with nonparametric methods: since the state space models are difference–stationary, forecast error variances will increase without bound as the lead time increases. This is perhaps unsurprising, since nonparametric methods are designed to make minimal assumptions about the nature of the trend and hence cannot be expected to deliver precise predictions far beyond the range of the data. However, the state space viewpoint provides additional insight into precisely what can and cannot be achieved by extrapolating nonparametric trend estimates. We have also seen that the distinction between 'deterministic' and 'stochastic' trends is not as clear as it initially appeared: the nonparametric estimation techniques of Chapter 4 were motivated by the idea of a deterministic trend, but we have seen that they can also be regarded as a means of estimating an unobserved realisation of an underlying stochastic process.

5.6   Nonlinear models

In the models discussed so far in this chapter, observations $\{Y_t\}$ are considered to depend linearly on previously occurring quantities. This may account adequately for the residual dependence in a regression model, which typically is not of primary interest. However, in other situations a better characterisation of potentially nonlinear dependence structure can

---

[7] The slope β_t in (5.55) represents a change per day; this has been multiplied by 365 to obtain an annual value.

lead to improved understanding of the system under investigation. One possibility is to use nonparametric techniques to estimate the dependence structure in a general autoregressive model of the form

$$Y_t = f(Y_{t-1}, \ldots, Y_{t-d}) + \delta_t, \tag{5.57}$$

where the function f is estimated nonparametrically; essentially this is the autoregressive equivalent of a nonparametric regression model (see Härdle, Lütkepohl and Chen, 1997, for a review of the relevant theory). As in the regression setting, nonparametric estimation of such a general model can be difficult if the dimension d is large (see Section 4.2); hence it is often useful to simplify (5.57), for example by assuming that the autoregression function is additive so that the model can be written as

$$Y_t = \sum_{j=1}^{d} f_j(Y_{t-j}) + \delta_t \tag{5.58}$$

for smooth functions $f_1(\cdot), \ldots, f_d(\cdot)$. In an environmental application of this type of model, Stenseth et al. (1997) analysed annual indices of Canadian lynx and snowshoe hare populations and, for each species, used the estimated functions $\{f_j(\cdot)\}$ to suggest ecological mechanisms governing the population dynamics.

Just as (5.58) can be considered as the autoregressive equivalent of an additive model, so the *functional-coefficient autoregressive (FCAR)* models of Chen and Tsay (1993) can be regarded as equivalent to varying coefficient models (Section 4.3.3). In an FCAR model, dependence on lagged values is modelled linearly as in a standard autoregression, but with autoregressive coefficients that are smooth functions of covariates. By allowing the autoregressive coefficients to depend on the time index, for example, it is possible to model gradual changes in the autocorrelation structure of a time series. There have been few, if any, applications of such models in the environmental sciences to date, however.

An alternative way of simplifying (5.57) is to approximate f with a piecewise linear function. For example, if $d = 1$ we might write

$$f(Y_{t-1}) = \begin{cases} c_1 + \phi_1 Y_{t-1} & \text{if } Y_{t-1} < \tau, \\ c_2 + \phi_2 Y_{t-1} & \text{if } Y_{t-1} \geq \tau, \end{cases}$$

for some threshold τ. This is an example of a *self-exciting threshold autoregressive* (SETAR) model: at any time, the process is in one of a number (here, two) of possible 'regimes' that are themselves determined by past values of the process. Variants on this exist, for example to ensure that $f(Y_{t-1})$ is a continuous function of Y_{t-1}. Environmental applications are limited, although the ideas have been applied by Stenseth et al. (1998a; 1998b) in ecological contexts and by Al-Awadhi and Jolliffe (1998) for the modelling of atmospheric pressure data. For further variants, and other nonlinear models for the function f in (5.57), see Chapter 3 of Tong (1990).

Another form of nonlinearity arises when the conditional variance of Y_t, given previous information, is not constant. The GARCH models described in Section 3.3 are designed to deal with this, in cases where changes in the variance cannot be linked directly either to external covariates or to changes in the mean. Alternatively, if the distribution of the data is such that the variance is related to the mean, analogues of generalised linear models may be used. Indeed, the inclusion of previous observations (possibly after transformation) as additional covariates in a GLM is a natural extension to the linear autoregressive model family. The resulting models are referred to by Fahrmeir and Tutz (2001, Section 6.2) as *generalised autoregressive models*. They also provide an escape from the assumption of Gaussianity: as discussed in Section 5.1.6, this assumption is not critical to the estimation of parameters in standard time series models, but it can affect model selection based on criteria such as the AIC that are calculated from the Gaussian likelihood used by most modern software packages. Moreover, prediction intervals calculated according to (5.20) may be seriously inaccurate in highly non-Gaussian settings.

Unfortunately, little is known at present about the properties of generalised autoregressive models except in a few special cases. For example, the conditions under which realisations from these models remain finite with probability 1 (which are analagous to the conditions for stationarity of standard time series models – compare the stationary and nonstationary simulations in Figure 5.5) have not been studied in general. Few alternatives are available, however. One exception, applicable to time series of counts, is the class of *integer-valued autoregressive* (INAR) processes, introduced independently by McKenzie (1985) and by Al-Osh and Alzaid (1987). A key challenge in modelling counts is to ensure that realisations of the models will themselves be integer-valued: it should be clear that standard autoregressive models cannot achieve this because, in general, quantities such as $\phi_1 Y_{t-1}$ will not be integer-valued. The INAR model class circumvents this using a binomial thinning operation: for example, the INAR(1) model can be written as

$$Y_t = X_{t-1} + \delta_t,$$

where $X_{t-1} \sim \text{Binomial}(Y_{t-1}, \phi)$ for some $\phi \in [0, 1)$ and (δ_t) is a sequence of independent, non-negative, integer-valued random variables. The properties of INAR models are given by Alzaid and Al-Osh (1990). As with the threshold models described above, environmental applications have so far been limited although the models have been applied in hydrological contexts by Thyregod *et al.* (1999) and by Pavlopoulos and Karlis (2008). The latter authors give a good review of the area. A promising alternative to INAR modelling is given by Cui and Lund (2010): the class of models introduced by these authors has some appealing features, such as the ability to represent long-memory structure in integer-valued time series, and has been applied to the study of precipitation occurrence.

Nonlinear and non-Gaussian versions of structural models, along with the corresponding extensions to the Kalman filter, are reasonably well developed, and are implemented in the `sspir` library in R (Dethlefsen and Lundbye-Christensen, 2006). In most cases, non-Bayesian solutions to the filtering and smoothing problem are approximate; see Fahrmeir and Tutz (2001, Chapter 8) and Durbin and Koopman (2001, Chapters 10 and 11), for example. The work of Naveau, Genton and Shen (2005) is an interesting exception to this general rule: these authors propose an exact extension to the Kalman filter that is designed to handle data from skewed distributions. In most cases, however, exact treatments are most easily accomplished within a Bayesian framework using Markov chain Monte Carlo methods. Nonlinear state space models are widely used in ecology: for clear and comprehensive introductions to the ideas involved, see Buckland *et al.* (2007, 2004), Newman *et al.* (2006), Thomas *et al.* (2005). Meyer and Millar (1999) and Brooks, King and Morgan (2004) present examples of nonlinear state space modelling using WinBUGS, including the code required to reproduce their analyses.

References

Al-Awadhi, S. and Jolliffe, I. (1998) Time series modelling of surface pressure data. *International Journal of Climatology*, **18**, 443–455.

Al-Osh, M. and Alzaid, A. (1987). First-order integer-valued (INAR(1)) process. *Journal of Time Series Analysis*, **8**, 261–275.

Alzaid, A. and Al-Osh, M. (1990) An integer-valued pth-order autoregressive structure (INAR(p)) process. *Journal of Applied Probability*, **27**, 314–324.

Banerjee, A., Dolado, J., Galbraith, J. W. and Hendry, D. F. (1993) *Co-integration, Error-Correction, and the Econometric Analysis of Non-stationary Data*. Oxford University Press, Oxford.

Barry, A. M., Burney, S. M. A. and Bhatti, M. I. (1997) Optimum influence of initial observations in regression models with AR(2) errors. *Applied Mathematics and Computation*, **82**, 57–65.

Beran, J. (1994) *Statistics for Long-Memory Processes*. Chapman & Hall/CRC, Boca Raton, Florida.

Beran, J. and Maechler, M. (2009) longmemo: *Statistics for Long-Memory Processes (Jan Beran)–Data and Functions*. R package version 0.9-6.

Besbeas, P., Freeman, S. N., Morgan, B. J. T. and Catchpole, E. A. (2002) Integrating mark–recapture–recovery and census data to estimate animal abundance and demographic parameters. *Biometrics*, **58**, 540–547.

Bloomfield, P. (1992) Trends in global temperature. *Climatic Change* **21**, 1–16.

Box, G. E. P., Jenkins, G. M. and Reinsel, G. C. (1994) *Time Series Analysis, Forecasting and control*, 3rd edition. Prentice-Hall International, New Jersey.

Brockwell, P. J. and Davis, R. A. (2002) *Introduction to Time Series and Forecasting*, 2nd edition. Springer-Verlag, New York.

Brooks, S. P., King, R. and Morgan, B. J. T. (2004) A Bayesian approach to combining animal abundance and demographic data. *Animal Biodiversity and Conservation*, **27**, 515–529.

Buckland, S. T., Newman, K. B., Fernández, C., Thomas, L. and Harwood, J. (2007) Embedding population dynamics models in inference. *Statistical Science*, **22**, 44–58.

Buckland, S. T., Newman, K. B., Thomas, L. and Koesters, N. B. (2004) State-space models for the dynamics of wild animal populations. *Ecological Modelling*, **171**, 157–175.

Chatfield, C. (2000) *Time-Series Forecasting*. Chapman & Hall/CRC, Boca Raton, Florida.

Chatfield, C. (2003) *The Analysis of Time Series–an introduction (sixth edition)*. Chapman & Hall/CRC Press, Boca Raton, Florida.

Chen, R. and Tsay, R. (1993) Functional-coefficient autoregressive models. *Journal of the American Statistical Association*, **88**(421), 298–308.

Conrad Lamon, E., Carpenter, S. R. and Stow, C. A. (1998) Forecasting PCB concentrations in Lake Michigan salmonids: a dynamic linear model approach. *Ecological Applications*, **8**, 659–668.

Conrad Lamon, E., Carpenter, S. R. and Stow, C. A. (2000) Depuration of PCBs in the Lake Michigan ecosystem. *Ecosystems*, **3**, 332–343. DOI:10.1007/s100210000030.

Cox, D. R. and Miller, H. D. (1965) *The Theory of Stochastic Processes*. Chapman & Hall, London.

Craigmile, P. F., Guttorp, P. and Percival, D. B. (2004) Trend assessment in a long memory dependence model using the discrete wavelet transform. *Environmetrics*, **15**, 313–335. DOI:10.1002/env.642.

Craigmile, P. F., Guttorp, P. and Percival, D. B. (2005) Wavelet-based parameter estimation for polynomial contaminated fractionally differenced processes. *IEEE Transactions on Signal Processing*, **53**, 3151–3161.

Cui, Y. and Lund, R. (2010). A new look at time series of counts. *Biometrika*, **96**, 781–792.

Davidson, R. and Mackinnon, J. G. (2004) *Econometric Theory and Methods*. Oxford University Press, New York.

Davison, A. C. (2003) *Statistical Models*. Cambridge University Press, Cambridge.

Dethlefsen, C. and Lundbye-Christensen S. (2006) Formulating state space models in R with focus on longitudinal regression models. *Journal of Statistical Software*, **16**, 1–15.

Draper, D. (1995) Assessment and propagation of model uncertainty (with discussion). *Journal of the Royal Statistical Society, Series B*, **57**, 45–97.

Durbin, J. and Koopman, S. J. (2001) *Time Series Analysis by State Space Methods*. Oxford University Press, Oxford.

Edwards, D. and Coull, B. C. (1987) Autoregressive trend analysis: an example using long-term ecological data. *Oikos*, **50**, 95–102.

Eubank, R. L. (2006) *A Kalman Filter Primer*. CRC Press, Boca Raton, Florida.

Fahrmeir, L. and Knorr-Held, L. (2000) Dynamic and semiparametric models. In *Smoothing and Regression: Approaches, Computation and Application* (ed. M. G. Schimek). John Wiley & Sons, Inc., New York. pp. 513–544.

Fahrmeir, L. and Tutz, G. (2001) *Multivariate Statistical Modelling Based on Generalized Linear Models*, 2nd edition. Springer-Verlag, New York.

Flury, B. (1997) *A First Course in Multivariate Statistics*. Springer, New York. xviii + 277 pp.

Fraley, C., Leisch, F., Maechler, M., Reisen, V. and Lemonte, A. (2006) *fracdiff: Fractionally Differenced ARIMA aka ARFIMA(p,d,q) Models*. R package version 1.3-1.

Gardner, G., Harvey, A. C. and Phillips, G. D. A. (1980) Algorithm AS154. an algorithm for exact maximum likelihood estimation of autoregressive-moving average models by means of Kalman filtering. *Journal of the Royal Statistical Society, Series C*, **29**, 311–322.

Granger, C. W. J. (1980) Long memory relationships and the aggregation of dynamic models. *Journal of Econometrics*, **14**, 227–238.

Granger, C. W. J. and Joyeux, R. (1980). An introduction to long-range time series models and fractional differencing. *Journal of Time Series Analysis*, **1**, 15–30.

Gudmundsson, G. (2004) Time-series analysis of abundance indices of young fish. *ICES Journal of Marine Science*, **61**, 176–183. DOI:10.1016/j.icesjms.2003.12.001.

Haddon, M. (2001) *Modelling and Quantitative Analysis in Fisheries*. Chapman & Hall/CRC, Boca Raton, Florida.

Härdle, W., Lütkepohl, H. and Chen, R. (1997) A review of nonparametric time series analysis. *International Statistical Review*, **65**(1) 49–72.

Harvey, A. C. (1989) *Forecasting, Structural Time Series Models and the Kalman Filter*. Cambridge University Press, Cambridge.

Haslett, J. and Raftery, A. E. (1989) Space–time modelling with long-memory dependence: assessing Ireland's wind power resource. *With discussion. Journal of the Royal Statistical Society, Series C*, **38**, 1–50.

Hosking, J. R. M. (1981) Fractional differencing. *Biometrika*, **68**, 165–176.

Hurst, H. E. (1951) Long-term storage capacity of reservoirs. *Transactions of the American Society of Civil Engineers*, **116**, 779–799.

ICES (2001) *Report of the Working Group on Assessment of Demersal Stocks in the North Sea and Skagerrak (ICES CM 2001: ACFM07)*. International Council for the Exploration of the Sea, Copenhagen.

Kendall, M. and Ord, J. (1990) *Time Series* 3rd edition. Edward Arnold.

Kot, M. (2001) *Elements of Mathematical Ecology*. Cambridge University Press, Cambridge.

Krebs, C. J. (2001) *Ecology: The Experimental Analysis of Distribution and Abundance*, 5th edition. Benjamin Cummings, San Francisco, California.

Krzanowski, W. J. (1988) *Principles of Multivariate Analysis*. Oxford University Press, Oxford.

McKenzie, E. (1985) Some simple models for discrete variate time series. *Water Research Bulletin*, **21**, 645–650.

McLeod, A. I. and Hipel, K. W. (1978) Preservation of the rescaled adjusted range. I. A reassessment of the Hurst phenomenon. *Water Resource Research*, **14**, 491–508.

Martín, M. L., Luna, M. Y., Morata, A. and Valero, F. (2004) North Atlantic teleconnection patterns of low-frequency variability and their links with springtime precipitation in the western Mediterranean. *International Journal of Climatology*, **24**, 213–230.

Meyer, R. and Millar, R. B. (1999) BUGS in Bayesian stock assessments. *Canadian Journal of Fishing and Aquatic Science* **56**, 1078–1086.

Milionis, A. E. and Davies, T. D. (1994) Regression and stochastic models for air pollution–I. Review, comments and suggestions. *Atmospheric Environment*, **28**(17), 2801–2810.

Mishra, A. K. and Desai, V. (2005) Drought forecasting using stochastic models. *Stochastic Environmental Research and Risk Assessment*, **19**(5), 326–339.

Monahan, J. F. (2001) *Numerical Methods of Statistics*. Cambridge University Press, Cambridge.

Moron, V., Vautard, R. and Ghil, M. (1998) Trends, interdecadal and interannual oscillations in global sea-surface temperatures. *Climate Dynamics*, **14**, 545–569.

Naveau, P., Genton, M. G. and Shen, X. (2005) A skewed Kalman filter. *J. Multivariate Analysis*, **94**, 382–400.

Newman, K. B., Buckland, S. T., Lindley, S. T., Thomas, L. and Fernández, C. (2006) Hidden process models for animal population dynamics. *Ecological Applications*, **16**, 74–86.

Pavlopoulos, H. and Karlis, D. (2008) INAR(1) modelling of overdispersed count series with an environmental application. *Environmetrics*, **19**, 369–393.

Percival, D. B., Overland, J. E. and Mofjeld, H. O. (2001) Interpretation of North Pacific variability as a short- and long-memory process. *Journal of Climate*, **14**, 4545–4559.

Petris, G. (2009) *dlm: Bayesian and Likelihood Analysis of Dynamic Linear Models*. R package version 1.0-2.

Pinheiro, J., Bates, D., DebRoy, S., Sarkar, D. and the R Core Team (2008) *nlme: Linear and Nonlinear Mixed Effects Models*. R package version 3.1-89.

Pole, A., West, M. and Harrison, J. (1994) *Applied Bayesian Forecasting and Time Series Analysis*. Chapman & Hall/CRC Press, Boca Raton, Florida.

Press, W., Teukolsky, S., Vetterling, W. and Flannery, B. (1992) *Numerical Recipes in FORTRAN*, 2nd edition. Cambridge University Press, Cambridge.

Priestley, M. B. (1981) *Spectral Analysis and Time Series*. Academic Press, New York.

Robeson, S. M. and Steyn, D. G. (1990) Evaluation and comparison of statistical forecast models for daily maximum ozone concentrations. *Atmospheric Environment Part B–Urban Atmosphere*, **24**(2), 303–312.

Ross, S. M. (2003) *Introduction to Probability Models*, 8th edition. Elsevier: Academic Press, Burlington.

Smith, R. L. and Chen, F. L. (1996) Regression in long-memory time series. In *Athens Conference on Applied Probability and Time Series*, vol. II (eds. P. Robinson and M. Rosenblatt). Springer-Verlag, New York. pp. 378–391. Lecture Notes in Statistics no. 115.

Stenseth, N., Falck, W., Bjørnstad, O. and Krebs, C. (1997) Population regulation in snowshoe hare and Canadian lynx: asymmetric food web configurations between hare and lynx. *Proceedings of the National Academy of Science USA* **94**, 5147–5152.

Stenseth, N., Chan, K., Framstad, E. and Tong, H. (1998a) Phase and density-dependent population dynamics in Norwegian lemmings: interaction between deterministic and stochastic processes. *Proceedings of the Royal Society, London, B*, **265**, 1957–1968.

Stenseth, N., Falck, W., Chan, K. S., Bjørnstad, O., O'Donoghue, M., Tong, H., Boonstra, R., Boutin, S., Krebs, C. and Yoccoz, N. (1998b) From patterns to processes: phase and density dependencies in the Canadian lynx cycle. *Proceedings of the National Academy of Science USA*, **95**, 15430–15435.

Sweeting, T. J. (1992) Asymptotic ancillarity and conditional inference for stochastic processes. *Annals of Statistics*, **20**, 580–589.

Thomas, L., Buckland, S. T., Newman, K. B. and Harwood, J. (2005) A unified framework for modelling wildlife population dynamics. *Australia and New Zealand Journal of Statistics*, **47**, 19–34.

Thyregod, P., Carstensen, J., Madsen, H. and Arnbjerg-Nielsen, K. (1999) Integer valued autoregressive models for tipping bucket rainfall measurements. *Environmetrics*, **10**, 395–411.

Tomassini, L., Reichert, P., Künsch, H. R., Buser, C., Knutti, R. and Borsuk, M. E. (2009) A smoothing algorithm for estimating stochastic, continuous-time model parameters and its application to a simple climate model. *Journal of the Royal Statistical Society, Series C*, **58**, 679–704.

Tong, H. (1990) *Non-linear Time Series: A Dynamical System Approach*. Oxford University Press, Oxford.

Toth, E., Brath, A. and Montanari, A. (2000) Comparison of short-term rainfall prediction models for real-time flood forecasting. *Journal of Hydrology*, 132–147.

Tsai, H. and Chan, K. (2005) Maximum likelihood estimation of linear continuous time long memory processes with discrete time data. *Journal of the Royal Statistical Society, Series B*, **67**, 703–716.

van der Wal, J. and Janssen, L. (2000) Analysis of spatial and temporal variations of PM10 concentrations in the Netherlands using Kalman filtering. *Atmospheric Environment*, **34**, 3675–3687.

Washington, R., Hodson, A., Isaksson, E. and MacDonald, O. (2000) Northern hemisphere teleconnection indices and the mass balance of Svalbard glaciers. *International Journal of Climatology*, **20**, 473–487.

Webster, R. and Oliver, M. A. (2001) *Geostatistics for Environmental Scientists*. John Wiley & Sons, Ltd, Chichester.

Young, P. C. (1998) Data-based mechanistic modelling of environmental, ecological, economic and engineering systems. *Environmental Modelling and Software*, **13**, 105–122.

Young, P. C., Lane, K., Ng, C. N. and Palmer, D. (1991) Recursive forecasting, smoothing and seasonal adjustment of nonstationary environmental data. *Journal of Forecasting*, **10**, 57–89.

Yule, G. U. (1927) On a method of investigating periodicities in disturbed series with special reference to Wolfer's sunspot numbers. *Philosophical Transactions of the Royal Society, London*, **A226**, 267–298.

Zuur, A. F. and Pierce, G. J. (2004) Common trends in northeast Atlantic squid time series. *Journal of Sea Research*, **52**, 57–72.

Zuur, A. F., Tuck, I. D. and Bailey, N. (2003) Dynamic factor analysis to estimate common trends in fisheries time series. *Canadian Journal of Fishing and Aquatic Science*, **60**, 542–552.

6

Other issues

In this final chapter of Part I we briefly review some other, more specialised, statistical techniques that may be useful for trend analyses in specific applications.

6.1 Multisite data

With the exception of the state space models of Section 5.5, the techniques discussed so far have mostly been designed for the analysis of a single series. In many applications, however, multiple series are routinely available, either as measurements on several variables (multivariate data) or as records from multiple spatial locations (multisite, or spatiotemporal, data), as with the Dutch wind speed data of Section 1.3.1. In this section we discuss multisite data; multivariate methods are considered later.

The simplest approach to a multisite analysis is to deal with each site separately, ignoring the information from all of the other sites. However, this does not make full use of the available information. It may be, for example, that none of the sites individually shows a statistically significant trend over the period of record but that the estimated trends from all sites are broadly similar. In this case, one might suspect that there is a genuine signal in the data, and the combined evidence for a trend is stronger than that from an individual site (it is also possible, of course, that neighbouring sites show similar trends simply because of correlation between them; this issue is discussed below). Multisite data therefore provide the opportunity of making more precise statements about series structure at individual locations. This is sometimes referred to in the literature as 'borrowing strength', because the analysis for an individual site is strengthened by including information from its neighbours, or as 'space-for-time', because the use of information from other sites can be regarded as a way of increasing the sample size, which would normally be possible only by collecting longer series. Furthermore, multisite data provide the opportunity to characterise the spatial variation of different aspects of time series structure; this can provide useful insights into the underlying mechanisms.

In some situations it may be appropriate to combine the various series, for example by averaging them, and then to analyse the combined series. However, this approach has

Statistical Methods for Trend Detection and Analysis in the Environmental Sciences, First Edition.
Richard E. Chandler and E. Marian Scott.
© 2011 John Wiley & Sons, Ltd. Published 2011 by John Wiley & Sons, Ltd.

its drawbacks if the substantive question of interest relates to individual spatial locations rather than to spatially aggregated summary measures. One such drawback is the implicit assumption that the individual series all have essentially the same structure (i.e. that there is no systematic variation over space): if this is not the case, then a single combined series may not be representative of any individual location. A further difficulty is that periods of record usually vary between sites: in this case, any combined series will contain artificial inhomogeneities caused by the different numbers of observations contributing at each time point. In general, then, it is necessary to extend our existing analysis techniques to handle multisite data.

In Section 3.4.3 it was noted that observational records may contain discontinuities due to changes in measurement practice or monitoring equipment. When dealing with multisite data, in addition the method of data collection may vary between sites: for example, there may be different types of measuring instruments, different calibration procedures or variation in observer practice. Such features can give rise to apparent spatial inconsistencies or inhomogeneities, with data from neighbouring sites showing less agreement than might be expected. This potentially complicates the analysis, with a risk of overinterpreting apparent spatial structure unless artefacts of the measurement process are acknowledged. Prior to any detailed analysis, therefore, it is worth taking care to understand the data-collection mechanism (including aspects that may not be documented) and its likely consequences. This is generally true, of course, but the potential for unwittingly producing sophisticated nonsense is perhaps greater when dealing with multisite data than in many other contexts.

6.1.1   Visualisation

At the exploratory stage of a multisite analysis, in addition to the techniques discussed in Chapter 2 it is useful to visualise the spatial structure of the data. One way to do this is to produce maps showing the spatial variation of summary statistics for each location. Statistics of interest may include the mean, variance, lag 1 autocorrelation coefficient and linear trend slope. For series at sub-annual timescales, it may be worth producing separate maps for each month of the year, or for each season. Care should be taken, however, to ensure that the periods of record from which such maps are constructed are at least broadly comparable: otherwise, apparent spatial structure may emerge as an artefact of differences in record lengths. If, for example, there is an increasing trend in a particular region and records from the eastern sites tended to start later than those from those in the west, a map of series means would give a potentially false impression of an east–west gradient.

One of the most appealing and readily interpretable methods for displaying spatial structure is the contour map: such maps can be produced easily in many software packages, for example using the `contour` command in R. However, note that contouring algorithms by definition produce a smooth representation of spatial variation: the effect of this may be to obscure important local features. Bearing in mind that one of the aims of an exploratory analysis is to obtain a preliminary understanding of the data structure (Section 2.1), it is worth considering alternative techniques that are as faithful to the data as possible. Grey-scale or colour images, with an appropriately chosen colour scale, can be extremely useful mapping tools if the sites under consideration form a regular grid; contours can be overlaid on such images to aid interpretation of the colour scale. Situations involving regularly gridded data tend to be fairly specialised, however. When sites are distributed irregularly in space, an alternative way of displaying spatial structure is using a 'bubble map' in which, at each site, a circle is drawn such that the size of the circle reflects the value of the quantity of interest. A variation on this is to use circles of different colours, rather than different sizes.

Example 6.1

Figure 6.1(a) is a bubble map of decadal trends in annual mean wind speed at each of the Dutch weather stations in Figure 1.1, for the period 1961–2000. At each station, annual values have been calculated in the same way as for the two stations considered in Section 2.1.1, discarding any year with fewer than 335 days' valid data. At IJmuiden, three annual values have been discarded; at Hoek van Holland, two values; at three further sites, one value; and the annual series from the remaining eight sites are complete. The small number of discarded values ensures that trend estimates from the different stations are broadly comparable. The quantities mapped in Figure 6.1(a) are the slopes of linear trend models fitted to the data from each site using least squares, with time measured in decades. Note that this analysis is purely exploratory: there are no diagnostics to determine the adequacy of the linear trend model and no attempt to determine whether any of the trends are significantly different from zero. However, the linear trend slope provides a simple and interpretable summary of overall change during the period of interest and, by mapping the slopes in this way, we can obtain preliminary insights into the spatial variation of changes in the wind speed regime.

(a)   (b)

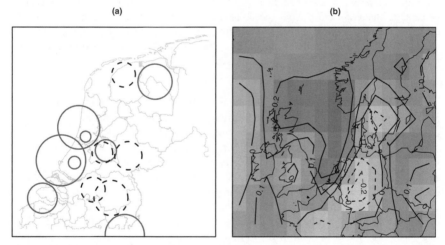

Figure 6.1   Trends (ms⁻¹ / decade) in annual mean wind speeds, 1961–2000 (a) for data from the 13 Dutch weather stations shown in Figure 1.1, (b) for gridded reanalysis data for northwestern Europe. Positive values are denoted by solid circles (contours); negative values by dashed circles (contours). Circle sizes in (a) are indicative only.

Figure 6.1(a) suggests that there have been increases in wind speed at some sites (represented by solid circles) and decreases at others. The largest increase is $0.27 \, \mathrm{m\,s^{-1}}$ per decade and the largest decrease is $-0.13 \, \mathrm{m \, s^{-1}}$ per decade. A possible interpretation of these mixed results is that, overall, there has been no change in wind speed over the Netherlands and that the differences between stations merely reflect random variation about zero. However, the locations of positive and negative values suggest some spatial organisation: with the exception of Beek (the most southerly station), there seems to be a tendency towards positive trends at 'coastal' stations and negative trends at 'inland' ones. A possible explanation for the anomalous behaviour at Beek is its increased exposure due to its altitude: it is 126 m above sea level, whereas with one exception the other stations are all below 20 m. Obviously, a formal analysis would be required to confirm the apparent spatial structure, but nonetheless the example illustrates the use of simple visualisation techniques to reveal potentially interesting structures in multisite data.

To explore further the apparent spatial structure in Figure 6.1(a), Figure 6.1(b) shows the corresponding trends for a large region of northwestern Europe. The data used here were derived from the NCEP-NCAR reanalysis data set (Kalnay *et al.*, 1996), which provides estimates of many climate variables on a regular grid. More details can be found in Yan *et al.* (2002); indeed, Figure 6.1(b) is identical to Figure 1(b) in that paper, except for the use of a slightly different 40-year period in the analysis. The regular grid provides the opportunity to visualise spatial variation using a grey-scale image with overlaid contours. Interestingly, the map shows a transition from negative trends of around -0.2 m s^{-1} per decade in continental Europe, to positive trends up to 0.3 m s^{-1} per decade in the northwest of the region, with a well-defined transition zone along the North Sea coast. Qualitatively, this agrees well with the apparent spatial organisation in Figure 6.1(a). To put these trends in context, the mean wind speed in Europe ranges from around 7 m s^{-1} on the continent to over 12 m s^{-1} in the northeast Atlantic Ocean. Over the course of half a century, therefore, a sustained change of 0.2 m s^{-1} per decade represents a substantial percentage change in mean wind speed. This could have significant implications for, for example, the risk of storm damage to buildings and oil rigs, and for the power output of wind farms. For more details of this data set and a formal analysis based on generalised linear models (Section 3.5), see Yan *et al.* (2002). ■

Arguably, maps such as those in Figure 6.1 suffer from the disadvantage that they do not show whether the trends are statistically significant. One possible way to overcome this is to use different line thicknesses or additional shading to indicate the areas where trends are 'formally' significant; another is to produce a map of p-values corresponding to tests of the null hypothesis that the underlying trend slope is zero. In general, however, such techniques are problematic for several reasons. An obvious one is that test results cannot be considered as reliable without carrying out appropriate diagnostics at each site to verify the underlying assumptions. Another is that such procedures lead to a partition of any spatial region on the basis of acceptance or rejection of the null hypothesis, and this may not be the most helpful way to think about the problem. Figure 6.1(b) illustrates the issue well: if the usual least squares assumptions hold, formal tests could be carried out to demonstrate that the largest (in absolute value) trends are significantly different from zero but that the smaller trends, such as those in the transition zone along the North Sea coast, are not. However, it would be wrong to conclude from this that 'there is no evidence for any trend' in the transition zone: it is clear that there are genuine changes in the wind speed regime and that the underlying trends vary smoothly in space, so that there are very few locations where they are exactly zero. A more constructive formal analysis, here and in many other environmental applications involving multisite data, would aim to characterise the spatial variation in trends using a spatial–temporal model. We now discuss some possible methods for achieving this.

6.1.2   Modelling

The spatial dimension of multisite data presents the modeller with two distinct challenges: first the need to account for systematic intersite differences and second the fact that data from neighbouring sites generally cannot be considered as independent. One natural framework for handling systematic differences between sites is to use regression techniques. To illustrate, suppose that a linear trend model is considered to be appropriate at each site:

$$Y_{st} = \beta_{0s} + \beta_{1s}t + \varepsilon_{st} = \mu_{st} + \varepsilon_{st}, \text{ say,} \tag{6.1}$$

where now Y_{st} denotes the response of interest at site s and time t and a subscript s has been added to the regression coefficients β_0 and β_1 to allow for the possibility that they vary between sites. In many cases, as in Figure 6.1(b), exploratory analysis suggests that the coefficients vary smoothly in space or that they depend on other spatially varying covariates such as altitude. It may therefore be plausible to represent them using a model such as

$$\beta_{js} = \beta_j + f_j\left(\text{Long}_s, \text{Lat}_s\right) \qquad (j = 0, 1), \tag{6.2}$$

where $f_0(\cdot, \cdot)$ and $f_1(\cdot, \cdot)$ are smooth functions centred on zero and Long_s, Lat_s are respectively the longitude and latitude of site s. The term β_j in (6.2) can be interpreted as an overall 'regional' value of the corresponding coefficient. Substituting (6.2) into (6.1), we obtain

$$Y_{st} = \beta_0 + f_0\left(\text{Long}_s, \text{Lat}_s\right) + \beta_1 t + f_1\left(\text{Long}_s, \text{Lat}_s\right)t + \varepsilon_{st}.$$

The first three terms on the right-hand side here are a constant β_0, a spatial component $f_0\left(\text{Long}_s, \text{Lat}_s\right)$, representing the expected variation of the response at time $t = 0$, and a trend component $\beta_1 t$. The fourth term is effectively an interaction (see Section 3.2.2) between the spatial covariates and the time index, and represents the spatial variation in trend slopes. This kind of approach can be applied to any regression based models, including GLMs.

In general, it may be difficult to specify a parametric form for functions such as $f_0(\cdot, \cdot)$ and $f_1(\cdot, \cdot)$ in (6.2) so that nonparametric methods are required. One possibility is to use smoothing techniques such as those described in Chapter 4 – in this case we have a varying coefficient model (see Section 4.3.3). An alternative is to try and approximate the $\{f_j(\cdot, \cdot)\}$ using simpler, appropriately chosen, parametric functions as building blocks; see Chandler (2005) for details. This makes the fitting of relatively complicated models feasible, although potentially at the expense of some lack of detail in the representation of regional variation.

Intersite differences do not always exhibit the degree of spatial organisation shown in Figure 6.1(b). In this case it is not appropriate to consider smooth representations of the coefficients in models such as (6.1), and it may be more plausible to consider them as realised values of random variables:

$$\beta_s = \begin{pmatrix} \beta_{0s} \\ \beta_{1s} \end{pmatrix} \sim \text{MVN}\left(\mu_\beta, \Sigma_\beta\right), \qquad \text{say,} \tag{6.3}$$

where now μ_β is the overall 'regional' value of the coefficients (corresponding to $(\beta_0 \ \beta_1)'$ in the smoothly varying model (6.2)) and the covariance matrix Σ_β reflects the magnitudes of differences between coefficient vectors of the individual sites and this regional value. In conjunction with (6.1), (6.3) defines a *random effects model* and the $\{\boldsymbol{\beta}_s\}$ are referred to as *random effects*. In the simplest case, the random effects at different sites may be considered as independent, leading to a complete lack of spatial structure: in fact, this is one possible way of accounting for potential differences between measurement processes at different sites (see the discussion at the beginning of Section 6.1.1). More generally, one might consider that the effects from neighbouring sites are likely to be correlated: this can be accommodated using *Gaussian process models*. These require the specification of a spatial correlation structure for the random effects, in which the correlations depend on intersite distances, (see Pinheiro and Bates, 2000, or Diggle and Ribeiro, 2007, for example). Indeed, some spatial correlation structures lead to smoothly varying effects, so that this class of model can be regarded as an alternative means of representing smooth spatial structure nonparametrically. Finally, there is the possibility that some of the regression coefficients in models such as (6.1) vary systematically in space whereas others are less structured: in this case it may be appropriate to model the unstructured coefficients using random effects and to retain a *fixed effects* representation such as (6.2) for the remainder. This leads to the class of *mixed effects models*.

The modelling strategies outlined above are designed to address the issue of systematic (but possibly unstructured) intersite differences, corresponding to spatial variation in $\{\mu_{st}\}$ in the context of model (6.1). They do not, however, address the issue of intersite dependence, which relates to the error component $\{\varepsilon_{st}\}$. A fully spatiotemporal model would represent the spatial correlation structure of this error component, perhaps via a stationary zero-mean Gaussian process. In conjunction with a spatial or spatiotemporal random effects model for $\{\mu_{st}\}$, such a model has a hierarchical structure with two 'levels' of randomness, sometimes referred to as the *data level* (reflecting the variation of the individual observations about their expected values) and the *population level* (reflecting the spatial variation of the expected values themselves). Hierarchical models are being used increasingly for the analysis of space–time data: see Wikle, Berliner and Cressie (1998), Banerjee, Carline and Gelfand (2004, Chapter 8) and Cressie and Wikle (2011), for example.

Although random effects, mixed effects and hierarchical models are conceptually appealing, in many cases model fitting and inference is far from trivial. For example, maximum likelihood estimates of variance parameters such as $\boldsymbol{\Sigma}_\beta$ in (6.3) can be severely biased; a modification of the usual likelihood criterion to obtain *restricted maximum likelihood* (REML) estimates of variance parameters is often advocated as a means of overcoming this, although the comparison of models involving different fixed and random effects is somewhat complicated (see Snijders and Bosker, 1999, Section 4.6 for an introduction to the concepts involved and Searle, Casella and McCulloch, 1992, Chapter 6 for a thorough account). In R, the `lme()` and `nlme()` commands in the `nlme` library (Pinheiro *et al.*, 2008) allow ML and REML fitting of such models. Another difficulty, however, is that maximising the (restricted) likelihood can be a formidable computational task except in relatively simple cases. This perhaps explains why currently there is considerable interest in the use of Bayesian methods, and the associated elegant computational machinery (see Gelman *et al.*, 1995, for example), to fit such models. For an application of these ideas to a problem involving environmental trend analysis, see Chapter 10 in the present volume.

In view of the complexity and computational challenges of hierarchical models, it is worth exploring alternative methods for dealing with intersite dependence that may be suitable for routine use. In the context of multisite data, perhaps the simplest approach is to fit models simultaneously to the data from all sites by maximising a log-likelihood function computed as though the sites were independent, and then to adjust standard errors and likelihood ratios for the dependence. It may be shown (Chandler *et al.*, 2007) that this procedure is justified so long as the fitted models contain an adequate representation of temporal dependence and that the adjustments to standard errors can be computed relatively easily (see Chandler, 2005, for details of the computations in the case of GLMs and Chandler *et al.*, 2007, for the general case). Adjustments to likelihood ratios for comparing nested models are slightly more difficult to compute; for recent developments see Chandler and Bate (2007) and references therein. An example of the application of these ideas in the context of environmental trend analysis can be found in Chapter 8 of the present volume.

In Section 3.3 we saw that generalised least squares (GLS) provides a means of accounting for temporal dependence in time series regression models and in Section 3.5 the extension to generalized estimating equations (GEEs) for GLMs was discussed. It is natural, then, to ask whether GLS and GEEs can be used to account for spatial as well as temporal dependence, and thereby improve on the techniques discussed in the previous paragraph. In principle this can be done; indeed, the pioneering work of Haslett and Raftery (1989) can be regarded as an application of this kind of idea to the modelling of multisite long-memory time series. However, subtle pitfalls await the uninitiated: in certain situations, as demonstrated by Sullivan Pepe and Anderson (1994), the use of such techniques can yield biased estimates of parameters of interest. This is particularly true in complex situations where the fitted models cannot capture fully the structure of a system, as is often the case in environmental applications. Thus the simpler approach, of fitting models as though sites are independent and then adjusting standard errors, may sometimes be preferred. A general lesson from this is that when dealing with complex data structures it is necessary to understand clearly the implications both of the model being fitted and the methodology for fitting it.

6.2   Multivariate series

Multivariate series arise when several variables are observed simultaneously at each time instant. If the variables are mutually independent, they can be analysed individually, but more often there are intervariable dependencies that are of interest to the analyst. In some settings, the nature of the dependence may suggest plausible strategies for modelling it. For example, in the multisite case considered above it is natural to consider the use of spatial correlation functions. More generally, however, the structure of the dependence is less clear and other techniques must be considered. Accessible introductions to 'classical' multivariate methods can be found in Manly (1994) and Everitt and Dunn (2001); Krzanowski (1988) gives a more detailed treatment.

6.2.1   Dimension reduction

By far the most common approach to multivariate problems is to start with some form of dimension reduction, of which principal components analysis (PCA – sometimes referred to as *empirical orthogonal function*, or EOF, analysis) is the best known. The idea behind

PCA is to transform a vector $\mathbf{Y} = (Y_1 \ldots Y_k)'$ of random variables into another vector $\mathbf{Y}^* = (Y_1^* \ldots Y_k^*)'$ of *principal components*, via a linear transformation:

$$Y_1^* = a_{11}Y_1 + a_{12}Y_2 + \cdots + a_{1k}Y_k$$
$$Y_2^* = a_{21}Y_1 + a_{22}Y_2 + \cdots + a_{2k}Y_k$$
$$\vdots \qquad\qquad \vdots$$
$$Y_k^* = a_{k1}Y_1 + a_{k2}Y_2 + \cdots + a_{kk}Y_k$$

or, in matrix form,

$$\mathbf{Y}^* = \mathbf{AY},$$

where A is a $k \times k$ matrix with (i, j)th element a_{ij}, chosen in such a way that:

- $\text{Var}(Y_1^*) \geq \text{Var}(Y_2^*) \geq \cdots \geq \text{Var}(Y_k^*)$. This is intended to ensure that the first component Y_1^* in some sense contains the most information about variation within the system as a whole, followed by the second component and so on.

- $Y_1^*, \ldots, Y_k^*$ are mutually uncorrelated random variables. This can be regarded as a means of avoiding the duplication of information between components.

- $A'A = I_k$, the $k \times k$ identity matrix. Without such a constraint, the variances of the components could be made arbitrarily large simply by scaling the elements of A.

It can be shown (Krzanowski, 1988, Section 2.2.4) that these criteria define a unique matrix A, that the rows of this matrix are the eigenvectors of the covariance matrix of $\mathbf{Y}$, and that the variances of the principal components are the corresponding eigenvalues.

In carrying out a PCA, the hope is firstly that most of the important variation is captured in a relatively small number of components (thereby achieving the goal of dimension reduction) and secondly that the *loadings* $\{a_{ij}\}$ are such that the components have a plausible subject-matter interpretation. For example, if Y_1 and Y_2 represent respectively the population sizes of predator and prey species in a particular area and a PCA yields components

$$Y_1^* = Y_1/\sqrt{2} + Y_2/\sqrt{2},$$
$$Y_2^* = Y_2/\sqrt{2} - Y_1/\sqrt{2},$$

then Y_1^* may be interpreted as an index of overall population abundance and Y_2^* as a measure of the excess of prey over predators (which may be interpreted as an index of the predators' food supply). The scaling by $\sqrt{2}$ here ensures that the loadings satisfy the constraint $A'A = I_2$.

In practice, of course, given only data $\mathbf{y}_1, \ldots, \mathbf{y}_T$ (each considered to be drawn from the distribution of $\mathbf{Y}$), the underlying covariance matrix is usually unknown and the principal components reported by software packages are the eigenvectors of the sample version

$$\frac{1}{T-1} \sum_{t=1}^{T} (\mathbf{y}_t - \overline{\mathbf{y}})(\mathbf{y}_t - \overline{\mathbf{y}})', \tag{6.4}$$

where $\bar{\mathbf{y}} = T^{-1}\sum_{t=1}^{T}\mathbf{y}_t$ is the sample mean of the observations.[1] If the data can be regarded as an independent random sample from the distribution of $\mathbf{Y}$, then (6.4) is just the usual estimator of the underlying covariance matrix. For stationary multivariate time series, the same estimator can be used: except for the use of divisor $T-1$ instead of T, the elements of this matrix are the zero-lag cross-covariances between each pair of series (see Section 2.1.5). In the stationary case, therefore, it may be useful to carry out a PCA and then, providing the components have a clear subject-matter interpretation, to compute the time series of the first few components (the value of the rth component at time t is $y_{rt}^{*} = a_{r1}y_{1t} + a_{r2}y_{2t} + \cdots + a_{rk}y_{kt}$, in an obvious notation) and analyse them separately using the methods for stationary series described in Chapter 5. Although this sounds straightforward, some care is required: for example, PCA yields different results depending on whether or not variables are standardised prior to analysis. For discussion of the issues involved, see Krzanowski (1988, Section 2.2.5), and Everitt and Dunn (2001, Chapter 3).

The technique of *factor analysis* is at some level similar to PCA, in that the aim is to derive a small number of interpretable new variables. A fundamental difference, however, is that factor analysis appeals to an underlying statistical model whereas principal components are defined implicitly via the criteria given above. For a vector $\mathbf{Y} = (Y_1 \ldots Y_k)'$ of random variables, the factor analysis model is

$$Y_1 = b_{11}F_1 + b_{12}F_2 + \cdots + b_{1p}F_p + \varepsilon_1$$
$$Y_2 = b_{21}F_1 + b_{22}F_2 + \cdots + b_{2p}F_p + \varepsilon_2$$
$$\vdots \qquad\qquad \vdots$$
$$Y_k = b_{k1}F_1 + b_{k2}F_2 + \cdots + b_{kp}F_p + \varepsilon_k,$$

where now $F_1, \ldots, F_p (p < k)$ are the values of p unobserved or latent *factors* and the $\{\varepsilon_i\}$ are uncorrelated random variables with zero mean. A factor analysis of data $\mathbf{y}_1, \ldots, \mathbf{y}_t$ will therefore produce a sequence of vectors $\mathbf{F}_1, \ldots, \mathbf{F}_T$ such that $\mathbf{F}_t = (F_{1t} \ldots F_{pt})'$ contains the values of each factor at time t. Although the precise definition of the factors here looks rather different from that of principal components, and although the precise implementation of factor analysis requires some skill (Manly, 1994, Section 7.6; Krzanowski, 1988, Section 16.2), operationally they are often interpreted in a very similar way and there are close connections between them (Everitt and Dunn, 2001, Section 3.3). For current purposes, however, the most important connection is that factor analysis, like PCA, relies implicitly on the sample covariance matrix (6.4) of the data.

Unfortunately, this reliance on the sample covariance matrix means that for nonstationary multivariate time series, it is generally inappropriate to use either PCA or factor analysis. The reason is that in the nonstationary case, by definition the $\{\mathbf{y}_t\}$ are drawn from different distributions so that in general they do not share the same covariance matrix; (6.4) is therefore meaningless as a measure of interseries association. Nonetheless, the idea of reducing a large number of interdependent series into a smaller number of uncorrelated ones is appealing, and some of the ideas can be extended to a time series context. For

---

[1] The components are not always calculated in this way, since it is computationally more stable to derive them from the singular value decomposition (SVD) of the $T \times k$ matrix with $\mathbf{y}_t$ in the tth row (Krzanowski, 1988, Section 4.1). This is the algorithm used in the prcomp() command in R, for example.

example, a natural extension of the factor analysis model is to replace the latent variables $F_1, \ldots, F_p$ with latent stochastic processes, which might be interpreted as trends that are common to some or all of the observed series. This idea leads to the technique of *dynamic factor analysis* (Molenaar, 1985; Molenaar, de Gooijer and Schmitz, 1992), which has strong connections with state space models (Section 5.5; see also Harvey, 1989, Section 8.5). Accessible introductions to the area, and environmental applications, can be found in Zuur, Tuck and Bailey (2003) and Zuur and Pierce (2004); see also Chapter 9 of the present volume. Another approach, which stands in a similar relation to dynamic factor analysis as does PCA to ordinary factor analysis, is the *min/max autocorrelation factor* (MAF) approach of Switzer and Green (1984). Here the aim is to transform the original series into a smaller number of component series, ordered according to the speed of decay of their sample autocorrelation functions. Since nonstationary processes, and in particular trends, lead to slowly decaying ACFs (see Sections 2.1.3 and 5.2.1), such a procedure should lead to information on trends being concentrated in the first few component series. The idea has been applied in the context of environmental trend analysis by Shapiro and Switzer (1989) but, perhaps due to a lack of widely available software, appears not to have been pursued: it would be interesting to explore its potential further.

6.2.2   Multivariate models

In addition to techniques that aim to reduce the dimensionality of multivariate series, there are methods that seek to model the interseries structure directly. The state space modelling framework of Section 5.5 is an obvious example of this. In addition, *vector ARMA* or *VARMA* models are multivariate analogues of ARMA models. For example, the model equation for a zero-mean vector AR(1) is

$$\mathbf{Y}_t = \boldsymbol{\phi}\mathbf{Y}_{t-1} + \varepsilon_t, \tag{6.5}$$

where $\boldsymbol{\phi}$ is a $k \times k$ matrix and (ε_t) is a sequence of uncorrelated zero-mean random vectors, each with the same covariance matrix. The process (6.5) is stationary only if the eigenvalues of $\boldsymbol{\phi}$ are all less than 1 in absolute value (Brockwell and Davis, 2002, Section 7.4); this is the multivariate extension of the AR(1) stationarity condition discussed in Section 5.1.1. It might be supposed that the individual series in a VAR(1) process would themselves have an AR(1) structure. However, the analysis of the predator–prey system in Example 5.6 shows that this is not the case. This system is a bivariate VAR(1) with nonzero mean and with

$$\boldsymbol{\phi} = \left(\begin{array}{cc} \phi_1 & \psi_1 \\ \psi_2 & \phi_2 \end{array} \right) ;$$

it was demonstrated earlier that the individual component series in fact behave as ARMA(2,1) processes. Clearly, therefore, the link between the behaviour of VARMA processes and their univariate counterparts is not straightforward. Moreover, the identification of appropriate models for multivariate series (which requires inspection of the sample cross-correlation functions, as well as the individual ACFs and PACFs) is rather more difficult than for univariate series. For an introduction to the problems, and some ways of overcoming them, see Chatfield (2003, Section 12.3).

In situations where the variables of interest depend upon covariates as well as upon each other, it is natural to consider the development of multivariate regression models that represent the simultaneous (inter-)dependence structure. In a multivariate regression model, the standard multiple regression equation (3.34) is replaced with a system of k equations, one for each of the response variables considered (note that the covariates are not necessarily restricted to be the same for each response variable, although this is of course a possibility). Such a model could in principle be fitted using ordinary least squares estimation separately for each response variable. However, such a procedure fails to account for potential correlations between the error terms for the different responses. Simultaneous estimation using generalised least squares (see page 99) overcomes this problem. It also allows the fitting of models in which constraints are imposed on the regression coefficients across two or more responses: this would be of interest, for example, if one wished to fit a model in which all of the responses shared a common linear trend. Multivariate regression models have been studied extensively in econometrics; for good introductory treatments, see, for example, Greene (2003, Chapter 14) and Davidson and Mackinnon (2004, Chapter 12).

Most of the methods reviewed so far in this section are most appropriate when the individual series are Gaussian or nearly so. Few models are currently available for multivariate time series in non-Gaussian settings; see, however, Fahrmeir and Tutz (2001) for a general review of methods for non-Gaussian multivariate data. Moreover, few models are available in settings where different variables have markedly different distributions. This arises, for example, in the analysis of daily rainfall and temperature series at a particular location: although in many parts of the world daily temperature distributions may be approximately Gaussian, rainfall distributions are usually highly skewed. In such situations, often the only viable method of analysis is to deal with each series in turn, at each stage conditioning on the series already analysed. The rationale for this is that any joint distribution can be factorised as a product of conditional distributions:

$$f(\mathbf{y}) = f(y_1) f(y_2|y_1) f(y_3|y_2, y_1) \cdots f(y_k|y_{k-1}, \ldots, y_1)$$

in an obvious notation. In the rainfall–temperature example, therefore, one might start by modelling the rainfall series, perhaps within the framework of a GLM (see Section 3.5) to account for covariates that are responsible for trends, and then build a time series regression model for temperature in which rainfall is considered as a covariate. For an illustration of this, see Furrer and Katz (2007).

6.3   Point process data

In Section 5.4 we discussed models for irregularly spaced time series where the irregularity is due to the measurement regime. In some situations, however, the process under investigation itself consists of a sequence of events (for example, earthquakes, tropical cyclones or sightings of a particular animal species) occurring at irregular time intervals: this gives rise to *point process data*.

Broadly speaking, there are two distinct approaches to the analysis of such data: these focus respectively upon the sequence of intervals between events and upon the rate of occurrence of the events themselves. The first approach reduces the data to a regularly spaced sequence of intervals that can be analysed using methods described earlier in this book – noting, however, that the interval lengths are necessarily non-negative and hence non-Gaussian. In this section, therefore, we briefly review methods that focus on the rate of occurrence. Davison (2003, Section 6.5) gives an accessible introduction to the area; for a more detailed treatment, see Cox and Isham (1980) or Daley and Vere-Jones (2003).

6.3.1 Poisson processes

The simplest model for point process data is the *homogeneous Poisson process*, which can be characterised as follows:

- No two events can occur exactly simultaneously.

- The number of events in any time interval has a Poisson distribution with mean proportional to the length of the interval: $N(t_1, t_2) \sim \text{Poi}(\lambda |t_2 - t_1|)$, say, where $N(t_1, t_2)$ is the number of events in the interval (t_1, t_2) and $\lambda > 0$.

- The numbers of events in nonoverlapping time intervals are independent random variables.

The parameter λ is known as the *rate* or *intensity* of the process, and can be interpreted as the mean number of events per unit time. If N events are observed over a time period of length T, then λ can be estimated as N/T.

The homogeneous Poisson process is rarely adequate in itself as a model for environmental data, but it forms the basis for many other point process models. In this respect it plays much the same role as does white noise in the context of time series models. Nonetheless, in the initial stages of an analysis it may be useful to determine whether the observed occurrence times can be regarded as a realisation of a homogeneous Poisson process. The following properties of the process can be used as the basis of diagnostics to assess this:

- If the period of record is split into nonoverlapping intervals of equal length, the numbers of events in each interval are independent and identically distributed Poisson random variables. In this case, the sample mean and variance of the event counts should be roughly equal.

- The intervals between events are independent random variables, each having an exponential distribution with parameter λ. The independence of successive intervals can be assessed using the sample autocorrelation function of the interval sequence, and a check for consistency with the exponential distribution can be made using a quantile–quantile plot (for example Rice, 2006, Section 10.2).

- Conditional on the total number of events N in an interval $(0, T)$, the event times form an independent random sample from the uniform distribution on $(0, T)$. Again, this could be checked using a quantile–quantile plot.

In the context of a trend analysis, perhaps the most obvious extension of the homogeneous Poisson process is to allow the rate to vary over time. Specifically, an *inhomogeneous Poisson process* (IPP) differs from the homogeneous version in that the constant λ is replaced with an *intensity function* $\lambda(\cdot)$ and the distribution of $N(t_1, t_2)$ is Poisson with mean

$$\int_{t_1}^{t_2} \lambda(t)\,dt.$$

In an IPP, the sample variance of counts in disjoint intervals is expected to be higher than the mean; this is sometimes referred to as *overdispersion*.

If inhomogeneity is suspected, it will often be of interest to estimate the intensity function. Perhaps the easiest way to achieve this is to exploit the fact that for an IPP with intensity function $\lambda(\cdot)$, conditional on the total number of events N in an interval $(0, T)$, the times of the events form an independent random sample from the distribution with density $f(\cdot)$ proportional to $\lambda(\cdot)$ over that interval:

$$f(x) = \begin{cases} \lambda(x) / \int_0^T \lambda(t)dt, & x \in (0, T), \\ 0, & \text{otherwise.} \end{cases}$$

The density $f(\cdot)$ can be estimated from the observed event times using well-established nonparametric procedures, as implemented in the density() command in R, for example. Denoting by $\hat{f}(t)$ the resulting density estimate at time t, the corresponding estimate of the intensity function is just $\hat{\lambda}(t) = N\hat{f}(t)$.

In some settings it may be of interest to model the intensity function of an IPP in a parametric form, thus opening up the possibility of introducing dependence on covariates. For example, an IPP analogue of the linear trend model (3.2) has

$$\log \lambda(t) = \beta_0 + \beta_1 t \quad \Rightarrow \quad \lambda(t) = \exp[\beta_0 + \beta_1 t] \tag{6.6}$$

(note that setting $\lambda(t) = \beta_0 + \beta_1 t$ would lead to problems since $\lambda(t)$ cannot be negative). Such models can be fitted using maximum likelihood: it can be shown (Davison, 2003, Section 6.5.1) that if an IPP is observed over the period $(0, T)$ to yield events at times $t_1, t_2, \ldots, t_N$ then the log-likelihood for parameters in $\lambda(\cdot)$ takes the form

$$-\int_0^T \lambda(t)dt + \sum_{j=1}^N \log \lambda(t_j). \tag{6.7}$$

For the 'loglinear trend' model (6.6), this reduces to

$$N\beta_0 + \beta_1 \sum_{j=1}^N t_j - e^{\beta_0} \left[e^{\beta_1 T} - 1 \right] / \beta_1. \tag{6.8}$$

This must be maximised numerically, for example using routines nlm() or optim() in R, to obtain estimates of the parameters β_0 and β_1.

More generally, if the intensity function of an IPP is considered to depend on covariates, then to evaluate the contribution $-\int_0^T \lambda(t)dt$ to the log-likelihood (6.7), it is necessary to know the covariate values throughout the time interval $(0, T)$. This is usually impractical: even if high-resolution covariate data are available (at times $0, \Delta t, 2\Delta t, \ldots$, say) the integral must of necessity be approximated by the corresponding sum:

$$\int_0^T \lambda(t)dt \approx \Delta t \sum_{k=1}^{T/\Delta t} \lambda(k\Delta t).$$

This approximation will be accurate if the intensity function can be considered as approximately constant within intervals of length Δt. In this case, the number of events (N_k, say) in the kth such interval has approximately a Poisson distribution with mean $\lambda(k\Delta t)\Delta t$. Denote this mean by μ_k, and notice that any dependence of $\lambda(\cdot)$ upon the underlying covariates will be shared by the sequence (μ_k). This suggests that, instead of working directly with the IPP log-likelihood (6.7), approximate inference for IPPs can be conducted by fitting generalised linear or generalised additive models (Sections 3.5 and 4.2) to the time series (N_k) of counts, considered as independent Poisson variables with covariate-dependent means (μ_k). The following example illustrates the procedure.

Example 6.2
Consider data from the loglinear trend model (6.6), aggregated as described so that N_k has approximately a Poisson distribution with mean

$$\mu_k = \lambda(k\Delta t)\Delta t = \exp(\beta_0 + \beta_1 k\Delta t)\,\Delta t.$$

Taking logarithms, we obtain

$$\log \mu_k = \log \Delta t + \beta_0 + \beta_1 x_k,$$

where $x_k = k\Delta t$. This shows that the approximation is effectively a Poisson GLM for the (N_k) with a logarithmic link function (see Section 3.5); it can therefore be fitted using standard software routines such as `glm()` in R. Notice, however, that the intercept in the linear predictor here is not β_0, but $\log \Delta t + \beta_0$. To force software routines to report an estimate of β_0, it is necessary to declare $\log \Delta t$ as a known component of the linear predictor; such a component is referred to as an *offset* (Dobson, 2001, Section 9.2).

It is instructive to demonstrate the connection between the approximating GLM and the IPP log-likelihood (6.7). Ignoring constant terms, the Poisson log-likelihood for the approximating GLM is

$$\ell(\beta_0, \beta_1) = \sum_{k=1}^{T/\Delta t} \left[N_k \log \mu_k - \mu_k \right]$$

$$= \sum_{k=1}^{T/\Delta t} \left[N_k (\log \Delta t + \beta_0 + \beta_1 k\Delta t) - \exp(\beta_0 + \beta_1 k\Delta t)\,\Delta t \right]$$

$$= N \left[\log \Delta t + \beta_0 \right] + \beta_1 \sum_{k=1}^{T/\Delta t} N_k k\Delta t - e^{\beta_0} \sum_{k=1}^{T/\Delta t} \exp(\beta_1 k\Delta t)\,\Delta t. \quad (6.9)$$

The term $N \log \Delta t$ can be ignored since it does not depend on the parameters. Consider next what happens to the remaining terms as we increase the accuracy of the approximating model by letting Δt tend to zero. In this case, the values of N_k will eventually become either one (for intervals containing the original occurrence times) or zero. The intervals for which $N_k = 1$ are defined by

$$\left\{ k : (k-1)\Delta t \le t_j < k\Delta t \text{ for some } j \in \{1, \ldots, N\} \right\}.$$

It follows that as $\Delta t \to 0$, the term $\beta_1 \sum_{k=1}^{T/\Delta t} N_k k\Delta t$ in (6.9) tends to $\beta_1 \sum_{j=1}^{N} t_j$. Simultaneously, the final term in (6.9) tends to $e^{\beta_0} \int_0^T \exp(\beta_1 t)\,dt$ (this being the standard definition of an integral via approximating sums). Combining these results, it is seen that the approximating log-likelihood (6.9) becomes equivalent to the exact IPP log-likelihood (6.8) as Δt tends to zero. ∎

This example demonstrates the intuitively obvious fact that when using GLMs or GAMs to fit an IPP model, the accuracy of the approximation increases as Δt becomes smaller. Of course, this attracts some computational cost as the 'sample size' $T/\Delta t$ increases; however, this is perhaps not a major issue with modern computing power. In practice, the temporal resolution of covariate data will often be the main limitation to the accuracy that can be achieved.

As with all statistical models, it is useful to check the adequacy of any IPP model after fitting. In this respect, a useful diagnostic makes use of the fact that a temporal IPP with intensity function $\lambda(\cdot)$ can be transformed into a homogeneous Poisson process with unit rate (i.e. for which $\lambda = 1$) by rescaling the time axis according to the transformation

$$\tau(t) = \int_0^t \lambda(u)\mathrm{d}u.$$

Thus, if an IPP model has been fitted to a sequence of events at times $t_1, \ldots, t_N$ to yield an estimated intensity function $\hat{\lambda}(\cdot)$, one can compute the rescaled occurrence time of each event as $\tau(t_j) = \int_0^{t_j} \hat{\lambda}(u)\mathrm{d}u$. The sequence $\tau(t_1), \ldots, \tau(t_N)$ should behave like a realisation of a homogeneous Poisson process with unit rate; this can be assessed using the methods outlined above. For more details, see Ogata (1988).

6.3.2 Other point process models

A key assumption in Poisson process models is that the numbers of events in disjoint time intervals are independent random variables. In many applications this is unrealistic: for example, major earthquakes are usually followed by several aftershocks so that earthquake occurrence times tend to be more clustered than under a Poisson process model. Many models are available for such clustered point process data; for the specific example of earthquake sequences, the class of self-exciting processes introduced by Hawkes (1971a, 1971b), has been applied extensively (e.g. Ogata, 1989; Ogata and Vere-Jones, 1984). An appealing feature of this class is that it can be fitted using likelihood methods; the fitting of other families of clustered point process models, such as the Neyman–Scott and Bartlett–Lewis families (Cox and Isham, 1980, Section 3.4), is rather more difficult. Furthermore, limited progress has been made in developing tractable nonstationary versions of these models, which would be necessary for trend analysis purposes. A substantial amount of effort has been devoted, however, to the analysis of spatial and spatiotemporal point processes, and it is possible that the ideas could be applied to the purely temporal setting without too much difficulty. For an accessible review of spatial point processes, see Diggle (2003); a more detailed account of modern developments is given by Møller and Waagepetersen (2003) and an application in which nonstationary clustered point process models are fitted to spatial data (the locations of trees in a forest) is given by Waagepetersen (2007).

The presence of clustering in point process data induces positive correlation between the counts in successive intervals and, like nonstationarity, is a possible cause of overdispersion (see Section 6.3.1). Underdispersion, corresponding to negative correlation at short timescales, is also possible; it is encountered less frequently, however. We have encountered the phenomenon in relation to the study of tropical cyclones: this is presumably because the occurrence of a cyclone in a particular region dissipates a huge amount of energy, as a result of which further cyclone formation tends to be inhibited for a short time afterwards. Another environmental example is provided by Ridout and Besbeas (2004), in relation to clutch sizes (i.e. numbers of eggs per nest) for a particular species of bird. Unfortunately, few models are available for underdispersed point process data. A pragmatic approach to their analysis may therefore be to discretise to a sufficiently fine timescale that at most one event occurs in any time interval, and to analyse the resulting zero–one series of counts as a binary time series, for example using GLMs (Section 3.5) in which lagged observations are used as covariates to account for the negative autocorrelation.

6.3.3   Marked point processes

Often, the occurrence of an event in a point process is accompanied by one or more items of other information (*marks*), such as the magnitude of an earthquake or the number of individuals seen at a particular sighting of a rare species. In this case we have a *marked point process*. The sequence of marks forms an irregularly spaced time series and could therefore be analysed in principle using methods such as those described in Section 5.4. However, those methods are designed primarily for situations in which an underlying continuous-time process is sampled at the observation times; in the case of marked point processes this is not always the case. When analysing earthquake sequences, for example, magnitudes can only be associated with earthquakes that actually occur. In this case it may be more natural to treat the marks themselves as a regularly spaced time series, possibly considering the interevent intervals as covariates.

6.4   Trends in extremes

It is not unusual to find applications in which the study of a particular system is motivated by the risk of adverse consequences from 'extreme' conditions: for example, abnormally high rainfalls lead to floods and abnormally high temperatures can be associated with increased human mortality. In particular, trends in extremes may be of interest, to determine whether the risk of such adverse events is changing.

In most of the methods discussed so far in this book, the observations are regarded as realised values of random variables: in the presence of trends, the associated probability distributions change through time. To study trends in extremes, therefore, one might choose to fit an appropriate statistical model to the data and to study the temporal changes in the tails of the fitted distributions. If the available data contain enough of the 'extreme' events of interest, this approach can be useful. However, if interest lies in much rarer events, the available data may not contain sufficient information to provide reassurance that the fitted model represents the extremal behaviour adequately. The following hypothetical example illustrates this.

Example 6.3

An engineering company is commissioned to design a structure with an intended lifetime of 20 years. However, extreme weather conditions will cause the structure to fail. The company wishes to know how strong the structure needs to be in order that the overall probability of failure, at any point during its intended lifetime, is less than 10 %.

This question can be considered by making two simplifying assumptions: first, that the probability p of failure due to extreme weather conditions in any given year is the same throughout the intended lifetime of the structure and, second, that the occurrence of such extreme conditions is independent from year to year. In this case the overall probability of failure during the 20-year lifetime is

$$P(\text{Failure at some point}) = 1 - P(\text{No failures in any of the 20 years})$$
$$= 1 - (1 - p)^{20}$$

under the simplifying assumptions given above. If the overall probability of failure is to be less than 0.1, we therefore require

$$1 - (1 - p)^{20} < 0.1 \Rightarrow (1 - p)^{20} > 0.9 \Rightarrow (1 - p) > 0.9^{1/20} = 0.9947,$$

and hence $p < 1 - 0.9947 = 0.0053$. Now the average time between successive events that occur with probability 0.0053 independently each year is $1/0.0053 = 189$ years: this is called the *return period* of the event. The engineering company therefore needs to design for conditions that occur roughly once every two hundred years. Clearly, few (if any) existing data sets will contain reliable direct information regarding such conditions. ∎

In reality, of course, the assumptions underlying the calculation in this example are unlikely to be met: in particular, in a changing climate the probability p will evolve over time. However, the example does illustrate that to limit the risk of some adverse event considered over even a relatively modest time horizon (20 years in this instance), it can be necessary to consider the likely magnitudes of very rare events and hence to extrapolate well beyond the range of the available data. To have any kind of confidence in such extrapolation, more is required than simply to show that a particular model fits the data well. If, for example, we could appeal to some generally applicable theory that suggests a particular class of distributions for 'extreme' observations, this would lend some credence to extrapolations based on this class. The approach is directly analogous to the use of the central limit theorem (for example Wackerly, Mendenhall and Scheaffer, 2007, Section 7.3) to motivate the normal distribution as a model for quantities (such as annually aggregated time series) that can be regarded as sums or averages.

6.4.1   Approaches based on block maxima

Fortunately, there is a well-developed body of theory relating to distributions for extremes. This dates back to the work of Fisher and Tippett (1928), who showed that if M_n denotes the largest value obtained from n independent, identically distributed random variables $X_1, \ldots, X_n$ then, if M_n has a limiting distribution after suitable normalisation, it must belong to one of three types: the Gumbel, Fréchet or (reverse) Weibull. Subsequently, von Mises (1954) and Jenkinson (1955) proposed combining the three types into a single parametric family, nowadays referred to as the *generalised extreme value (GEV)* distributions. A random variable Y has a GEV distribution with parameters $\mu \in \mathbb{R}$, $\sigma > 0$ and $\xi \in \mathbb{R}$ if its distribution function, defined for $\{y : 1 + \xi(y - \mu)/\sigma > 0\}$, is

$$G(y) = P(Y \le y) = \begin{cases} \exp\left\{-\exp\left[-(y - \mu)/\sigma\right]\right\}, & \xi = 0, \\ \exp\left\{-\left[1 + \xi\left(\dfrac{y - \mu}{\sigma}\right)^{-1/\xi}\right]\right\}, & \text{otherwise.} \end{cases}$$

The parameters μ, σ and ξ are referred to as *location*, *scale* and *shape* parameters respectively. The cases $\xi = 0$, $\xi > 0$ and $\xi < 0$ correspond to the Gumbel, Fréchet and Weibull distributions respectively. The distribution of the original $\{X_i\}$ determines which, if any, of these cases holds: in this case, the original distribution is said to be in the *maximum domain of attraction* of the corresponding GEV. For some discrete distributions, such as the Poisson and geometric, no limiting distribution for maxima exists (Embrechts, Klüppelberg and Mikosch, 1997, Section 3.1). However, for many standard continuous distributions including the normal, exponential and gamma families, the maximum has a limiting Gumbel distribution; for 'heavy-tailed' distributions the limit is Fréchet and for distributions with a finite upper endpoint the limit is Weibull.

The existence, in a wide range of settings, of the GEV limiting distribution for maxima suggests a strategy for the analysis of rare events in a time series: split the series into blocks of equal length, extract the largest value in each block, fit a GEV distribution to the resulting sample of 'block maxima' and use the fitted distribution to extrapolate beyond the range of the data. In Example 6.3 above, one might fit a GEV distribution to the observed annual maxima of the weather variable of interest: the 0.9947 quantile of the fitted distribution then provides an estimate of the size of event for which the company must design. The usual checks of modelling assumptions should, of course, be carried out but, providing these checks reveal no obvious problems with the model, one might feel more comfortable with the use of a model that has both theoretical and empirical justification. If interest lies in minima rather than maxima, the same procedure can be used: simply reverse the sign of all the data values prior to analysis so as to convert minima to maxima.

Initially, the conditions leading to the GEV distribution above may appear restrictive: it arises by considering the maximum of a large number of independent, identically distributed observations. Fortunately, it can be shown that the requirement of independence is not necessary: the GEV limit also applies to maxima of stationary sequences, at least in the absence of long memory which, in this context, has a slightly different technical definition from that given in Section 5.3; see Coles (2001, Section 5.2) and Embrechts, Klüppelberg and Mikosch (1997, Section 4.4). The requirement for the $\{X_i\}$ to be identically distributed is also potentially problematic: in many environmental applications, it is convenient to consider a year as the basic time unit and hence to fit distributions to annual maxima. However, environmental series often display seasonality, so that observations within a year cannot be considered to be identically distributed (this has led some authors to consider modelling monthly maxima instead; see, for example, Rust, Maraun and Osborn, 2009). Experience suggests, however, that the GEV often provides a good fit to samples of annual maxima. For a heuristic explanation of this, consider what would happen if we randomly reordered the entire time series and then split the result into yearly blocks. The reordered observations are now guaranteed to be identically distributed, because each of them is equally likely to be any one of the original observations: the GEV result therefore holds for the maxima of the reordered sequence. However, the sample of annual maxima in the reordered sequence will generally be rather similar to the sample of maxima from the original ordering, at least providing the series is long enough; thus the two samples will lead to similar fitted models.

We now consider how to exploit the results above to study trends in extremes. As discussed at the beginning of this section, such trends correspond to temporal changes in the underlying distributions. Since the GEV distribution is characterised by just three parameters, trends in extremes can be modelled by fitting GEV distributions to time series of block maxima and allowing one or more of the parameters to depend on covariates. Model fitting, and comparison of nested models to test for trends in one or more parameters, can be carried out using maximum likelihood (see Section 3.5). For example, Coles (2001, Section 6.3) fits GEV distributions to annual maximum sea levels at Fremantle, Western Australia, and finds evidence for a linear trend in the location parameter: $\mu_t = \mu_0 + \mu_1 t$, where the time index t is measured in years. The extension to nonparametric representations of trends in GEV parameters, using methods similar to those described in Chapter 4, has been considered by Davison and Ramesh (2000). In R, routines for maximum likelihood fitting of GEV distributions, both with and without covariates, are available in the ismev library (Coles and Stephenson, 2006).

In Example 6.3, the concept of a return period was introduced as the average time elapsing between events of a given magnitude. This concept clearly has limited validity in the presence of trends: in this case, it is necessary to identify the planning horizon of interest (for example 20 years in the example above) and to study directly the properties of the largest value experienced over this period. Cox, Isham and Northrop (2002) have done this analytically for the nonstationary Gumbel case. They show in particular that a simple approach, in which calculations are based on a stationary Gumbel distribution with location and dispersion parameters corresponding to the midpoint of the planning horizon, leads to a slight underestimation of the maxima of interest. The extension to the GEV distribution does not appear to have been attempted, however.

The use of block maxima to study extremes has the potential disadvantage of being wasteful of data: by discarding all but the largest observation in each block, one typically ends up with a relatively small sample of maxima. This limits the precision with which model parameters can be estimated, and hence also the ability to detect trends in extremes. One way around this is to extend the methodology to retain the $r > 1$ largest observations in each block: arguments similar to those justifying the GEV distribution as a model for maxima can be used to deduce a limiting joint distribution for these largest observations. For details, see Coles (2001, Section 3.5).

6.4.2   Approaches based on threshold exceedances

An alternative approach to the analysis of extremes is to consider all observations that exceed some high threshold. Specifically, let X be any random variable and consider the quantity

$$H_\tau(x) = P\left(X - \tau \le x | X > \tau\right),$$

for some threshold τ. Conditional on X exceeding τ, therefore, $H_\tau(x)$ is the distribution function of the magnitude of the threshold exceedance. To characterise the extremal behaviour of the distribution of X, we consider what happens as τ increases to infinity. In this case it can be shown (Coles, 2001, Section 4.2) that if the distribution of X is in the maximum domain of attraction of a GEV(μ, σ, ξ) for block maxima, $H_\tau(x)$ is increasingly well approximated by

$$H(x) = \begin{cases} 1 - \exp\left(-x/\sigma^*\right), & \xi = 0, \\ 1 - \left(1 + \xi x/\sigma^*\right)^{-1/\xi}, & \text{otherwise,} \end{cases} \tag{6.10}$$

where $\sigma^* = \sigma + \xi(\tau - \mu)$. In this case, the limiting distribution of the threshold exceedance $X - \tau | X > \tau$ is called the *generalised Pareto distribution* (GPD) with parameters σ^* and ξ. The expected value of the distribution is $\sigma^* / (1 - \xi)$. When $\xi = 0$, the GPD reduces to an exponential distribution with mean σ^*.

The existence of the GPD limit suggests that, in practice, a GPD distribution should provide a reasonable fit to exceedances of high but finite thresholds. Given an independent, identically distributed sequence $X_1, \ldots, X_n$ therefore, one could choose a high threshold τ and fit a GPD distribution (for example using maximum likelihood) to the exceedances $\{X_i - \tau : X_i > \tau\}$. The precision of the resulting parameter estimates will depend on the number of observations exceeding τ. Obviously, reducing τ will increase the number of such observations, but at the same time the use of the GPD is justified only if τ is 'large' in some sense. In practice, therefore, one will usually want to choose the smallest value of τ for which the GPD distribution appears to provide a reasonable fit to the data. One way to identify such a threshold (implemented in R via the routine gpd.fitrange in the ismev library) is based on the observation that if the GPD limit has effectively been achieved at some threshold τ then it will also hold for all higher thresholds. Examination of (6.10) shows that the shape parameter of the limiting GPD is the same for all thresholds. Moreover, the dispersion parameter σ^* increases linearly with the threshold in such a way that the quantity $\sigma^* - \xi\tau = \sigma - \xi\mu$ (in the parameterisation of the associated GEV distribution) is threshold-independent. Thus, to identify an appropriate threshold one could fit a GPD at a range of thresholds $\tau_1 < \tau_2 < \cdots < \tau_m$ and plot the estimates of ξ and $\sigma^* - \xi\tau_i$ at each threshold: these estimates will behave unpredictably at thresholds that are too low, but should stabilise at a roughly constant value once a high enough threshold has been achieved. For more details, see Coles (2001, Section 4.3).

Having fitted a GPD to threshold exceedances, estimates of extremal properties can be obtained by noting that for any $x > 0$, $P(X \leq \tau + x)$ can be written as

$$P(X \leq \tau + x \mid X > \tau)P(X > \tau) + P(X \leq \tau + x \mid X \leq \tau)P(X \leq \tau).$$

Now $P(X \leq \tau + x \mid X > \tau) = H_\tau(x)$, which, under the GPD, is modelled according to (6.10). Next, let $\pi = P(X > \tau) = 1 - P(X \leq \tau)$. Finally, notice that, for $x > 0$, $P(X \leq \tau + x \mid X \leq \tau) = 1$. Putting these results together, we have

$$P(X \leq \tau + x) = \pi H_\tau(x) + 1 - \pi.$$

The probability π can be estimated as the proportion of observations exceeding τ and $H_\tau(x)$ can be estimated using the fitted GPD. Thus the model can be used to estimate, for example, the return period of any event with magnitude greater than τ. In general, the number of threshold exceedances available for fitting GPD distributions is substantially greater than the number of, say, annual maxima in a time series, with a corresponding increase in the precision with which quantities of interest can be estimated.

The potential presence of autocorrelation in a time series is more problematic for threshold exceedance methods than for block maxima: extreme events may span several time points so that the assumption of independent threshold exceedances is not tenable. A simple, and widely used, solution to this problem is to 'decluster' the exceedances prior to analysis: some heuristic criterion is used to identify groups of one or more exceedances that are considered to belong to the same event, and the analysis is restricted just to the largest observations from each group (see, for example Davison and Smith, 1990 and Smith, 1989). More sophisticated alternatives, which aim explicitly to model the dependence structure, are also available; see Embrechts, Klüppelberg and Mikosch (1997, Section 8.1) and references therein.

In the presence of nonstationarity, by analogy with the methods discussed above for block maxima one could consider allowing the parameters of the GPD to be covariate-dependent. There are some additional complications when working with threshold exceedances, however. An obvious one is that the probability of exceeding some threshold τ will itself change over time. In principle, this can be dealt with by incorporating covariate information into the threshold exceedance rate as well as into the GPD parameters; this is discussed in the next subsection. Another difficulty is that without severe restrictions on the form of covariate dependence, models fitted at different thresholds are incompatible. This was pointed out by Eastoe and Tawn (2009), who suggested that one could alternatively study threshold exceedances in the residuals from a nonstationary model fitted to the entire data set. The hope is that by working with residuals, most of the nonstationarity will have been removed, thus leading to simpler models for the threshold exceedances. Eastoe and Tawn (2009) also show how to calculate the marginal probabilities of extreme events from threshold models in the presence of covariates. A third possibility, in some sense lying between the 'fixed threshold' and 'residual' approaches, is to model exceedances over a time-varying threshold, chosen to ensure that the exceedance rate is roughly constant. Quantile regression (Section 4.3.5) may be used to set such a threshold. For examples of this approach, see Kyselý, Picek and Beranová (2010) and Friedrichs (2010).

Usually, in statistical models where parameters are covariate-dependent, one can draw conclusions about the conditional distribution of the variable of interest given the covariate(s). When working with threshold exceedances, however, some care is required since by definition any relationships are conditional on the response variable being extreme. To illustrate the point, suppose that a GPD is fitted to daily threshold exceedance data and it is found that the dispersion parameter σ^* increases with some covariate. In this case, since the mean of the GPD is proportional to σ^* (see above), it is tempting to conclude that large values of the covariate are associated with higher extremes on average. However, this conclusion is unwarranted without making further assumptions. It could be, for example, that the covariate values were typically larger on days that were not used in the model fit because the response variable was below the threshold. This can easily be checked, however: divide the observations into two groups corresponding to threshold exceedances and nonexceedances, and compare the distributions of the covariate values in these two groups. If these distributions are similar or if covariate values tend to be higher in the 'exceedance' than the 'nonexceedance' group, then one can legitimately conclude that increases in the covariate are indeed associated with higher extremes. It is nonetheless slightly unsatisfactory that the models are fitted to data that are sampled conditional on values of the response variable. To some extent, the residual modelling approach of Eastoe and Tawn (2009) deals with the problem automatically since (hopefully) the residuals are largely free of covariate effects. As an alternative one might consider fitting models that consider both the extreme and nonextreme observations simultaneously, while retaining the asymptotic arguments that justify extrapolation into the tails of distributions. Such models are discussed in the next subsection.

6.4.3 Modern developments

The use of block maxima and threshold exceedance methods to analyse extremes has been widespread since at least the 1960s.[2] We have seen that exceedance methods are less wasteful of data, and hence have the potential for more precise inference, than methods based on block maxima. However, as noted above, when analysing threshold exceedances in nonstationary series, it is necessary to consider the effect of covariates upon exceedance rates as well as upon the distribution of the exceedances themselves. As a first step in this direction, one could start by observing that the times at which exceedances occur form a point process: if this process is an inhomogeneous Poisson process, then the methods of Section 6.3 can be used to model covariate effects and trends in the intensity function (which, in this context, represents the time-varying threshold exceedance rate). In principle, therefore, one could combine such an analysis with a GPD analysis of the exceedances themselves to build up a complete picture of the extremal behaviour.

The connection with point processes may be developed further. Consider plotting an observed time series $(y_t : t = 1, \ldots, T)$ with the time axis rescaled to the interval $(0,1)$ and with all observations below some threshold τ removed: specifically, the points $\{(t/(T+1), y_t) : y_t > \tau\}$ are plotted. This is illustrated in Figure 6.2: the data here are daily wind speeds at De Bilt (see Section 1.3.1) between 1981 and 2000. The threshold exceedances are shown in black. On the plot, these exceedances appear as a point process in two dimensions: the 'horizontal' dimension represents rescaled time in the range $(0, 1)$ and the 'vertical' dimension represents the magnitude of the exceedance in the range (τ, ∞). In fact, as shown by Pickands (1971), under the same conditions that gave rise to the GPD distribution in the previous subsection, for large values of T and τ this point process is approximately an inhomogeneous Poisson process.[3] At location (t^*, y) on the plot, the intensity function of this process is

$$\left[1 + \xi \left(\frac{y - \mu}{\sigma} \right) \right],$$

where (μ, σ, ξ) are the parameters of a GEV distribution for the largest of the T observations. As noted by Smith (1989), this point process characterisation of extremes provides an alternative means of fitting models using maximum likelihood, where now the likelihood is derived from the inhomogeneous Poisson process. Moreover, it is possible to reparameterise in such a way that the parameters of the point process correspond directly to GEV distributions for (say) annual maxima (Coles, 2001, Section 7.5): this aids interpretation, and also can be regarded as unifying the block maxima and threshold exceedance approaches. To accommodate trends and other forms of nonstationarity, the parameters can be made covariate-dependent in the usual way. In R, the ismev library contains

---

[2] The 1975 UK Flood Studies Report (NERC, 1975) gave an excellent summary of the state of the art as practised in hydrology at the time, and set out the rationale for using extreme value methods in a way that makes for interesting reading today. In addition to the theoretical arguments given above, the authors pointed out that, historically, extreme events were more likely to be recorded than unexceptional values so that relatively long and complete records are available for extremes; also where complete time series were available, discarding all but the extreme observations was potentially helpful at a time when the available computer power was insufficient to contemplate the routine analysis of large data sets.

[3] An inhomogeneous Poisson process in two dimensions is defined in exactly the same way as for the one-dimensional case, but with regions of the plane replacing intervals on the real line.

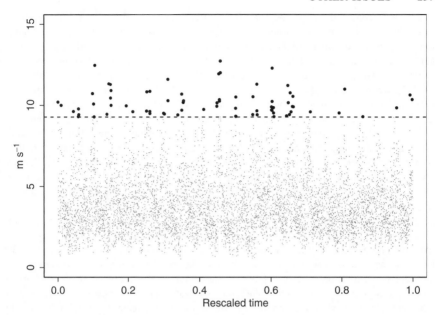

Figure 6.2   Two-dimensional point process of threshold exceedances for daily wind speed data from De Bilt, 1981–2000. The threshold used here is the 99th percentile of the overall wind speed distribution. Observations below the threshold are shown in grey.

routines for fitting and checking point process models for extremes. The methodology has also been extended by Chavez-Demoulin and Davison (2005) to allow nonparametric representation of trends in the model parameters.

Much current research into extremes centres on the development of tractable models for use in multivariate and spatial contexts; see De Haan and Pereira (2006), Butler *et al.* (2007), Cooley, Nychka and Naveau (2007), Eastoe (2009), Naveau *et al.* (2009) and references therein. In a temporal setting, an interesting development is the work of Toulemonde *et al.* (2009), who formulate autoregressive models that allow explicitly for autocorrelation in sequences of Gumbel-distributed maxima.

6.5   Censored data

In environmental applications as elsewhere, it is not always possible to record the exact values of the variable(s) of interest due to limitations in the measuring equipment. In pollution monitoring studies, for example, sensors may not be able to detect pollutant concentrations below some threshold (often referred to as the *limit of detection*, or LOD). At the other end of the scale, measuring equipment may fail at particularly high values of a variable such as wind speed. In such cases, values within the range of operation of the measuring equipment will be recorded exactly but values outside this range are often recorded merely as less than or greater than the appropriate threshold. Such data values are said to be *censored*. Notice that censoring is not the same as missingness: a censored value is known to lie in a particular interval, and thus provides information that can be incorporated into an analysis. Here we give a brief overview of methods that may be used to deal with censored data in the context of a trend analysis. We focus primarily on the case when small values of non-negative variables are censored, since this is the situation most commonly encountered in environmental applications. Most of the ideas can be applied, with obvious small modifications, to the case where large values are censored. For more details, see Helsel (2005) and Manly (2001, Chapter 10).

Where non-negative data are recorded as 'below LOD', one of the simplest approaches to the analysis is to substitute the censored values with some value between zero and the detection limit. Although common in practice, this approach has little to recommend it since it leads to underestimation of variability, as well as to potential biases when estimating trends and covariate effects. The approach is particularly problematic when, as is often done in the context of analyses of pollutant loadings, the data are log-transformed prior to analysis. Such a transformation maps the finite interval $(0, \text{LOD})$ to the range $(-\infty, \log \text{LOD})$: it should be clear that to assume any specific value in this semi-infinite range without good reason is foolhardy at best!

In situations where it is appropriate to assume a particular parametric form for the distribution of the response variable of interest, censored data can be handled easily in a likelihood based setting, at least in the absence of autocorrelation. Recall from Section 3.5 that if all data values are observed exactly, the likelihood function for the parameters of a model is defined, for each value of the parameter vector, as the joint density of the observations given the parameters. When the observations can be considered independent, this joint density is simply a product of contributions from each time point so that the log-likelihood is a sum. For censored data, the idea is easily modified: for each censored observation, the density of the observed value is replaced with the probability of the observation lying in the censoring interval so that the likelihood function is defined as

$$L\left(\boldsymbol{\theta}; \mathbf{y}\right) = \left[\prod_{t \in \mathcal{U}} f_t\left(y_t; \boldsymbol{\theta}\right)\right] \times \left[\prod_{t \in \mathcal{C}} P\left(Y_t \in \mathcal{I}_t; \boldsymbol{\theta}\right)\right],$$

where $\mathcal{U}$ denotes the set of times for which the observations are known exactly (i.e. are uncensored), $f_t(\cdot, \cdot)$ denotes the density of the observations, $\mathcal{C}$ denotes the set of times for which the observations are censored and, for $t \in \mathcal{C}$, $\mathcal{I}_t$ denotes the interval within which Y_t is known to lie. The log-likelihood is the logarithm of this quantity and, for any value of $\boldsymbol{\theta}$, can be calculated easily using modern software routines with built-in routines for calculating quantities such as $P\left(Y_t \in \mathcal{I}_t; \boldsymbol{\theta}\right)$. This log-likelihood can then be maximised using numerical techniques (for example using the routines nlm or optim in R) and, if necessary, the expected information matrix (and hence standard errors for the parameter estimates – see Section 3.1.6) can be estimated via numerical differentiation.

Example 6.4
Suppose that it is desired to fit the linear trend model (3.1) to the logarithms (assumed to be normally distributed) of a sequence $y_1, \dots, y_T$ of non-negative pollutant loading measurements, but that the LOD is $\tau > 0$ so that any values less than τ are censored. In the absence of censoring, the log-likelihood of the data would be

$$\ell\left(\beta_0, \beta_1; \mathbf{y}\right) = \sum_{t=1}^{T} \log \phi_{\mu_t, \sigma}\left(\log y_t\right) = -\frac{1}{\sigma^2} \sum_{t=1}^{T} \left(\log y_t - \mu_t\right)^2 + \text{constant},$$

where $\mu_t = \beta_0 + \beta_1 t$ is the trend function and $\phi_{\mu_t, \sigma}(\cdot)$ denotes the density of a normal distribution with mean μ_t and standard deviation σ. In the presence of censoring, let $\mathcal{C}$ be the set of times corresponding to censored observations. In this case, the log-likelihood for the parameters is

$$\ell(\beta_0, \beta_1; \mathbf{y}) = \sum_{t \notin C} \log \phi_{\mu_t, \sigma}(\log y_t) + \sum_{t \in C} \log \Phi_{\mu_t, \sigma}(\log \tau), \qquad (6.11)$$

where $\Phi_{\mu_t, \sigma}(\cdot)$ denotes the distribution function corresponding to $\phi_{\mu_t, \sigma}(\cdot)$: $\Phi_{\mu_t, \sigma}(z) = \int_{-\infty}^{z} \phi_{\mu_t, \sigma}(u) du$. Both $\phi_{\mu_t, \sigma}(\cdot)$ and $\Phi_{\mu_t, \sigma}(\cdot)$ are easily calculated using modern software packages, for example using routines dnorm and pnorm in R. Thus the log-likelihood can be calculated for any values of β_0, β_1 and σ. Notice that, in contrast to the uncensored case, the values of β_0 and β_1 maximising (6.11) will generally depend on the error variance σ^2. ∎

The problem of censoring occurs widely in the context of clinical trials where the response variable is the lifetime of a patient. Here, censoring occurs when a patient survives beyond the end of a trial so that the exact lifetime is known only to exceed the corresponding duration. In *survival analysis*, it is common to accommodate censoring in a likelihood based framework along the lines sketched above; see Davison (2003, Sections 5.4 and 10.8). The survival library in R (Therneau and Lumley, 2008) provides a comprehensive implementation of many of the techniques commonly used in survival analysis; such methods do not yet seem to have been widely adopted in the environmental sciences, however.

As usual, likelihood based methods have many advantages when their use can be justified. In practice, this means that the available data must be sufficiently complete that diagnostics can be carried out to check the distributional assumptions underlying any likelihood based analysis. If this is not the case then it may be preferable to use more robust methods for which fewer assumptions are required. One such approach, which may be appropriate in situations where there is a heavy degree of censoring, is to convert the time series into a binary sequence in which each observation is recorded merely as being above or below the censoring threshold: trends in this sequence can then be analysed using methods for binary data such as those described in Section 3.5. A more sophisticated approach derives from the observation that if the proportion of censored observations is high then, by definition, the uncensored observations must be extreme in some sense. This suggests that an extreme value analysis of the uncensored observations, using the threshold exceedance or point process methods of Section 6.4, may be profitable.

A more traditional method of dealing with censored data in environmental applications is to use rank based methods, on the basis that the data can still be ordered from smallest to largest without knowing the exact values of the censored observations: these may be treated as ties in the ranking. To test for the existence of trends, therefore, techniques such as the Mann–Kendall test (see Section 2.4), adjusted appropriately for ties, may be used. If interest lies rather in the slope of a linear trend, the Theil–Sen estimator may be adapted to deal with the censoring: see Akritas, Murphy and LaValley (1995).

The discussion above has implicitly assumed that there is only one censoring threshold in a time series. Matters are complicated if there are multiple thresholds – for example if the sensitivity of measuring equipment changes through time. For some of the approaches outlined above, small adjustments are required: for example, in a likelihood based approach, the contribution to the log-likelihood from each censored observation may still be computed using the appropriate censoring threshold. Rank based methods face the difficulty that observations that are censored at different thresholds cannot be ordered relative to each other. However, it may be argued (Helsel, 2005, page 189) that such pairs of observations where the ordering is indeterminate should be considered as ties and that the usual rank based procedures can be used on this basis.

Finally, we have not discussed the analysis of censored data in the presence of auto-correlation. Indeed, this issue does not seem to have received much attention in the literature. As noted in Section 2.4, most rank based test procedures implicitly assume that successive observations are uncorrelated, so that these should be treated with scepticism if autocorrelation is likely to be present. Likelihood based approaches can rapidly become intractable, particularly if a series contains successive observations that are censored. Perhaps the most straightforward methods for dealing simultaneously with autocorrelation and censoring are those based on converting the series into a binary sequence and the extension to use threshold exceedance methods from extreme value theory. In the former case, autocorrelation can be dealt with by including appropriate lagged values of the response variable into a model for the binary time series, and in the latter it can be dealt with by declustering the exceedances prior to analysis, as discussed in Section 6.4.

References

Akritas, M. G., Murphy, S. A. and LaValley, M. P. (1995) The Theil–Sen estimator with doubly censored data and applications to astronomy. *Journal of the American Statistical Association*, **90**, 170–177.

Banerjee, S., Carline, B. P. and Gelfand, A. E. (2004) *Hierarchical Modeling and Analysis for Spatial Data*. Chapman & Hall/CRC, Boca Raton, Florida.

Brockwell, P. J. and Davis, R. A. (2002) *Introduction to Time Series and Forecasting*, 2nd edition. Springer-Verlag, New York.

Butler, A., Heffernan, J. E., Tawn, J. A. and Flather, R. A. (2007) Trend estimation in extremes of synthetic North sea surges. *Journal of the Royal Statistical Society, Series C*, **56**, 395–414.

Chandler, R. E. (2005) On the use of generalized linear models for interpreting climate variability. *Environmetrics*, **16**, 699–715.

Chandler, R. E. and Bate, S. M. (2007) Inference for clustered data using the independence log-likelihood. *Biometrika*, **94**, 167–183.

Chandler, R. E., Isham, V., Bellone, E., Yang, C. and Northrop, P. J. (2007) Space–time modelling of rainfall for continuous simulation. In *Statistics of Spatial-Temporal Systems* (eds. B. Finkenstadt and V. Isham). CRC Press Boca Raton, Florida. pp. 169–207.

Chatfield, C. (2003) *The Analysis of Time Series–An Introduction*, 6th edition. Chapman & Hall/CRC Press, Boca Raton, Florida.

Chavez-Demoulin, V. and Davison, A. C. (2005) Generalized additive modelling of sample extremes. *Journal of the Royal Statistical Society, Series C*, **54**(1), 207–222.

Coles, S. (2001) *An Introduction to the Statistical Modelling of Extreme Values*. Springer Series in Statistics. Springer-Verlag, London.

Coles, S. and Stephenson, A. (2006) `ismev: An Introduction to Statistical Modeling of Extreme Values`. R package version 1.32.

Cooley, D., Nychka, D. and Naveau, P. (2007) Bayesian spatial modelling of extreme precipitation return levels. *Journal of the American Statistical Association*, **102**(479), 824–840.

Cox, D. R. and Isham, V. (1980) *Point Processes*. Chapman & Hall, London.

Cox, D. R., Isham, V. S. and Northrop, P. J. (2002) Floods: some probabilistic and statistical approaches. *Philosophical Transactions of the Royal Society London*, **A360**, 1389–1408.

Cressie, N. and Wikle, C. K. (2011) *Statistics for Spatio-temporal Data*. John Wiley & Sons, Inc., Hoboken, New Jersey.

Daley, D. J. and Vere-Jones, D. (2003) *An Introduction to the Theory of Point Processes*, vol. I, *Elementary Theory and Methods*, 2nd edition. Springer, New York.

Davidson, R. and Mackinnon, J. G. (2004) *Econometric Theory and Methods*. Oxford University Press, New York.

Davison, A. C. (2003) *Statistical Models*. Cambridge University Press, Cambridge.

Davison, A. C. and Ramesh, N. I. (2000) Local likelihood smoothing of sample extremes. *Journal of the Royal Statistical Society, Series B*, **62**, 191–208.

Davison, A. C. and Smith, R. L. (1990) Models for exceedances over high thresholds (with discussion). *Journal of the Royal Statistical Society, Series B*, **62**, 191–208.

De Haan, L. and Pereira, T. T. (2006) Spatial extremes: models for the stationary case. *Annals of Statistics*, **34**, 146–168.

Diggle, P. J. (2003) *Statistical Analysis of Spatial Point Patterns*, 2nd edition. Oxford University Press, Oxford.

Diggle, P. J. and Ribeiro, P. J. (2007) *Model-based Geostatistics*. Springer, New York.

Dobson, A. J. (2001) *An Introduction to Generalized Linear Models*, 2nd edition. Chapman & Hall, London.

Eastoe, E. F. (2009) A hierarchical model for non-stationary multivariate extremes: a case study of surface-level ozone and NO_x data in the UK. *Environmetrics*, **20**, 428–444. DOI: 10.1002/env.938.

Eastoe, E. F. and Tawn, J. A. (2009) Modelling non-stationary extremes with application to surface level ozone. *Journal of the Royal Statistical Society, Series C*, **58**, 25–45.

Embrechts, P., Klüppelberg, C. and Mikosch, T. (1997) *Modelling Extremal Events for Insurance and Finance*. Springer, Berlin.

Everitt, B. S. and Dunn, G. (2001) *Applied Multivariate Data Analysis*, 2nd edition. Arnold, London.

Fahrmeir, L. and Tutz, G. (2001) *Multivariate Statistical Modelling Based on Generalized Linear Models*, 2nd edition. Springer-Verlag, New York.

Fisher, R. A. and Tippett, L. H. C. (1928) On the estimation of the frequency distributions of the largest or smallest member of a sample. *Proceedings of the Cambridge Philosophical Society*, **24**, 180–190.

Friedrichs, P. (2010) Statistical downscaling of extreme precipitation events using extreme value theory. *Extremes*, **13**, 109–132. DOI: 10.1007/s10687-010-0107-5.

Furrer, E. M. and Katz, R. W. (2007) Generalized linear modeling approach to stochastic weather generators. *Climate Research*, **34**, 129–144.

Gelman, A., Carlin, J., Stern, H. and Rubin, D. (1995) *Bayesian Data Analysis*. Chapman & Hall, London.

Greene, W. H. (2003) *Econometric Analysis*, 5th edition. Prentice-Hall, New Jersey.

Harvey, A. C. (1989) *Forecasting, Structural Time Series Models and the Kalman Filter*. Cambridge University Press, Cambridge.

Haslett, J. and Raftery, A. E. (1989) Space–time modelling with long-memory dependence: assessing Ireland's wind power resource. With discussion. *Journal of the Royal Statistical Society, Series C*, **38**, 1–50.

Hawkes, A. G. (1971a) Point spectra of some mutually exciting point processes. *Journal of the Royal Statistical Society, Series B*, **33**, 438–443.

Hawkes, A. G. (1971b) Spectra of some self-exciting and mutually exciting point processes. *Biometrika*, **58**, 83–90.

Helsel, D. R. (2005) *Nondetects and Data Analysis – Statistics for Censored Environmental Data*. John Wiley & Sons, Inc., Hoboken, New Jersey. xv + 250 pp.

Jenkinson. A. F. (1955) The frequency distribution of the annual maximum (or minimum) values of meteorological events. *Quarterly Journal of the Royal Meteorological Society*, **81**, 158–172.

Kalnay, E., Kanamitsu, M., Kistler, R., Collins, W., Deaven, D., Gandin, L., Iredell, M., Saha, S., White, G., Woollen, J., Zhu, Y., Chelliah, M., Ebisuzaki, W., Higgins, W., Janowiak, J., Mo, K. C., Ropelewski, C., Wang, J., Leetmaa, A., Reynolds, R., Jenne, R. and Joseph, D. (1996) The NCEP/NCAR 40-year reanalysis project. *Bulletin of the American Meteorological Society*, **77**, 437–471.

Krzanowski, W. J. (1988) *Principles of Multivariate Analysis*. Oxford University Press, Oxford.

Kyselý, J., Picek, J. and Beranová, R. (2010) Estimating extremes in climate change simulations using the peaks-over-threshold method with a non-stationary threshold. *Global and Planetary Change*, **72**, 55–68.

Manly, B. F. J. (1994) *Multivariate Statistical Methods: A Primer*, 2nd edition. Chapman & Hall/CRC, Boca Raton, Florida.

Manly, B. F. J. (2001) *Statistics for Environmental Science and Management*. Chapman and Hall/CRC, Boca Raton, Florida.

Molenaar, P. C. M. (1985) A dynamic factor model for the analysis of multivariate time series, *Psychometrika*, **50**, 181–202.

Molenaar, P. C. M., de Gooijer, J. G. and Schmitz, B. (1992) Dynamic factor analysis of nonstationary multivariate time series. *Psychometrika*, **57**, 333–349.

Møller, J. and Waagepetersen, R. P. (2003) *Statistical Inference and Simulation for Spatial Point Processes*. Chapman and Hall/CRC, Boca Raton, Florida.

Naveau, P., Guillou, A., Cooley, D. and Diebolt, J. (2009) Modelling pairwise dependence of maxima in space. *Biometrika*, **96**, 1–17.

NERC (1975) Flood Studies Report, Natural Environment Research Council, London.

Ogata, Y. (1988) Statistical models for earthquake occurrences and residual analysis for point processes. *Journal of the American Statistical. Association*, **83**, 9–27.

Ogata, Y. (1989) Statistical model for standard seismicity and detection of anomalies by residual analysis. *Tectonophysics*, **169**, 159–174.

Ogata, Y. and Vere-Jones, D. (1984) Inference for earthquake models: a self-correcting model. *Stochastic Processes and Their Applications*, **17**, 337–347.

Pickands, J. (1971) The two-dimensional Poisson process and extremal processes. *J. Appl. Probab.* **8**, 745–756.

Pinheiro, J. C. and Bates, D. M. (2000) *Mixed-Effects Models in S and S-PLUS*. Springer-Verlag, New York.

Pinheiro, J., Bates, D., DebRoy, S., Sarkar, D. and the R Core Team (2008) *nlme: Linear and Nonlinear Mixed Effects Models*. R package version 3.1-89.

Rice, J. (2006) *Mathematical Statistics and Data Analysis*, 3rd edition. Duxbury Press, Belmont, California.

Ridout, M. S. and Besbeas, P. (2004) An empirical model for underdispersed count data. *Statistical Modelling*, **4**, 77–89. DOI: 10.1191/1471082X04st064oa.

Rust, H. W., Maraun, D. and Osborn, T. J. (2009) Modelling seasonality in extreme precipitation. *European Physics Journal, Special Topics*, **174**, 99–111.

Searle, S. R., Casella, G. and McCulloch, C. E. (1992) *Variance Components*. John Wiley & Sons, Inc., Hoboken, New Jersey.

Shapiro, D. E. and Switzer, P. (1989) Extracting time trends from multiple monitoring sites. Technical Report 132, Department of Statistics, Stanford University, Stanford, California.

Smith, R. L. (1989) Extreme value analysis of environmental time series: an application to trend detection in ground-level ozone. *Statistical Science*, **4**, 367–393.

Snijders, T. and Bosker, R. (1999) *Multilevel Analysis: An Introduction to Basic and Advanced Multilevel Modelling*. SAGE Publications, London.

Sullivan Pepe, M. and Anderson, G. (1994) A cautionary note on inference for marginal regression models with longitudinal data and general correlated response data. *Communications in Statistics*, **B23**, 939–951.

Switzer, P. and Green, A. A. (1984) Min/max autocorrelation factors for multivariate spatial imagery. Technical Report 6, Department of Statistics, Stanford University, Stanford, California.

Therneau, T. and Lumley, T. (2008) `survival`: *Survival Analysis, Including Penalised Likelihood*. R package version 2.34-1.

Toulemonde, G., Guillou, A., Naveau, P., Vrac, M. and Chevallier, F. (2009) Autoregressive models for maxima and their applications to CH_4 and N_2O. *Environmetrics*, **21**(2), 189–207. DOI: 10.1002/env.992.

von Mises, R. (1954) La distribution de la plus grande de n valeurs. In *Selected Papers*, Vol. II. American Mathematical Society, Providence, Rhode Island. pp. 271–294.

Waagepetersen, R. (2007) An estimating function approach to inference for inhomogeneous Neyman–Scott processes. *Biometrics*, **63**, 252–258.

Wackerly, D., Mendenhall, W. and Scheaffer, R. (2007) *Mathematical Statistics with Applications*, 7th edition. Duxbury Press, Belmont, California.

Wikle, C. K., Berliner, L. M. and Cressie, N. (1998) Hierarchical Bayesian space–time models. *Environmental and Ecological Statistics*, **5**(2), 117–154.

Yan, Z., Bate, S., Chandler, R. E., Isham, V. S. and Wheater, H. S. (2002) An analysis of daily maximum windspeed in northwestern Europe using Generalized Linear Models. *Journal of Climate*, **15**(15), 2073–2088.

Zuur, A. F. and Pierce, G. J. (2004) Common trends in northeast Atlantic squid time series. *Journal of Sea Research*, **52**, 57–72.

Zuur, A. F., Tuck, I. D. and Bailey, N. (2003) Dynamic factor analysis to estimate common trends in fisheries time series. *Canadian Journal of Fishing and Aquatic Science*, **60**, 542–552.

Part II

CASE STUDIES

7

Additive models for sulphur dioxide pollution in Europe

Marco Giannitrapani[1]*, Adrian Bowman[1], E. Marian Scott[1] and Ron Smith[2]

[1]*School of Mathematics and Statistics, University of Glasgow, Glasgow G12 8QW, UK*
[2]*CEH Edinburgh, Bush Estate, Penicuik, Midlothian EH26 0QB, UK*

7.1 Introduction

European governments were sufficiently convinced about the potential ecological harm of acid rain that they signed the Convention on Long-Range Transboundary Air Pollution in 1979. The convention facilitated the negotiation of concrete measures to control emissions of air pollutants, initially focused on sulfur dioxide (SO_2), and also the operation of networks to monitor air pollution across Europe under the EMEP programme (Co-operative Programme for Monitoring and Evaluation of the Long-Range Transmission of Air Pollutants in Europe). The monitoring networks run under the EMEP programme were all funded by national governments and, although there were continuing efforts to provide data sets to a uniform standard across the continent, there were inevitable inconsistencies. However, the data do provide a valuable record of change during a period when emissions of SO_2 decreased dramatically. The characteristics of the data sets are:

- Different methods of measurement are used by different countries and hence precision and limits of detection vary.

---

* Now at Novartis Farmaceutica S.A., E-08013 Barcelona, Spain.

Statistical Methods for Trend Detection and Analysis in the Environmental Sciences, First Edition.
Richard E. Chandler and E. Marian Scott.

- At many sites the measurement method has changed at some time during the series, sometimes as a result of changing science priorities.

- Numbers and locations of sites vary as national political policies and funding criteria alter.

- There are potentially long periods of missing data at some sites.

Initial inspection of the monitored concentrations indicated a decreasing trend over time, but certainly not generally a linear trend and not obviously directly linked to emission reductions. As the emitted sulfur in SO_2 can remain in the atmosphere for several days and travel about 5000 km before its deposit on the land surface, a simple direct link with emissions is unlikely. The underlying science suggests that the measured concentration observed at any site will be related to emissions, weather conditions and atmospheric chemistry. Giannitrapani *et al.* (2006) analysed the effects of pollutant emissions on monitored concentrations.

This chapter will look at the effects of meteorology on pollutant concentrations. The statistical framework of additive modelling provides one possible approach to study how covariates can influence the observed concentration profile. This case study was part of a larger study considering data on SO_2 gas concentrations. Here we will consider measurements made at a single site, at Kosetice in the Czech Republic, since the late 1980s. Data were collected as daily average concentrations and preliminary analyses showed that:

(a) The daily values had a skewed distribution and there were some outliers.

(b) There was high variability in the daily observations.

(c) There was evidence of nonlinear trends.

(d) There were seasonal effects.

(e) There was a 'day of the week' effect.

Skewness was addressed by applying a log transformation. In order to accommodate zero values, half of the smallest nonzero value was added to all data points before the log transformation was applied.

As the aim of the study is to investigate trends over periods of 10 or 20 years, the treatment of the variability at short time scales is not critical to the outcome as long as no important and unrecognised bias is introduced. The modified data used for the case

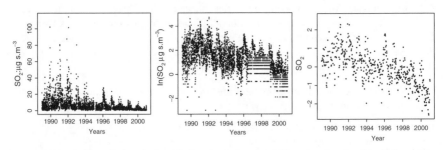

Figure 7.1   Left to right: daily SO_2 measurements at Kosetice in the Czech Republic; daily data on a log scale; weekly data, also on a log scale.

study were the weekly means of the log SO_2 values. However, there is clear evidence of a 'day of the week' effect and so, in order to protect against bias as a result of missing daily observations, the data were first adjusted by fitting a linear model with 'day of the week' as a factor. These adjusted data were then used to compute the weekly means, under the reasonable assumption that the 'day of the week' effect is similar over the years of interest. The weekly data showed less variability and, with fewer data points, the computations were considerably faster. Figure 7.1 shows the SO_2 concentrations on the original daily scale (left panel), on the natural log scale (middle panel) and on the weekly means of logged daily data (right panel). An additional advantage of the weekly data is a reduced effect of the change in resolution that is apparent in the late 1990s.

7.2   Additive models with correlated errors

7.2.1   An introduction to additive models

Nonparametric smoothing techniques aim to provide a means of modelling the relationships between variables without specifying any particular form for the underlying regression function (see Chapter 4). In the simplest setting, interest lies in estimating the regression function $m(\cdot)$ in a relationship between a response variable y and a covariate x, expressed in the model $y = m(x) + \varepsilon$, where ε denotes independent random error. When several covariates are present, Stone (1985), Buja, Hastie and Tibshirani (1989) and Hastie and Tibshirani (1986) proposed to extend the idea of multiple regression into a flexible form known as an additive model. For data $\{(y_i, x_{i1}, \ldots, x_{ip}); i = 1, \ldots, n\}$ the model can be represented as

$$y_i = \mu + \sum_{j=1}^{p} m_j(x_{ij}) + \varepsilon_i, \tag{7.1}$$

where the jth covariate has its own associated component m_j and the regression function is constructed from the combination of these components. All predictor effects are additive and can be examined separately, but there is no form specified for the explanatory variable functions $m_j(x_{ij})$. The remaining terms are a constant, μ, and an error term, ε, which in the simple case of independent errors is assumed to be distributed as $N(0, \sigma^2)$.

The assumption of independent errors implies that there is no systematic association between the errors (or discrepancies between the observations and predicted values) for consecutive observations. The series of SO_2 concentrations in Figure 7.1 clearly show patterns throughout the year: a high concentration is more likely to be followed by another high concentration than a very low concentration. If the explanatory variables give a sufficiently detailed explanation of why the concentrations form these patterns, then the assumption of independent errors is reasonable. However, a wide variety of applications generate data that are subject to correlations of various types, often temporal or spatial, and this is likely to be the case with the present data. A more general assumption that the error vector follows an $N(0, \sigma^2 R)$ distribution gives the flexibility to introduce temporal correlation in the residuals through the correlation matrix R. A simple approach, such as modelling residuals as a first-order autoregression (see Section 3.3.3), is often adequate.

The additive model is an extension of general linear regression, and this is the outcome if the functions $\{m_j(x_{ij})\}$ are the simple linear combinations $\{\beta_j x_{ij}\}$. When it is more helpful to allow the data to determine the shape of the trend, rather than using a predetermined function such as a linear or exponential form, then the explanatory variable functions $m_j(x_{ij})$ can be of arbitrary, but smooth, shape. Smoothing procedures can be used to estimate a trend for a response, y, which is less variable than y and is a function of explanatory variables $x_1, x_2, \ldots, x_p$.

7.2.2   Smoothing techniques

As discussed in Chapter 4, a wide variety of smoothing techniques (for example splines, lowess and local linear regression) is available, but the end result is usually relatively insensitive to the particular technique adopted. In this chapter, the local linear regression smoother (Cleveland, 1979) is used. This is a weighted regression of the response on the values of the explanatory variables in the neighbourhood. For observed data with a single explanatory variable, $\{(y_i, x_i); i = 1, \ldots, n\}$, the estimate $\hat{m}(x)$ is the least squares estimator $\hat{\alpha}$ from

$$\min_{\alpha, \beta} \sum_{i=1}^{n} \{y_i - \alpha - \beta(x_i - x)\}^2 w(x_i - x; h). \tag{7.2}$$

The weight function $w(.; h)$ should be a smooth, symmetric, unimodal function and the smoothing parameter or bandwidth, h, determines the size of the neighbourhood of x considered. The normal density function with mean 0 and standard deviation h,

$$w(x_i - x; h) = \exp\left[-\frac{(x_i - x)^2}{2h^2}\right], \tag{7.3}$$

is commonly used. Local linear regression has a number of attractive properties (Fan, 1993; Fan and Gijbels, 1992). Conceptually, it can be viewed as a relaxation of the usual linear regression model. As h becomes very large the weights attached to each observation by the kernel functions become more similar and the curve estimate approaches the least squares regression line. It is appealing to have this standard model within the nonparametric formulation. Approximate expressions for the bias and variance of the estimate are available from Ruppert and Wand (1994) as

$$\mathrm{E}\left[\hat{m}(x)\right] \approx m(x) + \frac{h^2}{2}\sigma_w^2 \frac{\mathrm{d}^2 m}{\mathrm{d}x^2} \tag{7.4}$$

and

$$\mathrm{Var}\left[\hat{m}(x)\right] \approx \frac{\sigma^2}{nh} \frac{a_w}{f(x)}, \tag{7.5}$$

where $\sigma_w^2 = \int z^2 w(z) \mathrm{d}z$, $a_w = \int w(z)^2 \mathrm{d}z$ and f denotes the local density on the x axis of the observations. Expression (7.4) shows that the larger the smoothing parameters and the degree of curvature ($\mathrm{d}^2 m/\mathrm{d}x^2$), the greater the bias. Conversely, expression (7.5) shows that the larger the smoothing parameters and the density of the observations close to x, the smaller the variance. A good choice of smoothing parameter achieves a suitable balance between bias and variance.

If the observed data are not linear, there is a trade-off between bias and variance as the bandwidth changes, as illustrated in the weekly mean log(SO_2) concentrations in Figure 7.2. With large values of h, the estimate approaches a linear trend. The variance of estimation is low across the range of the data but the estimate does not represent the underlying features of the data set well. The mean log SO_2 is overestimated at the start and end of the period and there are two periods (1991–1992 and 1996–1998) when it is underestimated, so it has high bias. A lower bandwidth gives a slowly varying curve more representative of the general shape in the data, while a very small bandwidth gives an estimate that follows the data very closely, and so has small bias, but has higher variance as the predicted values are based on smaller numbers of data values at any point in time. However, this estimate also shows that there are seasonal effects that should be explicitly modelled and it also raises the question of whether these seasonal effects are the same every year.

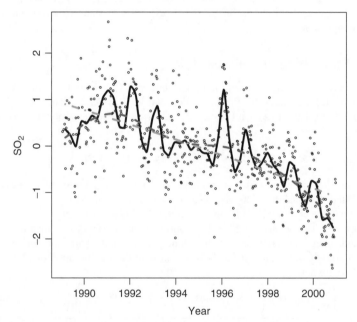

Figure 7.2   Smooth estimates of mean $\ln(SO_2)$ *over time, using the local linear approach. The full, dashed and dashed–dotted lines correspond to estimates constructed with smoothing parameters* $h = 0.1, 1, 10$ *respectively.*

Covariates that are defined on a cyclical scale, referring, for example, to seasonal information, require a different treatment. As the smooth estimate at any point is a weighted average of the values of the explanatory variable in the neighbourhood, a cyclical pattern of weights can be used to reflect the annual seasonal effect. Since linear regression is not an appropriate model on a cyclical scale, a local mean estimator can be formed from

$$\min_{\alpha} \sum_{i=1}^{n} \{y_i - \alpha\}^2 w(x_i - x; h), \tag{7.6}$$

where the weight function is now defined using a von Mises density as described in Section 4.2. For the weekly data considered here, the length of the seasonal cycle in the von Mises density was set to 53 weeks. The bandwidth h again sets the degree of smoothing applied to the data, with large h allowing influence from observations across a wide range while small values of h restrict attention only on those observations that lie close to the point of estimation. Loader (1999) describes an alternative approach based on a suitably scaled sine function. The cyclical pattern can be adjusted in various ways, most flexibly by introducing a bivariate term for year and week simultaneously, therefore allowing interaction between them.

These forms of smoothing are based on weighted least squares criteria and so the resulting estimates can be expressed as linear combinations of the elements of the vector of response data $\mathbf{y}$. It is convenient to define a *smoothing matrix* S whose rows contain the weights that are appropriate for particular estimation points on the scale of the covariate. The vector of estimated values $\hat{\mathbf{m}}$ at these points then has the convenient representation $\hat{\mathbf{m}} = S\mathbf{y}$. This is particularly useful for the construction of standard errors and methods of model comparison.

7.2.3   Smoothing correlated data

When the errors are correlated with joint distribution $N(0, \sigma^2 R)$, it is useful to consider the Choleski decomposition for the correlation matrix, $R = LL'$. Writing $\Gamma = L^{-1}$ as in Section 3.3.3, this leads to a local least squares criterion, which incorporates the correlation structure directly as

$$(\mathbf{y} - \alpha \mathbf{1}_n - X\beta)' \Gamma' W \Gamma (\mathbf{y} - \alpha \mathbf{1}_n - X\beta). \tag{7.7}$$

Here, X denotes a vector with ith element $(x_i - x)$, $\mathbf{1}_n$ is a vector of n ones and the matrix W has diagonal elements $w(x_i - x; h)$ with zeros elsewhere.

In solving the criterion (7.2) for the standard local linear estimator, an explicit solution can be derived. The details are given by Wand and Jones (1995), Bowman and Azzalini (1997) and other authors. In a similar manner, an explicit expression for the case of correlated errors can be derived for the value of $\hat{\alpha}$ in the minimisation of (7.7): $\hat{m}(x) = \sum_{i=1}^{n} a_i(x) y_i$, where

$$a_i(x) = \frac{\left(\sum_{l,j}^{n} \gamma_{li} \gamma_{lj} w_l\right)\left(\sum_{l,g,j}^{n} \gamma_{lg} \gamma_{lj} w_l x_g x_j\right) - \left(\sum_{l,j}^{n} \gamma_{li} \gamma_{lj} w_l x_j\right)\left(\sum_{l,g,j}^{n} \gamma_{lg} \gamma_{lj} w_l x_j\right)}{\left(\sum_{l,g,j}^{n} \gamma_{lg} \gamma_{lj} w_l\right)\left(\sum_{l,g,j}^{n} \gamma_{lg} \gamma_{lj} w_l x_g x_j\right) - \left(\sum_{l,g,j}^{n} \gamma_{lg} \gamma_{lj} w_l x_j\right)^2}. \tag{7.8}$$

Here, γ_{lj} indicates the (l, j)th element of Γ and $w_j = w(x - x_j; h)$. This reduces to the appropriate expression in the case of independent errors. The local mean estimator is a special case of the local linear estimator, and its explicit representation follows as

$$\hat{m}(x) = \sum_{i=1}^{n} \left\{ \frac{\sum_{j,l} \gamma_{ji} \gamma_{jl} w_j}{\sum_{g,j,l} \gamma_{jg} \gamma_{jl} w_j} \right\} y_i.$$

Local linear regression smoothing can easily be extended to the case of two covariates. With independent errors, the estimate can be defined as the least squares solution $\hat{\alpha}$ in the problem

$$\min_{\alpha, \beta_1, \beta_2} \sum_{i=1}^{n} \left\{ \left[y_i - \alpha - \beta_1(x_{i1} - x_1) - \beta_2(x_{i2} - x_2) \right]^2 \right.$$

$$\left. \times w(x_{i1} - x_1; h_1) w(x_{i2} - x_2; h_2) \right\}. \tag{7.9}$$

With correlated data, formulation (7.7) applies with X denoting a matrix whose ith row is $(x_{i1} - x_1, x_{i2} - x_2)$ and β denotes the vector $(\beta_1, \beta_2)'$. An explicit solution for $\hat{\alpha}$ is again available by performing the detailed algebra, along the lines described by Bowman and Azzalini (2003) in the independent case.

In all cases, the vector of fitted values at a set of estimation points of interest continues to have the vector–matrix representation $\hat{\mathbf{m}} = S\mathbf{y}$. However, S now involves the elements of the correlation matrix R as well as the weights w_j and the covariate values x_i. For consistency with Section 4.2.2, the development below focuses on the situation when the estimation points of interest are the observed covariate values. The corresponding S matrix is denoted by $\tilde{S}$: thus $\tilde{S}\mathbf{y}$ is the vector of fitted values corresponding to the available observations. For a summary of differences that may arise when considering alternative sets of evaluation points, see Section 4.2.2.

7.2.4   Fitting additive models

Univariate and bivariate smoothing procedures can be used as building blocks in the estimation of individual components of the additive model (7.1). The standard approach employs the backfitting algorithm, which iteratively applies smoothing with respect to each covariate, using as response the residuals based on the other model components. The details are described by Hastie and Tibshirani (1990) but the heart of the process is expressed in the iterative scheme

$$\hat{\mathbf{m}}_j^{(l)} = (I_n - P_0) S_j^* \left(\mathbf{y} - \hat{\mu} - \sum_{k<j}^{j} \hat{\mathbf{m}}_k^{(l)} - \sum_{k>j}^{p} \hat{\mathbf{m}}_k^{(l-1)} \right) \quad (j = 1, \ldots, p) \tag{7.10}$$

where $\hat{\mathbf{m}}_j^{(l)}$ indicates the estimate of component j at iteration l, I_n represents the identity matrix of order n, P_0 is the $n \times n$ matrix filled with the value $1/n$ and S_j^* denotes the fundamental smoothing matrix associated with this component, as defined in the section above on single components. To ensure unique definitions of the estimators, the intercept term is held at $\hat{\mu} = \bar{y}\mathbf{1}_n$, where $\bar{y}$ is the sample mean, and the estimator is premultiplied by the factor $(I_n - P_0)$ at each step to ensure that $\sum_i \hat{m}_j^{(l)}(x_{ij}) = 0$; this works because for any $n \times 1$ vector $\mathbf{x}$, $P_0\mathbf{x}$ is a vector filled with the mean of the elements of $\mathbf{x}$. The initial values of $\hat{\mathbf{m}}_j$ can be set to $\mathbf{0}$ and the process continues until convergence.

For a variety of reasons, such as the calculation of standard errors and the implementation of model comparison tests, it is very useful to have available the projection matrices that create the final estimates of the additive components, after convergence. Explicit expressions can be derived for the case of two components. More generally, Hastie and Tibshirani (1990) explain how the projection matrices can be recovered retrospectively by repeated use of indicator variables. However, it is straightforward simply to keep track of the relevant matrices as the iterations proceed. If $\tilde{S}_j^{(l)}$ denotes the matrix producing the current estimate of component $\mathbf{m}_j^{(l)}$ as $\hat{\mathbf{m}}_j^{(l)} = \tilde{S}_j^{(l)}\mathbf{y}$, then the backfitting scheme (7.10) can be expressed as

$$\tilde{S}_j^{(l)} = (I_n - P_0) S_j^* \left(I_n - \sum_{k<j} \tilde{S}_k^{(l)} - \sum_{k>j} \tilde{S}_k^{(l-1)} \right). \tag{7.11}$$

The recursions start with $\{\tilde{S}_j^{(0)} : j = 1, \ldots p\}$ defined as $T \times T$ matrices of zeros. At each stage, the updated projection matrix $\tilde{S}_j^{(l)}$ remains independent of the data $\mathbf{y}$. The result after convergence is a set of projection matrices $\{\tilde{S}_j; j = 1, \ldots, p\}$ that create the estimates of the individual components and the fitted values as $\hat{y} = \tilde{S}y$, where $\tilde{S} = \sum_{j=0}^{n} \tilde{S}_j$.

The connection between (7.10) and (7.11) should be clear, except that the final bracketed component of (7.11) contains no term corresponding to subtraction of the sample mean $\hat{\mu} = \bar{y}\mathbf{1}_n$. This is because for all the smoothers considered here, the rows of the smoothing matrices sum to 1 (i.e. each smoothed value is a weighted average of the observations) so that $S_j^*\mathbf{1}_n = \mathbf{1}_n$, whence

$$(I_T - P_0)\,S_j^*\hat{\mu} = \bar{y}\,(I_T - P_0)\,\mathbf{1}_n = \bar{y}\,(\mathbf{1}_n - \mathbf{1}_n) = \mathbf{0}.$$

Thus the subtraction in (7.10) is not strictly necessary; it is included there merely because most treatments of the backfitting algorithm suggest subtracting the sample mean as a first step.

In the discussion above it has been assumed that the correlation matrix R is known. In practice, it will often be required to estimate this. As in the case with linear models, an effective strategy is to fit an independence model and use the residuals from this to identify a suitable structure for the error component. This follows the approach of Niu (1996).

However, there is an intrinsic difficulty here in separating systematic trend from fluctuations due to correlation, when the data consist of a single series. The degree of smoothing used in the estimation of trend sets the balance between these two effects. There are suggested methods of bandwidth choice, such as that of Francisco-Fernández and Opsomer (2005), which aim to incorporate the presence of correlation into the selection process. However, these cannot provide a robust and reliable solution under all circumstances. The approach adopted here is based on judgement of the smoothness of trend expected, expressed through the concept of approximate degrees of freedom, as discussed below. The sensitivity of results to these judgements is also explored. An alternative is to base the choice on the frequency considerations discussed in Chapter 2, using the squared gain functions for the filter weights.

An additional issue arises as a result of the bias that is inevitably present in the estimation of the regression function. This bias will be transferred to the residuals, leading to inflated residual autocorrelations at short time lags. This is most easily seen by considering the residuals $r_i = y_i - \hat{m}(x_i)$, where $\hat{m}(x_i)$ denotes the fitted additive model $\hat{\mu} + \sum_j \hat{m}_j(x_{ij})$, evaluated at the ith observation. Estimates of correlation are derived from products of the form $r_i r_j$, which has mean value with principal terms $E\{\varepsilon_i\varepsilon_j\} + b_i b_j$, where b_i denotes the bias $m(x_i) - E\{\hat{m}(x_i)\}$. Since the bias is smooth in the covariates, then b_i and b_j will have the same sign where observations i and j are close. Where the covariates are themselves time related, this therefore inflates the estimates of correlation. However, the effect is therefore a conservative one, leading to a more cautious interpretation of the regression components in the fitted model.

With the SO_2 data, an additive model was fitted using the smoothing parameters $h_1 = 0.7$ for years and $h_2 = 0.3$ for seasonality. These values were chosen to reflect the substantial degrees of smoothness expected in the trend and seasonal curves. Hastie and Tibshirani (1990) use an analogy with linear models to describe the concept of *approximate degrees of freedom* in the context of smoothing techniques. For the jth smooth component this can be easily calculated as $\mathrm{tr}(\tilde{S}_j)$, where $\tilde{S}_j$ is the corresponding projection matrix as described above (see also Sections 4.1.4 and 4.2). The chosen values for h_1 and h_2 correspond to approximately 8 and 4 degrees of freedom. This allows substantial flexibility in the estimate of trend across the years, as this is of particular interest, and modest flexibility for the seasonal effect.

The residuals from this independent additive model were used to construct an autocorrelation function and this showed that a simple AR(1) process provides a good description for the error terms. As in most applications, principal interest lies in the mean structure and so it is sufficient to adopt a simple model for the errors, which allows the principal effects of correlation to be incorporated without undue complexity. Where there is strong evidence that more complex time series models are required, these can be adopted without difficulty, as long as the appropriate matrix inversions are feasible. For the SO_2 data monitored at Kosetice, the estimated correlation parameter, using the residuals from the independence model, is $\hat{\rho} = 0.366$, indicating a modest degree of correlation. An estimate of the correlation matrix R is then available by setting the (i, j)th entry to $\hat{\rho}^{|i-j|}$. Niu (1996) proposes iteration between estimation of the mean function using the current estimation of the correlation structure and estimation of the error structure for the current residuals. In the work described in this chapter a simple one-step approach was adopted, mirroring common practice with generalised least squares for linear models. With an AR(1) model, as described in Section 3.3.3, an explicit form is available for the matrix Γ in the modified local least squares criterion (7.7); this can be used to reduce the computational complexity.

Figure 7.3 illustrates the estimated components of the additive model which expresses the trend and seasonal effects on SO_2 simultaneously. The steady decline over the years and the marked seasonal pattern are both clear. The estimates shown were obtained using the independent-data criterion expressed in (7.2), rather than the correlation-adjusted method described in (7.7) and (7.8). Estimates obtained using the correlation-adjusted method were very similar to those shown. This corresponds to the ability of ordinary least squares procedures in linear models with correlated data to produce valid, if not fully efficient, estimators (see Section 3.1.1).

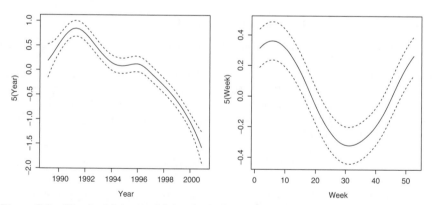

Figure 7.3   Fitted additive model for the $\ln(SO_2)$ *data. The left- and right-hand panels show the estimated components for the trend over years and and the seasonal effect of weeks respectively. The dashed lines identify a distance of two standard errors from the estimate, based on an assumption of AR(1) errors.*

However, it is also important to indicate the variability associated with these estimates through their standard errors and here it becomes important to incorporate the underlying correlation appropriately. An assessment of variability can easily be derived from the representation of the component estimates as $\hat{\mathbf{m}}_j = \tilde{S}_j \mathbf{y}$ through the expression $\text{Var}(\hat{\mathbf{m}}_j) = \tilde{S}_j R \tilde{S}_j' \sigma^2$. An estimate of σ^2 can be constructed from the dependence-adjusted residual sum of squares, which can be written as $\mathbf{y}'\mathbf{V}\mathbf{y}$, where $\mathbf{V} = (\mathbf{I} - \tilde{\mathbf{S}})'\mathbf{R}^{-1}(\mathbf{I} - \tilde{\mathbf{S}})$. This has mean value $E\{\mathbf{y}'\mathbf{V}\mathbf{y}\} = E\{\boldsymbol{\varepsilon}'\mathbf{V}\boldsymbol{\varepsilon}\} + \mathbf{m}'(\mathbf{I} - \tilde{\mathbf{S}})'\mathbf{R}^{-1}(\mathbf{I} - \tilde{\mathbf{S}})\mathbf{m} = \mathrm{tr}\{\mathbf{R}\mathbf{V}\}\sigma^2 + \mathbf{b}'\mathbf{R}^{-1}\mathbf{b}$, where $\mathbf{b}$ denotes the vector of bias terms at the observed data points. This therefore leads to an estimate of σ^2 as

$$\hat{\sigma}^2 = \frac{\mathbf{y}'\mathbf{V}\mathbf{y}}{\mathrm{tr}\{\mathbf{R}\mathbf{V}\}}.$$

Since this has mean value $\sigma^2 + \mathbf{b}'\mathbf{R}^{-1}\mathbf{b}/\mathrm{tr}\{\mathbf{R}\mathbf{V}\}$, the effect of bias again has a conservative effect by inflating the estimate of σ^2.

Following the terminology of Hastie and Tibshirani (1990), the normalising constant $\mathrm{tr}\{\mathbf{R}\mathbf{V}\}$ is referred to as the *approximate degrees of freedom for error*. The corresponding approximate degrees of freedom for the model is $\nu = n - \mathrm{tr}\{\mathbf{R}\mathbf{V}\}$. When $\mathbf{R}$ is replaced by the identity matrix this reduces to one of the standard definitions of approximate degrees of freedom described by Hastie and Tibshirani (1990) in the independent errors case.

Estimated standard errors for the model component $\hat{\mathbf{m}}_j = \tilde{\mathbf{S}}_j\mathbf{y}$ are then available as the square root of the diagonal entries of the matrix $\tilde{\mathbf{S}}_j\mathbf{R}\tilde{\mathbf{S}}_j'\hat{\sigma}^2$. The dashed lines in Figure 7.3 illustrate a band corresponding to ± 2 estimated standard errors around the estimate. The high precision of estimation is apparent. However, it is of interest to consider the effect of failing to incorporate the correlation structure in the estimation process. This can easily be done by replacing the matrix $\mathbf{R}$ in the expression above by the identity matrix. This leads to a reduction in standard errors of around 25 % over those that incorporate the correlation structure. This marked effect indicates the importance of adopting a model that properly incorporates correlated errors and therefore gives a more realistic assessment of precision in estimation.

7.2.5   Comparing nonparametric models

An essential part of using additive models is the ability to compare different candidate models and hence to assess the evidence for the retention or omission of particular components. Some terms may also be well described by parametric rather than nonparametric forms. Hastie and Tibshirani (1990) advocated the use of an F statistic, analogous to the standard procedure for linear models. Specifically, evidence against the null hypothesis that the data were generated by a model M_0, as opposed to the larger model M_1, is contained in the test statistic

$$F = \frac{(\mathrm{RSS}_0 - \mathrm{RSS}_1)/(\nu_0 - \nu_1)}{\mathrm{RSS}_1/\nu_1}, \tag{7.12}$$

constructed from the residual sum of squares (RSS) and approximate degrees of freedom for error (ν) of the two models. Hastie and Tibshirani (1990) proposed that this statistic should be compared to an F distribution with $\nu_0 - \nu_1$ and ν_1 degrees of freedom, again by analogy with linear models. The presence of bias in the residual sum of squares and the absence of the required properties in the underlying projection matrices mean that the test statistic will not follow an F distribution under the null hypothesis, M_0. However, this distribution does provide a helpful benchmark.

When dealing with correlated data, the dependence-adjusted residual sum of squares can be expressed as $\mathrm{RSS} = \mathbf{y}'\mathbf{V}\mathbf{y}$, where $\mathbf{V}$ denotes the matrix $(\mathbf{I} - \tilde{\mathbf{S}})'\mathbf{R}^{-1}(\mathbf{I} - \tilde{\mathbf{S}})$. This can be conveniently extended here to express the residual sum of squares for models M_0 and M_1 as $\mathrm{RSS}_k = \mathbf{y}'\mathbf{V}_k\mathbf{y}$, for the models indexed by $k = 0, 1$. It follows that

$E\{RSS_0 - RSS_1\} = E\{\varepsilon'(V_0 - V_1)\varepsilon\} + \mathbf{b}_0' R^{-1}\mathbf{b}_0 - \mathbf{b}_1' R^{-1}\mathbf{b}_1$, where $\mathbf{b}_0$ and $\mathbf{b}_1$ denote the bias vectors under the two models. It is very difficult to derive explicit expressions for the two bias terms, although Opsomer (2000) derived recursive expressions. The guiding principle that bias is controlled by curvature and the parameters h^2 and σ^2 remains true, but the situation is complicated by the interactions between the covariates as the backfitting algorithm progresses. However, in order to ensure that the bias terms cancel one another out as far as possible, it is clear that the same smoothing parameters should be used when fitting both models.

As discussed in Section 4.1.7, an alternative approach to the distributional calculations arises by expressing the F statistic in terms of quadratic forms as

$$F = \mathbf{y}'A\mathbf{y}/\mathbf{y}'B\mathbf{y},$$

where A is the matrix $(V_0 - V_1)/(v_1 - v_0)$ and B is the matrix $V_1/(n - v_1)$. The significance of the statistic F can then be expressed through its p-value, as

$$p = P\left\{\frac{\mathbf{y}'A\mathbf{y}}{\mathbf{y}'B\mathbf{y}} > F_{obs}\right\} = P\{\mathbf{y}'C\mathbf{y} > 0\},$$

where F_{obs} denotes the value calculated from the observed data and C is the matrix given by $(A - F_{obs}B)$. In Section 4.1.7 it was shown that in the absence of bias the quantity $\mathbf{y}'C\mathbf{y}$ reduces to $\varepsilon'C\varepsilon$. Following the discussion on bias cancellation above, the same replacement can be made as an approximation in the present setting. Johnson and Kotz (1972) summarise general results about the distribution of a quadratic form in normal variables. Specifically, the jth cumulant of the distribution of $\varepsilon'C\varepsilon$ is available as $2^{j-1}(j - 1)!\mathrm{tr}\{(VC)^j\}$. This allows the moments to be calculated and matched to a convenient distributional approximation, such as a shifted and scaled chi-squared distribution, $a + b\chi_c^2$. Bowman and Azzalini (1997) give all the details of this process.

7.3   Models for the SO_2 data

The SO_2 data from Kosetice which are analysed in this section are available on the web at www.emep.int. Meteorological data that were also available at the same station were hourly values of precipitation, temperature, humidity, wind speed and wind direction. These variables have been aggregated to the weekly scale, as described in Section 7.1. The wind information was combined into a mean direction, weighted by speed, to reflect the dominant direction of air flow. As a result of the skewness of SO_2 and rainfall, these variables were analysed on a log scale. As in the case of SO_2, a small constant was added to rainfall before the application of the log transformation, in order to accommodate zero values.

In broad terms, the reduction of SO_2 concentrations across many areas of Europe since 1980 is primarily linked to a decline in emissions, but changes in weather patterns also affect concentrations and are a confounding factor in the analysis. For example, an increasing frequency of westerly winds bringing relatively clean air from the Atlantic across Europe would reduce the average observed SO_2 concentration at many sites. Therefore, to evaluate the effectiveness of government policies on reductions of emissions or to estimate the possible future effects of climate change, it is important to identify these meteorological effects. Two specific questions of interest are identified for this case study:

(a) Does meteorology explain some of the variability of SO_2?

(b) Has the seasonal pattern of SO_2 changed over the years?

In order to answer question (a), the following two nested additive models were fitted:

$$\text{Model 1}: \ \log(SO_2) = \mu + m_y(\text{year}) + m_w(\text{week}) + \varepsilon,$$

$$\text{Model 2}: \ \log(SO_2) = \mu + m_y(\text{year}) + m_w(\text{week}) + m_r(\text{rain})$$
$$+ m_t(\text{temp}) + m_h(\text{hum}) + m_a(\text{air}) + \varepsilon.$$

Both models present the trend ('year', including weeks on a decimal scale) and the seasonality ('week', as week of the year), but model 2 also includes the meteorological variables. Therefore comparisons of these two models will allow a significant effect of meteorological variables to be identified.

The fit of each component of model 2 is shown in Figure 7.4, using the smoothing parameters equivalent to approximately 8 degrees of freedom for years and approximately 4 degrees of freedom for each of the other terms. The full lines show the estimates of the components of the additive model. The dashed lines identify a distance of two standard errors from the estimates.

Figure 7.5 shows the fitted values of the models as a function of time. Model 2 captures many more of the high concentration events than model 1 is able to fit. This is an indication that the peaks are mainly due to the meteorological effects. However, it would be useful to have a formal comparison of the models, to weigh the evidence more effectively. This was done using both approximate F and quadratic form tests, producing a p value lower than 0.001 in both cases. This gives clear evidence that, at Kosetice, meteorological variables give significant extra information in explaining the variability of SO_2.

Some of the panels in Figure 7.4 also raise the issue of whether the covariate effects may be linear rather than nonparametric in nature. For example, the trend with time may be linear, allowing a much simpler model structure to describe the changes. The adoption of a linear effect for one or more of the terms in model 2 creates a semiparametric model. The estimation process for models of this type has been thoroughly discussed by Hastie and Tibshirani (1990), Green and Silverman (1994) and other authors. In the present context, semiparametric models can be fitted simply by replacing the smoothing matrices for the appropriate terms by the usual matrix expressions for the fitted values in a linear regression. The model comparison methods follow immediately, using the projection matrix $\tilde{S}$ produced by the backfitting algorithm. When model 2 is compared with the restricted models that constrain each term in turn to be linear, all the nonparametric terms are found to be highly significant, with the exception of rain, where the p-value for a nonparametric effect is 0.379. Indeed, if rain is omitted altogether, the p-value is 0.134, suggesting that, given the information provided by the other meteorological variables, there is little to be gained by including rain.

In order to assess whether there has been a significant change in seasonality across years, model 3 includes the same meteorological variables as model 2, but with an additional bivariate term to describe the interaction between trend and the seasonality:

$$\text{Model 3}: \ \log(SO_2) = \mu + m_{yw}(\text{years, weeks}) + m_r(\text{rain})$$
$$+ m_t(\text{temp}) + m_h(\text{hum}) + m_a(\text{air}) + \varepsilon.$$

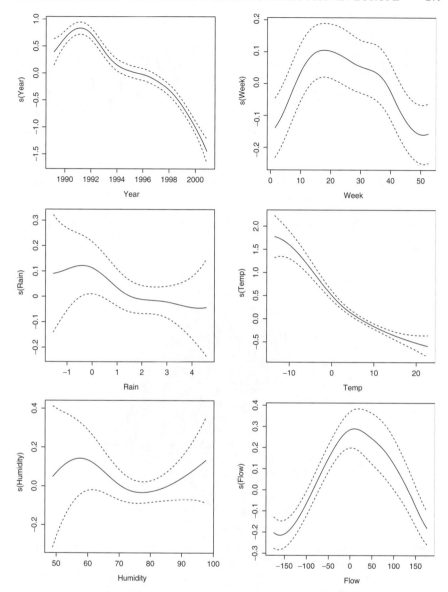

Figure 7.4   The estimated components of an additive model for the log SO_2 data at Koset-ice, adding meteorological information to the earlier model based on year and week.

In model 2, the seasonality is assumed constant across years, while with model 3 the shape of the seasonal term changes with year, as shown in Figure 7.6. A nonparametric term for rain is retained here. However, the results described below are essentially unchanged if rain is omitted.

A comparison of models 2 and 3 in a hypothesis test produces the p values 0.003 and < 0.001, using the F distribution and quadratic form calculations respectively. There is

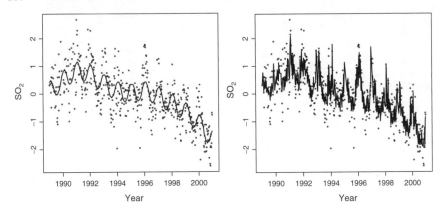

Figure 7.5   Fitted values of mean log(SO$_2$) *for model 1 (left) and model 2 (right).*

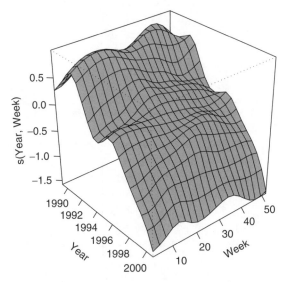

Figure 7.6   The estimated component from model 3, which expresses the interaction between year and week, through a bivariate surface.

therefore a strong indication that, at Kosetice, the seasonality of log(SO$_2$) changes across the years. This is consistent with the shape of the estimate of the bivariate component shown in Figure 7.6, where there seem to be differences in the patterns of decline in summer and winter.

It can therefore be concluded that at Kosetice the meteorology has a significant effect in explaining the variability of the SO$_2$ and that the nature of the seasonal effect changes from 1989 to 2000.

7.4 Conclusions

Techniques for fitting and testing additive models with correlated data have been described. Although the backfitting algorithm in its usual form, designed for independent data, can be applied to good effect, it is important to incorporate a suitable correlation structure into subsequent analysis. This places the results on a firmer methodological footing and gives confidence in the conclusions reached. In particular, it provides a more realistic assessment of the variability of the estimates and a more accurate method of model comparison through the distributional calculations involving the F statistic.

The selection of a suitable smoothing parameter is an important issue in nonparametric regression, although Bowman and Azzalini (1997) point out that the sensitivity of this choice is often less in issues of inference than in estimation. When the response data are correlated, the choice of smoothing parameter has the effect of setting the balance between the patterns in the data that are attributed to trend in the mean and those attributed to correlated errors. In the application discussed in this chapter, the smoothing parameters have been selected through judgement based on background knowledge of the nature of the trends and correlations to be expected in this setting. The concept of approximate degrees of freedom also provides a very useful and interpretable scale on which choices can be expressed.

As discussed earlier, it would be wise to check whether the qualitative results are sensitive to the degree of smoothing used. The test to compare models 1 and 2 was applied using 3, 4 and 6 degrees of freedom, which represents quite a wide range. In each case, the degrees of freedom for the trend over the years was doubled, to match the pattern used earlier in the paper. The p-values were all less than 0.001, which confirms that the evidence for the explanatory power of the meteorological information is very strong at all levels of smoothing. The test for the effect of rainfall was also repeated at these different degrees of smoothing, producing p-values of 0.120, 0.135 and 0.052 respectively. Again, these consistently indicate a lack of convincing evidence for the effect of rainfall. Finally, models 2 and 3 were also compared at the same levels of smoothing, producing p-values of 0.194, 0.003 and < 0.001 respectively. On this occasion, the evidence for interaction disappears when a very large degree of smoothing is applied. However, a reasonable interpretation is simply that the estimate is constrained to lie relatively close to a linear model, which lacks sufficient flexibility to express the interaction structure.

The analysis of trends in pollutants is an important environmental problem, both as a driver of policy and as a test of the success or failure of environmental initiatives. Weather is always important, as changes in the climate (both long term and from year to year) can change (a) the amount of pollutant emitted, (b) the transport of pollutant across the landscape and (c) the chemistry and other factors in the atmosphere that determine the fate of the pollutant. For example, recent warm summers have increased the energy requirements to run air conditioning, hot stationary air masses increase ozone and particulate production in the lower atmosphere from chemical reactions, and with less wind dispersal concentrations increase in the source regions and decline in remote areas. When SO_2 concentrations were very high over Europe in the 1980s, periods of sustained westerly winds from the Atlantic would significantly reduce concentrations in Western Europe. The methods described in this chapter allow a much more flexible approach to analysis of pollutant trends than has often been used in the past. There is a flexible fitting of seasonal components, meteorology can be explicitly modelled when data are available, adjustments for correlated data can be included directly in the modelling and the data can determine the shape of the trend.

Acknowledgement

Marco Giannitrapani acknowledges support from the Chancellor's fund of the University of Glasgow and the Centre for Ecology and Hydrology, Edinburgh.

References

Bowman, A. W. and Azzalini, A. (1997) *Applied Smoothing Techniques for Data Analysis – The Kernel Approach with S-Plus Illustrations*, vol. 18, of Oxford Statistical Science Series. Oxford University Press, Oxford.

Bowman, A. W. and Azzalini, A. (2003) Computational aspects of nonparametric smoothing with illustrations from the sm library. *Computational Statistics and Data Analysis*, 545–560.

Buja, A., Hastie, T. and Tibshirani, R. (1989) Linear smoothers and additive models. *Annals of Statistics*, **17**(2), 453–510.

Cleveland, W. S. (1979) Robust locally weighted regression and smoothing scatterplots. *Journal of the American Statistical Association*, **74**(368), 829–836.

Fan, J. (1993) Local linear regression smoothers and their minimax efficiencies. *Annals of Statistics*, **21**(1), 196–216.

Fan, J. and Gijbels, I. (1992) Variable bandwidth and local linear regression smoothers. *Annals of Statistics*, **20**(4), 2008–2036.

Francisco-Fernández, M. and Opsomer, J. D. (2005) Smoothing parameter selection methods for nonparametric regression with spatially correlated errors. *Canadian Journal of Statistics*, **33**, 279–295.

Giannitrapani, M., Bowman, A., Scott, M. and Smith, R. (2006) Sulphur dioxide in Europe: statistical relationships between emissions and measured concentrations. *Atmospheric Environment*, **40**, 2524–2532.

Green, P. G. and Silverman, B. W. (1994) *Nonparametric Regression and Generalized Linear Models: A Roughness Penalty Approach*. Chapman & Hall, London.

Hastie, T. and Tibshirani, R. (1986) Generalized additive models (with discussion). *Statistical Science*, **1**(398), 297–318.

Hastie, T. and Tibshirani, R. (1990) *Generalized Additive Models*. Chapman & Hall, London.

Johnson, N. L. and Kotz, S. (1972) *Distributions in Statistics: Continuous Univariate Distributions*, Vol. II. John Wiley & Sons, Inc., New York.

Loader, C. (1999) *Local Regression and Likelihood*. Springer.

Niu, X. F. (1996) Nonlinear additive models for environmental time series, with applications to ground-level ozone data analysis. *Journal of the American Statistical Association*, **91**(435), 1310–1321.

Opsomer, J. D. (2000) Asymptotic properties of backfitting estimators. *Journal of Multiple Analyses*, **73**(2), 166–179.

Ruppert, D. and Wand, M. P. 1994 Multivariate locally weighted least squares regression. *Annals of Statistics*, **22**, 1346–1370.

Stone, C. J. (1985) Additive regression and other nonparametric models. *Annals of Statistics*, **13**(2), 689–705.

Wand, M. P. and Jones, M. C. (1995) *Kernel Smoothing*. Chapman & Hall, London.

8

Rainfall trends in southwest Western Australia

Richard E. Chandler[1], Bryson C. Bates[2] and Stephen P. Charles[3]

[1]*Department of Statistical Science, UCL, Gower Street, London WC1E 6BT, UK*

[2]*Climate Adaptation Flagship, CSIRO Marine and Atmospheric Research, Wembley, Western Australia, Australia*

[3]*Climate Adaptation Flagship, CSIRO Land and Water, Wembley, Western Australia, Australia*

8.1 Motivation

The southwest of Western Australia (SWA) has been in a state of hydrological drought since the mid 1970s (Bates *et al.*, 2008; Power, Sadler and Nicholls, 2005). This has major implications for the management of water resources in the region, as exemplified by Figure 8.1(a), which shows total annual volumes of water entering 11 reservoirs to the east of the city of Perth. These reservoirs provide 25–45 % of the water for the region's Integrated Water Supply Scheme (IWSS), which caters for 84 % of the 1.9 million population of the state of Western Australia. The series shows a clear downward trend since the 1960s. As a result, despite the extensive use of demand management strategies, the region's water supply infrastructure has required substantial investment in recent decades (Bates and Hughes, 2009). For example, the state Water Corporation spent almost $A1 billion in source development between 1996 and 2006; a new seawater desalination

Statistical Methods for Trend Detection and Analysis in the Environmental Sciences, First Edition.
Richard E. Chandler and E. Marian Scott.

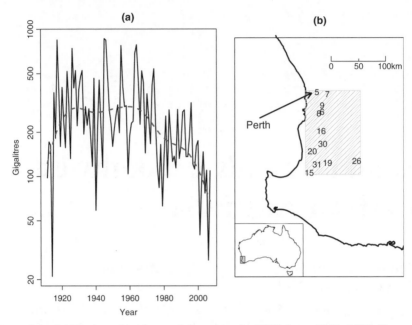

Figure 8.1   (a) Total combined annual flows into the IWSS dams, 1911–2007. The vertical axis is on a log scale. The dashed line is a trend estimate obtained using local linear regression with a Gaussian kernel and a bandwidth of h = 8 years. (b) Map of study area, with locations of stations supplying daily rainfall data used in this study (stations are numbered following Charles et al., 2007). Data from station 8 were subsequently discarded (see Section 8.3). Hatching shows the area used for basis function representation of regional variation (see Section 8.4); inset shows location of main map within Australia.

plant started operation in November 2006 and a further desalination plant is due for completion in 2011.

Against this background, it is of interest both to characterise the trend in reservoir inflows and to understand the reasons for it. Until recently, this trend was often seen to consist of periods of roughly constant levels separated by abrupt breaks, notably in 1975 (for example, Hennessy *et al.*, 2007, Section 11.6). However, using the test for discontinuities in an otherwise smooth regression function developed by Bowman, Pope and Ismail (2006), Bates *et al.* (2010) show that the series is in fact better described by a smooth trend such as that shown in Figure 8.1(a). This has major implications for water resource management in the region: under the 'abrupt breaks' scenario, inflows would be expected to fluctuate around their current level in the short to medium term, whereas under the 'smooth trend' scenario a continued decline is likely.

The most obvious potential cause of reduced reservoir inflows is a corresponding decline in the region's rainfall. However, this is not the only possible explanation: changes in forest management and land use (including open-cast bauxite mining) may also be partly responsible since these can affect both runoff processes and groundwater recharge. Research suggests that increases in stream yields follow the temporary removal of vegetation during mining and decreases follow mine restoration (e.g. Croton and Reed, 2007), although the net contribution of such effects to the decline in inflows is a matter of some controversy. Unfortunately, to resolve this would require extensive modelling of surface and subsurface hydrology, and the hydrogeological data needed to support such an exercise are not available. It is nonetheless possible to examine climate data from the region, to obtain at least a partial understanding of climatic factors that may be driving the change in inflows. Bates *et al.* (2010) reported the results of such an exercise, in which changes in precipitation were related to changes in regional atmospheric circulation. The aim of that paper was to summarise a suite of analyses linking the inflow series indirectly to large-scale climatic changes; as such, the details of each individual analysis were necessarily brief. The present contribution gives a more complete description of precipitation trends in the region.

To place the ensuing analysis in context it is helpful to understand the topography, climatology and human geography of the region; relevant aspects are summarised in the next section. Section 8.3 gives details of the data used in the study, along with quality checks carried out to ensure that the modelled trends were free of artefacts. Sections 8.4 and 8.5 describe the methodology and results respectively; and conclusions are given in Section 8.6.

8.2   The study region

The study region is an area of roughly $100 \times 150 \, \text{km}^2$, lying to the south and east of Perth in SWA, as shown in Figure 8.1(b). The Darling Range, an escarpment 320 km in length and with an average height of around 300 m, runs parallel to the coast at a distance of 25–50 km. The IWSS catchments are all located in the Darling Range, with the reservoirs at or near its west-facing scarp slope. The human population is concentrated on the coastal plain in the west; to the east lies Western Australia's 'wheat belt', with large but scattered farms and few centres of population.

SWA experiences a Mediterranean climate, with hot dry summers (December to February) due to the influence of the subtropical belt of high pressure. In winter, this belt moves northwards and airflow is predominantly westerly; this brings in moisture from the Indian Ocean. Thus, most of the annual precipitation (typically in excess of 70 % – see Bates *et al.*, 2010) falls between May and October. Previous studies (for example, Bates *et al.*, 2008, IOCI, 2002) suggest that substantial precipitation declines have occurred during early winter (May to July): for example, over the period 1975–2004 the mean regionally averaged rainfall for these months was reduced by 14 % compared with the previous 70 years.

The eastward movement of moist air from the Indian Ocean results in a west–east rainfall gradient over the region, which is exaggerated by the rain shadow effect of the Darling Range. Rainfall also decreases from south to north, reflecting the transition between subtropical weather systems to the north and low-pressure, mid-latitude systems to the south; see, for example, Bates *et al.* (2008, Figure 3).

8.3   Data used in the study

Previous studies of precipitation in SWA (for example, Charles *et al.*, 2007) have used daily rainfall data from a network of 31 measuring stations, 12 of which fall within the study area indicated in Figure 8.1(b). Data from these 12 stations, obtained from the Bureau of Meteorology and from the Department of Water, Western Australia, were initially considered for this analysis. The station locations are indicated by numbers in Figure 8.1(b); the numbering is the same as in the reference above. There are few stations in the east of the study area, due to its being thinly populated: this makes it difficult to characterise precipitation trends reliably in this eastern region.

Data for the winter half-year only (May–October) have been considered, because these months contribute most of the region's rainfall and have been associated with precipitation declines as noted above. This gives a total of 184 days' data each year. All available winter data for the period 1958–2007 have been used, although not all stations were operational throughout this period and most had some missing observations (see Table 8.1).

Table 8.1 Details of records used for precipitation trend analysis. All records start in 1958 and end in 2007 except stations 30 (start 1972, end 1999) and 31 (start 1974, end 2002). 'Long.', 'Lat.' and 'Alt.' denote longitude, latitude and altitude respectively. Italicised values are 'effective altitudes' used in the modelling (see Section 8.4) and '% missing' denotes the percentage of missing days during the years of observation.

Station	Long. (°E)	Lat. (°S)	Alt. (100 m)		% missing
5	115.97	31.93	0.20	*0.0*	0
6	116.07	32.33	2.40	*3.0*	10
7	116.17	31.97	3.00	*3.0*	2
8	116.00	32.37	0.32	*1.5*	37
9	116.07	32.20	2.50	*3.0*	6
15	115.82	33.57	0.63	*1.0*	1
16	116.05	32.72	2.67	*2.5*	1
19	116.17	33.37	1.90	*2.0*	17
20	115.88	33.13	1.16	*1.0*	5
26	116.73	33.33	2.62	*2.5*	2
30	116.08	32.98	2.80	*3.5*	3
31	115.97	33.40	1.40	*2.0*	4

Location details (latitude and longitude) were available for each station. Station altitudes were also available for 10 of the 12 stations; approximate altitudes were read from maps for stations 30 and 31.

The data from most stations are recorded to a nominal resolution of 0.1 mm. However, inspection reveals a strong preference for even-numbered decimal digits except at stations 30 and 31 where, for most years in the record, the observations are given to two decimal places. Apart from these two stations, the tendency to record even digits does not seem to be systematic (it does not occur predominantly in some years or at some stations, for example). In other studies (for example Yang et al., 2006) it has been found that differences in recording resolution can lead to spurious structures that are detectable by the methods used below. To avoid this, therefore, all data were rounded to a resolution of 0.2 mm prior to analysis.

An initial analysis revealed further problems associated with changes in recording practice at weekends. These are exemplified by Figure 8.2, which compares the distributions of rainfall recorded at station 20 from 2001 onwards with those recorded prior to this date. The distributions of cube-rooted rainfall amounts are shown for each day of

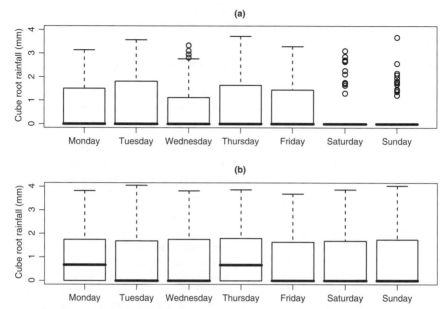

Figure 8.2   Distributions of cube-rooted daily rainfall recorded at station 20 on each day of the week (a) from 2001–2007, (b) prior to 2001.

the week: the cube-root transformation compresses the upper tail of the distribution, so that the structure in the remaining part is more clearly visible. The figure shows that the majority of post-2000 weekend readings are zero, whereas this was not the case in earlier years. This is clearly an artefact of the data collection mechanism rather than a genuine feature of the precipitation. In fact, many weekend readings are missing from this station from 2001 onwards. We conjecture that since late 2000, the gauge has rarely been read at weekends and that the observer has adopted a practice of recording zeros when the gauge is empty on a Monday morning, but missing values otherwise (since in this case it is not possible to determine when the rain fell). The effect is to underreport the number of wet days at this station in the latter part of the record, and hence to exaggerate any drying trends.

Similar problems have been identified at some other stations. These have all been referred back to documentation held by the Bureau of Meteorology. Unfortunately, this documentation is often sparse, but in a few cases there was confirmation that, for example, weekend monitoring of a station had ceased on a particular date. To avoid bias in our trend assessments, therefore, all suspect observations (including those, such as at station 20, where there is evidence of selective reporting) have been discarded from the subsequent analyses. Table 8.2 gives a summary. A comparison with Table 8.1 shows that the high percentages of missing data at stations 6 and 19 are largely associated with selective reporting of zeros as at station 20. Station 8 did not display such clear evidence of selective reporting. However, the extremely high percentage (37 %) of missing observations here, without any supporting documentation, casts doubt on the quality of the entire record, which therefore has not been considered further. The analysis below is based on data from the 11 remaining stations: there are roughly 92 000 observations in total.

Table 8.2   Rainfall observations discarded from the trend analysis.

Station	Obs. discarded	Reason
6	All Sunday observations, 1969–1985	Selective reporting of zeros on Sundays. There are also occasional problems with Sunday readings outside the 1969–1985 period, but there are sufficiently few of these not to affect the analysis to any appreciable degree
8	All	High proportion of missing observations throughout, leading to suspicions of operational difficulties or selective reporting and casting doubt on the quality of the available data
16	All observations from 1964 to 1973	Differences between observer practice at weekends and on weekdays. Weekday observations are recorded to a coarser resolution than weekends, and the proportion of recorded wet days in particular is lower on weekdays than at weekends. Weekend observations are consistent with those from other stations, and with post-1973 observations from this station
19	All weekend observations, 1975 onwards	Selective reporting of zeros at weekends
20	All weekend observations, 2001 onwards	Selective reporting of zeros at weekends (see Figure 8.2)
26	All weekend observations, 2001 onwards	Selective reporting of zeros at weekends

Apart from specific problems relating to this particular data set, daily precipitation data often exhibit apparent spatial inconsistencies as a result of variation in observer practice at different stations (see, for example, Yang *et al.*, 2006). These inconsistencies are typically associated primarily with the treatment of very small rainfall amounts and, to reduce the effect of this, it is common to threshold the data prior to analysis. Specifically, if y_{st}^* denotes the recorded precipitation at station s on day t, then the thresholded observation is defined as

$$y_{st} = \max\left(y_{st}^* - \tau, 0\right)$$

for some small threshold $\tau > 0$. Previous authors (for example Charles, Bates and Hughes, 1999; Charles *et al.*, 1999) have used a threshold of $\tau = 0.3$ mm for the data used here. For the present study, a threshold of 0.25 mm has been used: since the data are rounded to a resolution of 0.2 mm, exactly the same days will be classified as dry or wet after thresholding as in previous work.

8.4   Modelling methodology

In the analysis below we fit statistical models simultaneously to data from the remaining 11 stations, and use these models to infer the nature and extent of trends in both rainfall occurrence and intensity. Daily rainfall data are usually highly variable, so that weak signals can be difficult to detect in individual records by conventional standards of statistical significance. Fitting simultaneously to data from all stations is one way to overcome this, since the evidence for a signal is strengthened if it is present in several independent series. Of course, since the stations are located in a relatively small geographical area, on any given day the observations from different stations are unlikely to be independent and this must be accounted for in the analysis. However, unless all the records are completely interdependent, extra information can always be gained by simultaneous fitting (see Section 6.1).

8.4.1   Generalised linear models for daily rainfall

Our analysis is based on generalised linear models (GLMs), which were introduced in Section 3.5. GLMs were first used to analyse individual daily rainfall time series by Coe and Stern (1982) and Stern and Coe (1984), and more recently by Fealy and Sweeney (2007) and Furrer and Katz (2007). For more flexible representation of structure, generalised additive models (GAMs; see Section 4.2) have also been used; see Hyndman and Grunwald (2000), Beckmann and Buishand (2002) and Underwood (2009). Chandler and Wheater (2002) considered the use of GLMs for simultaneous modelling of rainfall data from multiple locations; the analysis below follows similar lines, but also incorporates more recent developments.

In most applications of GLMs to daily rainfall, two separate models are used: logistic regression to model the probability of rainfall occurrence and gamma distributions to model the amount of rain if nonzero. Logistic regression specifies the probability π_{st} of rain at station s on day t via the relationship

$$\log\left(\frac{\pi_{st}}{1 - \pi_{st}}\right) = \beta_0 + \sum_{j=1}^{p} \beta_j x_{st}^{(j)} \, , \tag{8.1}$$

where $\mathbf{x}_{st} = \left(x_{st}^{(1)} x_{st}^{(2)} \cdots x_{st}^{(p)}\right)'$ is a covariate vector and $\beta_0, \beta_1, \ldots, \beta_p$ are coefficients to be estimated. Subsequently, if station s experiences rain on day t then the rainfall amount is taken to have a gamma distribution with mean μ_{st}, where

$$\log \mu_{st} = \gamma_0 + \sum_{j=1}^{q} \gamma_j \xi_{st}^{(j)} \tag{8.2}$$

for covariate vector $\boldsymbol{\xi}_{st} = \left(\xi_{st}^{(1)} \xi_{st}^{(2)} \cdots \xi_{st}^{(q)}\right)'$ and coefficients $\gamma_0, \gamma_1, \ldots, \gamma_q$.

Since a gamma distribution has two parameters, (8.2) does not fully specify the rainfall amounts model: a dispersion parameter is also required (see Section 3.5). As usual in GLMs, the dispersion parameter is assumed constant for all observations: for a gamma GLM this means that the gamma distributions have a common shape parameter or, equivalently, that the coefficient of variation is constant (McCullagh and Nelder, 1989, Chapter 8). This assumption plays the same role as the assumption of constant variance in a Gaussian multiple regression.

8.4.2   Temporal and spatial dependence

GLMs for mutually independent observations are usually fitted using maximum likelihood. However, in the present application there is likely to be dependence between observations in both space and time.

Temporal dependence can be handled by including functions of previous days' rainfalls as additional covariates in the models; see Section 3.5. Spatial (i.e. interstation) dependence is more difficult and, in our work, models are fitted as though observations from different stations are independent. As discussed in Section 6.1, this procedure delivers valid estimates of the coefficients describing the marginal time series structure at each station, but the standard errors of the parameter estimates must be adjusted appropriately. The required adjustments follow from the theory of estimating functions (see, for example, Davison, 2003, Section 7.2). This theory deals with very general situations where a statistical model with parameter vector θ is fitted to data $\mathbf{y}$ by solving a set of equations of the form

$$\mathbf{U}(\theta; \mathbf{y}) = \mathbf{0} \, ,$$

where the vector $\mathbf{U}(\theta; \mathbf{y})$ has the same number of elements as θ. For large samples in regular problems, any estimator defined in this way has approximately a multivariate normal distribution: the target of the estimation procedure is the mean vector of this distribution, θ_0, say. The covariance matrix of the distribution is $H^{-1}VH^{-1}$, where $H = \mathrm{E}\left[\partial\mathbf{U}/\partial\theta|_{\theta=\theta_0}\right]$ is the matrix of partial derivatives of $\mathbf{U}$ evaluated at θ_0 and V is the covariance matrix of $\mathbf{U}(\theta_0; \mathbf{y})$. In the present setting, denoting by $\ell(\theta; \mathbf{y})$ the log-likelihood for θ under the (incorrect) assumption that the observations are mutually independent, the parameter estimates solve $\partial\ell/\partial\theta = \mathbf{0}$; thus the theory of estimating functions can be applied by setting $\mathbf{U}(\theta; \mathbf{y}) = \partial\ell/\partial\theta$. In this case, the matrix H is given by $H = \mathrm{E}\left[\partial^2\ell/\partial\theta\partial\theta'\right]$, which, apart from a change of sign, is just the expected information for θ (Section 3.1.6). If the log-likelihood corresponds to the joint density from which the data were generated, then it can be shown that $V = -H = J$, say (Davison, 2003, Section 4.4). In this case, therefore, the asymptotic covariance matrix $H^{-1}VH^{-1}$ reduces to J^{-1}, in agreement with the classical theory for maximum likelihood estimation.

To apply this theory, it is necessary to estimate the matrices H (or equivalently J) and V. J is unaffected by interstation dependence so the usual estimator, $\hat{J}$, say, can be used; see Section 3.5 for the estimator usually used in GLMs. For V, notice that the log-likelihood $\ell(\theta; \mathbf{y})$ can be written as a sum of contributions from each time point and hence $\mathbf{U}(\theta; \mathbf{y}) = \partial\ell/\partial\theta$ is also a sum of contributions: $\mathbf{U}(\theta; \mathbf{y}) = \sum_{t=1}^{T} \mathbf{U}_t(\theta)$, say. If the model accurately reflects the dependence of each observation upon preceding values (achieved here by including previous days' rainfalls as covariates) then, under general conditions, these contributions are uncorrelated (Chandler *et al.*, 2007). Thus

$$V = \mathrm{Var}\left[\sum_{t=1}^{T} \mathbf{U}_t(\theta_0)\right] = \sum_{t=1}^{T} \mathrm{Var}\left[\mathbf{U}_t(\theta_0)\right] = \sum_{t=1}^{T} \mathrm{E}\left[\mathbf{U}_t(\theta_0)\,\mathbf{U}_t'(\theta_0)\right]$$

and the empirical counterpart of this expression can be used to estimate V:

$$\hat{V} = \sum_{t=1}^{T} \mathbf{U}_t(\hat{\theta})\,\mathbf{U}_t'(\hat{\theta})$$

where $\hat{\boldsymbol{\theta}}$ denotes the estimated parameter vector. For GLMs, the quantities $\left\{ U_t \left(\hat{\boldsymbol{\theta}} \right) \right\}$ have a convenient algebraic form: see Chandler (2005), for example. The covariance matrix of $\hat{\boldsymbol{\theta}}$ can now be estimated as $\hat{J}^{-1} \hat{V} \hat{J}^{-1}$, whence standard errors can be obtained.

Although standard errors provide a means of assessing the significance of individual coefficients in a model, it is often of interest to examine multiple terms simultaneously. If observations are independent and models are all fitted to the same set of observations (which may require that fitting is restricted to the subset of observations for which all potential covariate values are available), this can be done using likelihood ratio tests as described in Section 3.5.2. Denoting by ℓ_1 and ℓ_0 the maximised log-likelihoods for models with and without the terms of interest then, in large samples and under the null hypothesis that the data were generated from the simpler of the two models, the likelihood ratio statistic $\Lambda = 2 (\ell_1 - \ell_0)$ has approximately a chi-squared distribution with degrees of freedom equal to the number of coefficients being tested. If the observed value of Λ exceeds an appropriate percentage point of this distribution, the null hypothesis can be rejected.

Once again, however, in the current application it is necessary to consider the implications of potential dependence between the observations. If the dependence is purely temporal and is adequately represented by the models, then the usual procedures remain valid; thus the only real problem is caused by interstation dependence. We handle this using an empirical adjustment to the likelihood ratio statistic (Chandler and Bate, 2007). Unfortunately, for GLMs involving a dispersion parameter (such as the model for rainfall amounts here) and for which model comparison would normally be carried out using a deviance based F test, this adjustment can be used for the numerator of the F statistic but not for the denominator. For the purposes of model comparison, therefore, we have used maximum likelihood estimators of dispersion parameters throughout and have used likelihood ratio tests based on the chi-squared distribution.

The theory outlined above is implemented in the GLIMCLIM software package (Chandler, 2002), which has been used for all of our modelling.

8.4.3   Covariates considered

Prior to modelling, a small number of potential covariates were identified as candidates for a parametric representation of spatial and temporal structure in both precipitation occurrence and amounts. The fitted models are likely to be oversimplified because of the small number of covariates considered: this is a deliberate decision that acknowledges the relative sparsity of information about, for example, regional variation that can be obtained from a network of 11 stations. An alternative approach would be to represent the structure nonparametrically, for example using GAMs. However, nonparametric modelling can be computationally costly for extremely large data sets and when (as seen below) many interactions are present. Moreover, the adjustment of standard errors and model comparison statistics for intersite dependence is an area that has so far received relatively little attention in the nonparametric modelling literature; see, however, Bowman, Giannitrapani and Scott (2009). Thus a parsimonious parametric approach, although imperfect, is intended to provide a reasonable approximation to the main features of interest. The adequacy of this approximation can be checked using appropriate residual analyses.

The potential covariates fall into several distinct categories. First, to account for temporal dependence, functions of previous days' rainfalls were considered. These include indicator variables taking the value 1 if rainfall occurred previously and 0 otherwise, as well as log-transformed rainfall amounts (in fact, the transformation used was $\log(1 + y)$ rather than $\log y$, to avoid problems when previous days were dry). Network-averaged values of these variables were considered, as well as at-site values. Both weighted and unweighted network averages were tried: in the weighted case, the weights were taken to decay exponentially with distance from the site of interest and the decay rate was estimated by maximum likelihood.

Seasonality was represented via Fourier covariates (see Section 3.2), specifically $\cos(2\pi k \times \text{day of year}/365)$ and $\sin(2\pi k \times \text{day of year}/365)$, for $k = 1$ and 2. The annual cycle and its first harmonic should provide a reasonable representation of intraseasonal variation within the winter half-year considered here.

Systematic regional variation, for example the west–east and south–north rainfall gradients (see Section 8.2) along with the effects of topography, was represented by considering station altitude as a covariate along with functions of latitude and longitude. With regard to station altitude, residuals from a preliminary modelling exercise showed some unexpected structure suggesting that in some cases the recorded station altitudes were incorrect. Further investigation revealed that some stations are located in areas of heterogeneous topography so that their altitudes may not be representative of the rainfall they receive. For example, one station is 32 m above sea level, but is situated at the bottom of a narrow valley whose walls rise steeply to an altitude of over 150 m on both sides. We therefore defined an 'effective altitude' for each station, that could be considered more representative of the rainfall it receives. This was determined subjectively by inspection of maps. Since any such exercise is necessarily imprecise, effective altitudes were defined to the nearest 50 m; the values for each station are given in Table 8.1.

Any regional structure that is not accounted for by dependence on (effective) altitude is likely to vary smoothly in space. Thus, a representation using smooth basis functions was considered: bases of Legendre polynomials were used for both latitude and longitude. Polynomial bases are likely to provide a reasonable approximation to the regional structure when, as here, the west–east and south–north rainfall gradients are anticipated to be essentially monotonic; see Chandler (2005). The degree of polynomials considered was restricted to a maximum of two, however, due to the relatively limited information available about regional variation. A quadratic surface has the flexibility to represent moderate curvature, but by restricting the complexity of the fitted surfaces we hope to avoid the risk of overfitting.

Finally, a linear trend covariate was considered. According to models (8.1) and (8.2), this will lead to modelled precipitation changes that are linear on the log-odds scale for occurrence, and on the log scale for amounts. Although this could be seen as unrealistic, it is primarily intended to provide a convenient and interpretable measure of overall change during the period of record.

8.4.4   Modelling strategy

The basic approach to the modelling was, separately for occurrence and amounts, to start by building a 'baseline' model representing the overall intraseasonal, regional and autocorrelation structure in the data. This was done in stages, at each stage adding one or more terms to the model and using a dependence-adjusted likelihood ratio statistic to test for the significance of the new terms. Some care is required in interpreting the results of significance tests in large data sets such as this one, since they may detect genuine effects that are nonetheless too small to be of any practical relevance (Chandler, 2005).

Therefore, terms were only retained if they appeared significant at the 1 % level. At each stage, the decision regarding which term or group of terms to consider next was informed by physical understanding and an examination of model diagnostics (see below). As usual in this kind of modelling (see Section 3.2), sine and cosine terms were always considered in pairs and no covariate representing precipitation at a lag of $k > 1$ days was considered without including lags $1, \ldots, k - 1$ as well. Furthermore, no interactions were considered without including the corresponding main effects and no higher-order interactions were considered without including all of the corresponding lower-order interactions.

After acceptable baseline models had been found for both occurrence and amounts (judged on the basis of diagnostics, along with a lack of further significant covariates or interactions among the candidates considered), they were extended by adding the trend and its interactions with the existing covariates. The interactions were built up in stages, testing for significance at each stage as before. Finally, the overall significance of the trends was assessed by comparing the 'trend' and 'baseline' models using likelihood ratio statistics adjusted for interstation dependence.

8.5   Results

Table 8.3 summarises the final models, both with and without the trend and its interactions. The adjusted likelihood ratio tests indicate overwhelming evidence for trends in both occurrence and amounts. The remainder of the discussion therefore focuses on the models containing trend terms.

Table 8.3   Summary of fitted models for rainfall occurrence and amounts: 'df' is number of coefficients; 'log L' maximised log-likelihood ignoring intersite dependence; 'Λ_{ADJ}' dependence-adjusted likelihood ratio statistic for comparing 'trend' and 'baseline' models; 'p' associated p-value.

	Model	df	log L	Λ_{ADJ}	p
Occurrence	Baseline	21	−46 364.54	–	–
	Linear trend	24	−46 330.51	156.11	$< 10^{-16}$
Amounts	Baseline	13	−69 789.60	–	–
	Linear trend	16	−69 765.80	56.31	3.6×10^{-12}

In both of these models, regional variation is represented using a combination of effective altitude and a quadratic function of latitude and longitude; moreover, both models contain terms representing intraseasonal variation, although there are more of these for occurrence than for amounts. The occurrence model also contains terms representing autocorrelation; such terms were found not to be statistically significant in the amounts model. Autocorrelation in rainfall occurrence is represented by covariates defined as the proportion of stations experiencing precipitation at lags of one and two days, with an additional indicator for precipitation occurrence at the location of interest at a lag of one day. When computing the proportion of stations experiencing precipitation, the option of giving more weight to stations that are close to the location of interest was explored, but found not to be justified on the basis of likelihood ratio tests.

The occurrence model contains interactions between previous days' rainfalls and both seasonal and regional components: this indicates intraseasonal and regional variation in the strength of temporal autocorrelation. There are no interactions between seasonal and regional components; thus the intraseasonal variation in precipitation occurrence is the same at all locations. Nor are there any interactions between the trend and previous day's precipitation occurrence; if present, these would relate to the probability of a wet day following another wet day and hence to the relative clustering of wet and dry days. There are, however, interactions between the trend and regional components; these are explored below.

The amounts model contains interactions between seasonal components and both altitude and longitude; thus intraseasonal variation is altitude- and longitude-dependent. Another way of viewing the seasonality–longitude interaction is that the west–east precipitation gradient is seasonally varying. The amounts model also contains interactions between the trend function and regional components; again, these are explored below.

Neither model contains any interactions between the trend and seasonal components: these have been explored, but found to be insignificant. Thus there is no direct evidence for changes in intraseasonal structure.

8.5.1   Diagnostics

Before examining the fitted models in detail, in particular to determine the implied structure of precipitation trends in the region, it is worth checking the modelling assumptions. The first check is that the models account for all of the systematic intraseasonal and interannual structure in the data. This can be checked by examining the mean and standard deviation of Pearson residuals (see Section 3.5) for each month and year of the record. Figure 8.3 shows the results for the amounts model. The corresponding plots for precipitation occurrence are very similar.

The dashed lines in the top two plots of Figure 8.3 are approximate 95 % limits about zero, computed under the model and adjusted for intersite dependence as described by Chandler (2002, Appendix D.3). If the model has captured the systematic seasonal and interannual structure, then approximately 95 % of the mean residuals should lie between these lines. The plots of residual standard deviations can be used to diagnose problems with the assumption of a common dispersion parameter for the gamma distributions; see the discussion after Equation (8.2).

Except for a slightly higher residual standard deviation in October, the monthly plots in Figure 8.3 reveal little structure. Thus the model has successfully captured any seasonality in the data. The uniformity of standard deviations from May through September suggests that the gamma dispersion parameter is remarkably constant throughout this period. Similarly, the annual plots reveal no obvious trends in either the mean or standard deviation of residuals: we conclude that the linear trend covariate is adequate to describe the long-term variation in precipitation amounts.

In a similar way, the representation of regional variation can be checked by examining maps of mean Pearson residuals at each station: if the models are adequate, there should be no obvious spatial structure in such maps. The results are not shown here; they are indeed spatially unstructured however. As with other applications of this methodology (e.g. Chandler and Wheater, 2002; Yang et al., 2005), the models appear systematically to over- or underpredict at some stations, but not in a spatially organised way. For example, the occurrence model has a large positive mean residual, indicating underprediction, at station 9; however, at neighbouring station 6, the same model has a significant negative mean residual. Such results are associated either with local-scale variations in rainfall that

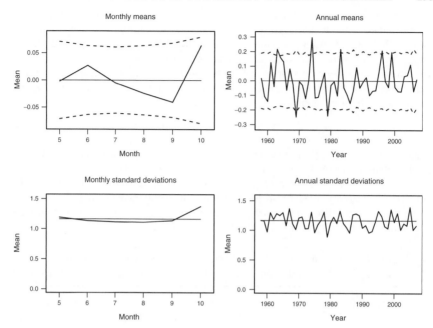

Figure 8.3   Monthly and annual summaries of Pearson residuals from the amounts model including linear trend. Dashed lines in mean plots are 95 % pointwise limits about zero, adjusted for intersite dependence.

the models are not designed to reproduce or with variation in observer practice at different stations that has not been eliminated by the thresholding described in Section 8.3.

The final check is for the assumption of gamma distributions in the amounts model. For this we have used the fact that if Y has a gamma distribution with mean μ and shape parameter ν, then Y/μ has a gamma distribution with mean 1 and shape parameter ν. The fitted model provides an estimate of μ for each observation, so that the observations can be scaled to have the same gamma distribution under the model. To check this, we plot the observed quantiles of these scaled observations against the quantiles of the expected distribution. The result reveals a satisfactory fit. It is not shown here, however, since for present purposes the precise distributional assumptions are less important than the assumed variance structure. Our checks of residual standard deviations suggest that the variance structure is indeed adequate so that, regardless of whether or not the observations follow gamma distributions, the results can be justified as coming from a quasi-likelihood fit (Section 3.5).

Overall then, these diagnostics suggest that the modelling assumptions are broadly satisfied and that the models do a good job of representing the systematic structure in the data. We can therefore examine the fitted models in some detail, and have some confidence in any conclusions that emerge.

8.5.2   Trends in wet-day precipitation amounts

A detailed interpretation of the fitted models is complicated, both because of the nonlinear link functions and because of the presence of so many interaction terms (10 for the occurrence model and seven for amounts). Since interest here is primarily in characterising the precipitation trends, however, we concentrate on the contributions of the trend component. Table 8.4 shows an edited extract of GLIMCLIM output for the fitted amounts model. This shows that if all other covariates are unchanged (for example by considering the same calendar day each year so that the seasonal covariates are effectively fixed) then a unit increase in the trend covariate, which is in units of decades, changes the log mean wet-day rainfall by an amount

$$-0.0289 - (0.0222 \times \text{Lat1}) - (0.0498 \times \text{Long1}) \ . \tag{8.3}$$

Here, 'Lat1' and 'Long1' are first-order Legendre polynomial transformations of latitude and longitude respectively: these are linear functions taking the value -1 at the southern (respectively western) edge of the study area and increasing to $+1$ at the northern (respectively eastern) edge. At all locations for which this quantity is positive (respectively negative), therefore, the model indicates an overall increase (respectively decrease) in precipitation intensity over the study period.

Table 8.4   Edited extract from GLIMCLIM output for the final amounts model, showing contributions of the trend covariate to the linear predictor.

Main effect:	Coefficient	Std Err
-----------	-----------	-------
...		
Trend 1 - Linear (0.1 per year)	-0.028895	0.0110
...		
2-way interactions:	Coefficient	Std Err
------------------	-----------	-------
...		
Trend 1 - Linear (0.1 per year) with Legendre poly 1 for Latitude	-0.022237	0.0058
Trend 1 - Linear (0.1 per year) with Legendre poly 1 for Longitude	-0.049816	0.0077

Perhaps surprisingly, (8.3) yields a positive value at stations 15, 20 and 31 (see Figure 8.1(b) for locations); it is negative at all other stations, however. Its actual value ranges from -0.057 to $+0.036$, at stations 26 and 15 respectively. Since these quantities are on a log scale due to the use of a logarithmic link function in model (8.2), they correspond to multiplicative decadal changes in mean precipitation intensity of $\exp(-0.056) = 0.94$ and $\exp(0.036) = 1.04$. These results indicate that wet-day amounts have decreased except in the southwest corner of the region.

8.5.3   Trends in precipitation occurrence

For the occurrence model, the analogue of (8.3) is

$$-0.0747 + (0.0312 \times \text{Long1}) + (0.0338 \times \text{Altitude}) \ , \tag{8.4}$$

where 'Altitude' here denotes the effective station altitude given in Table 8.1. Unfortunately, however, a straightforward interpretation of (8.4) is not possible, due to the presence of covariates representing previous days' rainfalls in the model (recall that the amounts model contains no such covariates). This complicates matters to a surprising degree, as we now demonstrate.

Notice first that for station s and day t our occurrence model (8.1) has the form

$$\pi_{st} = g^{-1}\left[\alpha_{st} + \sum_{j=1}^{k} f_{stj}\left(\mathbf{y}_{t-j}\right)\right]. \tag{8.5}$$

Here, $g^{-1}(\cdot)$ is the inverse of the logistic link function, $\mathbf{y}_{t-j}$ is the vector of all binary rainfall occurrence indicators on day $t - j$ and α_{st} denotes the contributions of all other covariates to the linear predictor. Any interactions involving previous days' occurrence are subsumed into the functions $\left\{f_{stj}\left(\cdot\right)\right\}$ so that previous occurrences make no contribution to α_{st}. For the particular model fitted here, $k = 2$ and the functions $f_{st1}(\cdot)$ and $f_{st2}(\cdot)$ are linear: both contain a contribution from the overall proportion of wet stations on the respective days, and f_{st1} has an additional contribution from the binary occurrence indicator at station s itself. These functions are also time-varying, because of interactions between previous days' occurrences and the seasonal covariates. Contributions to α_{st} come from the regional, seasonal and trend covariates and their mutual interactions, along with the intercept. Therefore α_{st} varies slowly with t: the values of the trend and seasonal covariates change little from day to day. As a result we can approximate $\alpha_{s(t-1)}, \ldots, \alpha_{s(t-k)}$ with α_{st}. Although this depends on the values of covariates at time t, these covariates play no further role from here on; we therefore treat α_{st} as a constant, and all subsequent calculations are considered as conditional upon its value.

Denoting by S the number of stations considered, (8.5) in fact defines a system of S equations for day t. It is convenient to stack these into a single vector equation:

$$\boldsymbol{\pi}_t = h\left[\boldsymbol{\alpha}_t + \sum_{j=1}^{k} \mathbf{f}_{tj}\left(\mathbf{y}_{t-j}\right)\right],$$

where $h(\cdot) = g^{-1}(\cdot)$ is considered as acting componentwise on the right-hand side. For our model, the linearity of the $\left\{f_{stj}\left(\cdot\right)\right\}$ means that $\mathbf{f}_{tj}\left(\mathbf{y}_{t-j}\right)$ can be written as $\boldsymbol{\Phi}_{tj}\mathbf{y}_{t-j}$ for some $S \times S$ matrix $\boldsymbol{\Phi}_{tj}$. Notice also that $\boldsymbol{\pi}_t = \mathrm{E}\left(\mathbf{y}_t|\mathbf{y}_{t-1}, \ldots, \mathbf{y}_{t-k}\right)$.

Suppose now that we want to find the vector of unconditional occurrence probabilities on day t, $\bar{\boldsymbol{\pi}}_t$, say. By the law of iterated expectations we have $\bar{\boldsymbol{\pi}}_t = \mathrm{E}\{\mathbf{y}_t\} = \mathrm{E}\left\{\mathrm{E}\left[\mathbf{y}_t|\mathbf{y}_{t-1}, \ldots, \mathbf{y}_{t-k}\right]\right\} = \mathrm{E}\{\boldsymbol{\pi}_t\}$. Combining this with the previous expression for $\boldsymbol{\pi}_t$, we obtain

$$\bar{\boldsymbol{\pi}}_t = \mathrm{E}\left\{h\left[\boldsymbol{\alpha}_t + \sum_{j=1}^{k} \mathbf{f}_{tj}\left(\mathbf{y}_{t-j}\right)\right]\right\}, \tag{8.6}$$

where the expectation is over the joint distribution of $\mathbf{y}_{t-1}, \ldots, \mathbf{y}_{t-k}$.

In general, the functions $h(\cdot)$ and $\left\{\mathbf{f}_{tj}\left(\cdot\right)\right\}$ will be nonlinear. Thus the expectation on the right-hand side of (8.6) depends on the full joint distribution of $\mathbf{y}_{t-1}, \ldots, \mathbf{y}_{t-k}$. A possibly unexpected consequence is that the marginal occurrence probabilities $\bar{\boldsymbol{\pi}}_t$ are affected by interstation dependence (which is not explicitly modelled by the GLM): a single GLM could thus yield different expectations depending on the assumed dependence structure. To understand this more clearly, it may be helpful to note that our occurrence model can be regarded as partially defining a Markov process with state space consisting of the 2^{kS} possible configurations of rainfall occurrence at each of the S stations over a consecutive k-day period. The properties of such processes are usually studied via the matrix of interstate transition probabilities from one time point to the next (see for example Ross, 2003, Chapter 4); however, in the current context, to specify this matrix in full would require a model for the joint probabilities of precipitation occurrence simultaneously at all stations, conditional on previous days' configurations.

Dependence of (8.6) upon the full joint distribution can be avoided if both $h(\cdot)$ and $\{\mathbf{f}_{tj}(\cdot)\}$ are linear or if, for each $s \in \{1, \ldots, S\}$ and $j \in \{1, \ldots, k\}$, the sth component of $\{\mathbf{f}_{tj}(\cdot)\}$ acts only on the sth component of $\mathbf{y}_{t-j}$. In our case, the latter condition does not hold since the conditional occurrence probability π_{st} depends on the proportion of stations experiencing rainfall on each of the previous days; hence each component of $\mathbf{f}_{tj}(\cdot)$ acts on all of the components of $\mathbf{y}_{t-j}$.

Fortunately, as already noted, for our occurrence model the functions $\{\mathbf{f}_{tj}(\cdot)\}$ are linear so that (8.6) can be written as

$$
\bar{\boldsymbol{\pi}}_t = \mathrm{E}\left\{h\left[\boldsymbol{\alpha}_t + \sum_{j=1}^{k} \boldsymbol{\Phi}_{tj}\mathbf{y}_{t-j}\right]\right\} .
$$

The function $h(\cdot)$ is not linear; however, if it is approximately linear over the range of its argument then we can write

$$
\mathrm{E}\left\{h\left[\boldsymbol{\alpha}_t + \sum_{j=1}^{k} \boldsymbol{\Phi}_{tj}\mathbf{y}_{t-j}\right]\right\} \approx h\left[\mathrm{E}\left\{\boldsymbol{\alpha}_t + \sum_{j=1}^{k} \boldsymbol{\Phi}_{tj}\mathbf{y}_{t-j}\right\}\right]
$$

$$
= h\left[\boldsymbol{\alpha}_t + \sum_{j=1}^{k} \boldsymbol{\Phi}_{tj}\mathrm{E}\left\{\mathbf{y}_{t-j}\right\}\right] . \qquad (8.7)
$$

A further approximation, again using the fact that day-to-day changes in marginal occurrence probabilities are due solely to seasonality and long-term trends and hence are small, is to replace $\mathrm{E}\{\mathbf{y}_{t-j}\}$ with $\mathrm{E}\{\mathbf{y}_t\} = \bar{\boldsymbol{\pi}}_t$. Thus we have $\bar{\boldsymbol{\pi}}_t \approx h\left[\boldsymbol{\alpha}_t + \sum_{j=1}^{k} \boldsymbol{\Phi}_{tj}\bar{\boldsymbol{\pi}}_t\right]$, and the vector of marginal occurrence probabilities can be calculated approximately by solving the equation

$$
\bar{\boldsymbol{\pi}}_t - h\left[\boldsymbol{\alpha}_t + \sum_{j=1}^{k} \boldsymbol{\Phi}_{tj}\bar{\boldsymbol{\pi}}_t\right] = \mathbf{0} . \qquad (8.8)
$$

Our function $h(\cdot)$ is the inverse logistic: $h(x) = \left[1 + \exp(-x)\right]^{-1}$. Thus an analytical solution to (8.8) does not exist. However, numerical techniques can be used. We find that a straightforward Newton–Raphson scheme works well; this starts with an initial value $\bar{\boldsymbol{\pi}}_t^{(0)}$ and then iterates

$$
\bar{\boldsymbol{\pi}}_t^{(\ell+1)} = \bar{\boldsymbol{\pi}}_t^{(\ell)} - \left(\boldsymbol{I} - \boldsymbol{H}^{(\ell)}\sum_{j=1}^{k}\boldsymbol{\Phi}_{tj}\right)^{-1}\left(\bar{\boldsymbol{\pi}}_t^{(\ell)} - h\left[\boldsymbol{\alpha}_t + \sum_{j=1}^{k}\boldsymbol{\Phi}_{tj}\bar{\boldsymbol{\pi}}_t^{(\ell)}\right]\right)
$$

to convergence. Here, I is the $S \times S$ identity matrix and $H^{(\ell)}$ is the diagonal $S \times S$ matrix with sth element $h'\left[\alpha_{st} + \sum_{j=1}^{k} \phi_{stj} \bar{\pi}_t^{(\ell)}\right]$, where ϕ_{stj} is the sth row of Φ_{tj}. For the inverse logistic function, we have $h'(x) = e^{-x} h^2(x)$.

Since the primary aim of this study is to learn about precipitation trends, it is arguably less relevant to study the occurrence probabilities themselves than their changes over time. Equation (8.8) can be used to explore this for our model. The absence of interactions between the trend and previous days' covariates ensures that if the trend covariate changes with all others held constant (so that one is effectively looking at the same calendar day in a different year), α_t will be affected but the $\{\Phi_{jt}\}$ will not. Suppose then that α_t changes to $\alpha_t + \delta$. Then the new marginal occurrence probability vector, $\bar{\pi}_t^*$, say, approximately satisfies

$$\bar{\pi}_t^* = h\left[\alpha_t + \delta + \sum_{j=1}^{k} \Phi_{tj} \bar{\pi}_t^*\right] .$$

Row s of this vector equation reads

$$\bar{\pi}_{st}^* = h\left[\alpha_{st} + \delta_s + \sum_{j=1}^{k} \phi_{stj} \bar{\pi}_t^*\right]$$

$$= h\left[\alpha_{st} + \sum_{j=1}^{k} \phi_{stj} \bar{\pi}_t + \delta_s + \sum_{j=1}^{k} \phi_{stj}\left(\bar{\pi}_t^* - \bar{\pi}_t\right)\right]$$

$$\approx \bar{\pi}_{st} + \left[\delta_s + \sum_{j=1}^{k} \phi_{stj}\left(\bar{\pi}_t^* - \bar{\pi}_t\right)\right] h'\left[\alpha_{st} + \sum_{j=1}^{k} \phi_{stj} \bar{\pi}_t\right] ,$$

the last line following from a first-order Taylor expansion and the approximation $\bar{\pi}_{st} \approx h\left[\alpha_{st} + \sum_{j=1}^{k} \phi_{stj} \bar{\pi}_t\right]$, which is row s of (8.8). We thus have

$$\bar{\pi}_{st}^* - \bar{\pi}_{st} \approx \left[\delta_s + \sum_{j=1}^{k} \phi_{stj}\left(\bar{\pi}_t^* - \bar{\pi}_t\right)\right] h'\left[\alpha_{st} + \sum_{j=1}^{k} \phi_{stj} \bar{\pi}_t\right] .$$

Reassembling the S resulting equations into vector form now yields

$$\bar{\pi}_t^* - \bar{\pi}_t \approx H\left[\delta + \sum_{j=1}^{k} \Phi_{tj}\left(\bar{\pi}_t^* - \bar{\pi}_t\right)\right] ,$$

where H is a diagonal matrix of derivatives as before, now evaluated at $\bar{\pi}_t$. Finally, after some rearrangement, we obtain

$$\bar{\pi}_t^* \approx \bar{\pi}_t + \left(I - H\sum_{j=1}^{k} \Phi_{tj}\right)^{-1} H\delta . \tag{8.9}$$

This shows that for the model fitted here, it is far from straightforward to translate changes in the linear predictor to the corresponding changes in occurrence probabilities;

the latter changes are not even guaranteed to have the same sign as the former at all stations. In particular, it is not possible to draw any conclusions directly from the coefficients of the trend covariate in the model. Essentially, this is because the precipitation occurrence at station s on day t depends on the occurrences at all stations on the previous two days. Thus the modelled trend at station s has a direct component from the trend covariate, along with an indirect component from trends at the other stations; if these components have opposite signs then the net effect cannot be determined without evaluating the probabilities explicitly.

The accuracy of the fundamental relationship (8.8) depends on the adequacy of the approximations involved in deriving it. Chief among these is (8.7), which is justified providing $h(\cdot)$ is reasonably linear throughout the range of its argument. This argument is just the linear predictor in the GLM. For the model fitted to our data set, 90 % of these linear predictors lie in the range $(-1.75, 1.40)$. A plot of $h(x) = [1 + \exp(-x)]^{-1}$ shows that it is very close to linear over this interval; thus we might expect that (8.8) will be accurate in this particular application. Notice also that if $k = 0$ then (8.8) is exact: thus, even for nonlinear $h(\cdot)$, the approximation should be reasonable for models with weak or moderate temporal dependence structure.

To check the adequacy of (8.8) we have compared the results with those obtained via simulation. For the period 1960–2005, 100 multistation precipitation time series were generated from the fitted models. For a given station and date, the proportion of simulations for which precipitation occurred can be used as an estimate of the corresponding probability. As we have seen, marginal occurrence probabilities are partially determined by the interstation dependence structure and, in the present application, this dependence is strong: on any given day, stations tend to be either mostly wet or mostly dry. To capture this in the simulations, we have used the dependence structure proposed by Yang *et al.* (2005) in which the number of wet stations on any given day has a beta-binomial distribution with the location parameter determined by the occurrence model (8.1). For our data, the shape parameter of this beta-binomial distribution is estimated as 0.57; this corresponds to 'U'-shaped distributions for numbers of wet stations and hence to strong interstation dependence (see Yang *et al.*, 2005, for more details).

Specimen results from this simulation exercise, showing marginal probabilities of precipitation occurrence at stations 20 and 26 on 1 July and 1 October each year, are shown in Figures 8.4(a) and (b). For these stations, (8.4) yields values of -0.064 and 0.037 respectively. 1 July corresponds to the wettest part of the year, whereas 1 October is towards the end of the winter period when precipitation is less frequent.

The solid lines in Figures 8.4(a) and (b) are the marginal probabilities of rainfall occurrence computed from (8.8): these show that the modelled trends in rainfall occurrence appear very close to linear. They also show that the decline at station 20 is substantially greater than the increase at station 26, which might be expected given the values of (8.4) at the two stations. The trend slopes in the two plots seem very similar, but the difference between the two stations is reduced in October relative to July. This is to be expected: the model contains interactions between covariates representing seasonal and regional variation, but no seasonal–trend interactions. For the remainder of our discussion, therefore, we focus on the results for 1 July since the results for other calendar dates are similar.

The dashed lines in Figures 8.4(a) and (b) show the simulation based estimates of the same marginal probabilities. These are quite variable, since each estimate is based on just 100 daily values. Overall, however, the theoretical approximations seem to match the simulations well. As a further check, the plots also show empirically derived estimates

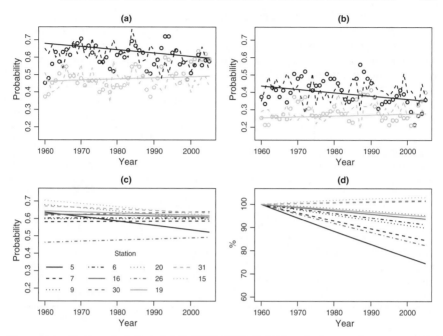

Figure 8.4   Changes in daily rainfall, 1960–2005. Solid lines in the top plots are marginal probabilities of rainfall occurrence at stations 20 (black) and 26 (grey) on (a) 1 July and (b) 1 October each year, obtained from (8.8); dashed lines are estimates obtained from 100 simulations of the fitted model and points are estimates obtained directly from observations (see text for details). Bottom plots: (c) shows modelled marginal occurrence probabilities at all stations on 1 July each year and (d) shows the expected rainfall on 1 July each year as a percentage of expected rainfall on 1 July 1960.

of these occurrence probabilities, obtained directly from the data without reference to the fitted models. Each empirical estimate is computed using data from a 5-year, 15-day window centred on the calendar date of interest. Thus, for example, the estimate at station 20 for 1 July 1960 is computed from all observations at that station between 24 June and 8 July, for the years 1958–1962. These empirical estimates suggest that the model does a good job at reproducing trends in precipitation occurrence: although they tend to cluster in runs above or below the solid trend lines, this is largely due to the fact that successive empirical estimates are strongly interdependent because they are computed from overlapping observation windows. It is perhaps worth noting that although the models were fitted to these same data, they are calibrated to reproduce conditional rather than marginal relationships: the indirect relationship between the two, approximated by (8.8), demonstrates that a comparison of observed and modelled trends in the marginal occurrence structure is quite a stringent test of the internal model structure.

Having established that (8.8) provides a good approximation to the observed occurrence probabilities, Figure 8.4(c) plots these approximated probabilities for each year at all stations. This plot shows substantial variation between stations: some show little change or small increases over the period of record, while others show marked decreases. To assist interpretation of this plot, Table 8.5 shows the changes in marginal probabilities

Table 8.5   Summary of modelled precipitation changes on 1 July, 1960–2005. δ_{OCC} is the change in conditional log odds of precipitation occurrence; $\bar{\pi}_{2005} - \bar{\pi}_{1960}$, is the change in marginal occurrence probabilities derived from (8.8); 'Approximation', the value of (8.9); δ_{AMT}, the change in log mean wet day rainfall amount. Stations are listed in order from north to south (see Figure 8.1).

| | Station | | | | | |
	5	7	9	6	16	30
δ_{OCC}	−0.418	0.092	0.065	0.065	−0.016	0.144
$\bar{\pi}_{2005} - \bar{\pi}_{1960}$	−0.114	0.005	0.000	0.000	−0.017	0.021
Approximation	−0.107	0.007	0.002	0.002	−0.016	0.023
δ_{AMT}	−0.096	−0.177	−0.107	−0.092	−0.038	−0.020

| | Station | | | | |
	20	26	19	31	15
δ_{OCC}	−0.290	0.166	−0.060	−0.114	−0.306
$\bar{\pi}_{2005} - \bar{\pi}_{1960}$	−0.083	0.029	−0.026	−0.039	−0.087
Approximation	−0.077	0.031	−0.024	−0.036	−0.079
δ_{AMT}	0.083	−0.256	−0.012	0.077	0.161

between 1960 and 2005 at all stations. The three lowest-lying stations (5, 15 and 20 – see Table 8.1) all show substantial decreases in precipitation occurrence over this period, whereas stations with effective altitudes of 300 m or above (numbers 6, 7, 9 and 30) show little change or, in some cases, a slight increase. This is perhaps to be expected given the decadal contributions (8.4) from the trend covariate to the occurrence model. The values of these contributions over the 45-year period considered here are given as δ_{OCC} in Table 8.5. A comparison of the first two rows in the table confirms our earlier assertion that it is difficult to translate changes from one scale to the other: although there are no stations here where δ_{OCC} and $\bar{\pi}_{2005} - \bar{\pi}_{1960}$ have different signs, it is clear that the marginal probabilities would decrease at any station with $0 < \delta_{OCC} < 0.05$.

The third row of Table 8.5 shows the value of the approximation (8.9) to $\bar{\pi}_{2005} - \bar{\pi}_{1960}$. This is computed using the values of δ_{OCC} given in the table, along with the marginal probabilities for 1 July 1960 from (8.8). The approximation is seen to be generally accurate at all stations.

8.5.4   Combined trends in occurrence and amounts

The final row of Table 8.5 shows the modelled change, again over the period 1960–2005, in log mean wet-day rainfall amounts on 1 July. This is straightforward to obtain from (8.3) (simply multiply by 4.5, which is the period length in decades), because the amounts model contains no covariates representing lagged rainfalls. The tabulated station ordering brings out the north–south pattern of trends in wet-day rainfall amounts: by and large, negative changes are most pronounced at the most northerly stations with positive changes in the south. The only two stations that deviate from this overall pattern are 20 and 26; these both fall slightly off the main north–south line of stations (see Figure 8.1) and the deviations here are due to the contribution of station longitude to (8.3).

Having established the trends in occurrence and amounts separately, we can combine these to obtain an overall picture of changes in precipitation over the region. If $\bar{\pi}_{st}$ is the probability of occurrence at station s and time t, and μ_{st} is the expected rainfall amount conditional on occurrence, then the overall expected rainfall is $\bar{\pi}_{st}\mu_{st}$. For each station, Figure 8.4(d) shows the overall percentage change in this quantity on 1 July each year between 1960 and 2005. Unsurprisingly, there is once again considerable variation between stations. However, decreases in overall mean precipitation are now apparent almost everywhere. These decreases are in excess of 10 % over the 45-year period at the three most northerly stations (5, 7 and 9), along with the easternmost station 26. Of the remaining stations, numbers 6, 16, 20 and 19 show overall decreases of between 5 % and 10 %, and stations 30, 31 and 15 show small increases of 1 %, 2 % and 3 % respectively. An examination of Table 8.5 suggests that these changes are primarily driven by changes in precipitation intensity on wet days, although changes in the frequency of occurrence play an important role, notably at stations 5 (where a reduction in occurrence frequencies compounds the changes in intensity) and 30 (where slight increases in occurrence frequencies, associated with the high altitude of this station, offset the reduced intensities).

8.6 Summary and conclusions

The analysis above has revealed substantial trends in both the frequency and intensity of precipitation in the study area over the last half-century. A linear representation of these trends appears to provide a useful summary of overall change, although there is pronounced regional variation in the trend slopes. By fitting models simultaneously to data from all stations, this regional variation has been represented explicitly. The results show that the patterns of change in occurrence and intensity are different: reductions in occurrence are greatest at low-altitude locations in the west of the study area, whereas reductions in intensity are greatest in the north and east. The net effect is generally a reduction in midwinter precipitation of between 5 % and 25 %, except in the southwest corner of the study area. Although the models indicate increases in frequency or intensity at a few stations, these increases are all small and no station has a modelled increase in both frequency and intensity.

The differences between the regional structures of change in frequency and intensity suggest that the physical mechanisms behind the precipitation decline are complex. They are, however, consistent with a southward shift of the mid-latitude weather systems that are responsible for most of the winter precipitation in the region (Alexander *et al.*, 2010; Hope, Drosdowsky and Nicholls, 2006): this would weaken the strength of westerly flow at the study latitudes and hence reduce the transport of moisture inland.

Our results agree in broad terms with the conclusions from previous studies, although there are some discrepancies in the detail. The most notable of these is the absence of interaction between trend and seasonal components in our models, suggesting that (at least on the scale of the linear predictors) there is no intraseasonal variation in the magnitude of precipitation trends. IOCI (2002) and Bates *et al.* (2008), for example, concluded that over a much larger region, the precipitation decline has been strongest during the months of May to July. However, the apparent discrepancy is probably due mainly to differences in study domains. For example, the results in Bates *et al.* (2008) relate to regional averages over almost all of the land area shown in Figure 8.1(b) whereas the present work provides no basis for extrapolation outside the shaded rectangle. Our model based approach does, however, provide a more formal means of testing for intraseasonal variation than has previously been attempted, and it would be of interest to extend it to the larger region and compare the results with those from earlier studies. The absence of intraseasonal variation in our particular study area seems quite clear; this implies that the physical mechanisms responsible for the precipitation changes here are essentially the same throughout the winter half-year.

Data quality problems may also be responsible for some differences between the present results and those of previous studies. The problems noted in Section 8.3 have been uncovered here for the first time: failure to address these would lead to exaggerated estimates of drying trends. Indeed, one of the key messages from this work is the importance of checking the data carefully prior to any trend analysis.

The present study was primarily motivated by a desire to understand the decline in IWSS inflows shown in Figure 8.1(a). The results suggest that reduced precipitation is likely to be a major contributor to this decline. Indeed, over the 1958–2007 period covered by the precipitation data, the trend in inflows is monotonically decreasing, which is entirely consistent with our modelled precipitation changes over this period. Since most of the water supply catchments are located within the Darling Range, and the average height of this range is around 300 m, our analysis implies that reduced inflows are probably associated with declines in precipitation intensity rather than frequency: at this altitude, precipitation frequency has changed relatively little over the last half-century.

A final contribution of this paper has been the development of simple approximate expressions for some of the model properties; theoretical properties are difficult to derive for nonlinear, nonstationary and multivariate time series models, and as far as we are aware this is the first time that such a derivation has been attempted for the class of models considered here. For our occurrence model, the approximations have been shown to provide a quick, computationally efficient and apparently accurate alternative to simulation if interest is in the marginal occurrence probabilities at any specific time point. However, it is possible that this example is particularly well suited to the approximations because the lagged terms enter linearly into the linear predictor, and the inverse link function $h(\cdot)$ is itself approximately linear throughout most of our data set. More experience is therefore required to understand better how accurate and useful such approximations may be in general, and an alternative strategy, of fitting models that represent the marginal probabilities directly, may also be worth considering. An example of this latter approach is given by Kneib and Fahrmeir in Chapter 10 of the present volume.

References

Alexander, L. V., Uotila, P., Nicholls, N. and Lynch, A. (2010) A new daily pressure dataset for Australia and its application to the assessment of changes in synoptic patterns during the last century. *Journal of Climate*, **23**(5), 1111–1126. Doi: 10.1175/2009JCLI2972.1.

Bates, B., Hope, P., Ryan, B., Smith, I. and Charles, S. (2008) Key findings from the Indian Ocean Climate Initiative and their impact on policy development in Australia. *Climatic Change*, **89**, 339–354. DOI: 10.1007/s10584-007-9390-9.

Bates, B. C. and Hughes, G. (2009) Adaptation measures for metropolitan water supply for Perth, Western Australia. In *Climate Change Adaptation in the Water Sector* (eds F., Ludwig P., Kabat H. van Schaik and M. van der Valk). Earthscan, London. pp. 187–204.

Bates, B. C., Chandler, R. E., Charles, S. P. and Campbell, E. P. (2010) Assessment of apparent non-stationarity in time series of annual inflow, daily precipitation and atmospheric circulation indices: a case study from southwest Western Australia. *Water Resources Research* **46**, W00H02, doi:10.1029/2010WR009509.

Beckmann, B. R. and Buishand, T. A. (2002) Statistical downscaling relationships for precipitation in the Netherlands and north Germany. *International Journal of Climatology*, **22**, 15–32.

Bowman, A., Giannitrapani, M. and Scott, E. M. (2009) Spatiotemporal smoothing and sulphur dioxide trends over Europe. *Journal of the Royal Statistical Society, Series C* **58**, 737–752.

Bowman, A. W., Pope, A. and Ismail, B. (2006) Detecting discontinuities in nonparametric regression curves and surfaces. *Statistical Computing*, **16**, 377–390.

Chandler, R. E. (2002) GLIMCLIM: Generalized linear modelling for daily climate time series (software and user guide). Technical Report 227, Department of Statistical Science, University College London, London WC1E 6BT. Available at http://www.ucl.ac.uk/Stats/research/Resrpts/abstracts.html.

Chandler, R. E. (2005) On the use of generalized linear models for interpreting climate variability. *Environmetrics*, **16**, 699–715.

Chandler, R. E. and Bate, S. M. (2007) Inference for clustered data using the independence log-likelihood. *Biometrika*, **94**, 167–183.

Chandler, R. E. and Wheater, H. S. (2002) Analysis of rainfall variability using Generalized Linear Models – a case study from the West of Ireland. *Water Resources Research*, **38**(10). Doi: 10.1029/2001WR000906.

Chandler, R. E., Isham, V., Bellone, E., Yang, C. and Northrop, P. J. (2007) Space–time modelling of rainfall for continuous simulation. In *Statistics of Spatial-Temporal Systems* (eds B. Finkenstadt and V. Isham). CRC Press, Boca Raton, Florida. pp. 169–207.

Charles, S. P., Bates, B. C. and Hughes, J. P. (1999) A spatio-temporal model for downscaling precipitation occurrence and amounts. *Journal of Geophysical Research*, **104**(D24), 31657–31669.

Charles, S. P., Bari, M. A., Kitsios, A. and Bates, B. C. (2007) Effect of GCM bias on downscaled precipitation and runoff projections for the Serpentine catchment, Western Australia. *International Journal of Climatology*, **27**, 1673–11690.

Charles, S. P., Bates, B. C., Whetton, P. H. and Hughes, J. P. (1999) Validation of a downscaling model for changed climate conditions in southwestern Australia. *Climate Research*, **12**, 1–14.

Coe, R. and Stern, R. D. (1982) Fitting models to daily rainfall. *Journal of Applied Meteorology*, **21**, 1024–1031.

Croton, J. T. and Reed, A. J. (2007) Hydrology and bauxite mining on the Darling Plateau. *Restoration Ecology*, **15**(4), S40–S47.

Davison, A. C. (2003) *Statistical Models*. Cambridge University Press, Cambridge.

Fealy, R. and Sweeney, J. (2007) Statistical downscaling of precipitation for a selection of sites in Ireland employing a generalised linear modelling approach. *International Journal of Climatology*, **15**, 2083–2094.

Furrer, E. M. and Katz, R. W. (2007) Generalized linear modeling approach to stochastic weather generators. *Climate Research*, **34**, 129–144.

Hennessy, K., Fitzharris, B., Bates, B., Harvey, N., Howden, S., Hughes, L., Salinger, J. and Warrick, R. (2007) Australia and New Zealand. In *Climate Change 2007: Impacts, Adaptation and Vulnerability. Contribution of Working Group II to the Fourth Assessment Report of the Intergovernmental Panel on Climate Change* (eds M. L., Parry O. F., Canziani J. P., Palutikof P. J. van der Linden and C. E. Hanson). Cambridge University Press, Cambridge. pp. 507–540.

Hope, P., Drosdowsky, W. and Nicholls, N. (2006) Shifts in the synoptic systems influencing southwest Western Australia. *Climate Dynamics*, **26**(7–8), 751–764. Doi: 10.1007/s00382-006-0115-y.

Hyndman, R. J. and Grunwald, G. K. (2000) Generalized additive modelling of mixed distribution Markov models with application to Melbourne's rainfall. *Australia and New Zealand Journal of Statistics*, **42**, 145–158.

IOCI (2002) Towards understanding climate variability in south western Australia – research reports on the first phase of the Indian Ocean Climate Initiative. Technical Report, Indian Ocean Climate Initiative. Available from http://www.ioci.org.au/publications/ pdf/IOCI_FPR_1.pdf.

McCullagh, P. and Nelder, J. A. (1989) *Generalized Linear Models*, 2nd edition. Chapman & Hall, London.

Power, S., Sadler, B. and Nicholls, N. (2005) The influence of climate science on water management in Western Australia. *Bulletin of American Meteorological Society*, **86**, 839–844.

Ross, S. M. (2003) *Introduction to Probability Models*, 8th, edition. Elsevier: Academic Press, Burlington.

Stern, R. D. and Coe, R. (1984) A model fitting analysis of rainfall data (with discussion). *Journal of the Royal Statistical Society, Series C*, **A147**, 1–34.

Underwood, F. M. (2009) Describing long-term trends in precipitation using generalized additive models. *Journal of Hydrology*, **364**, 285–297.

Yang, C., Chandler, R. E., Isham, V. S., Annoni, C. and Wheater, H. S. (2005) Simulation and downscaling models for potential evaporation. *Journal of Hydrology*, **302**, 239–254.

Yang, C., Chandler, R. E., Isham, V. S. and Wheater, H. S. (2006) Quality control for daily observational rainfall series in the UK. *Water and Environment Journal*, **20**(3), 185–193. DOI: 10.1111/j.1747-6593.2006.00035.x.

9

Estimation of common trends for trophic index series

Alain F. Zuur[1], Elena N. Ieno[2], Cristina Mazziotti[3], Giuseppe Montanari[3], Attilio Rinaldi[3] and Carla Rita Ferrari[3]

[1]*Highland Statistics Ltd, 6 Laverock Road, Newburgh AB41 6FN, UK*

[2]*Highland Statistics, Suite N 226, Av Finlandia 21, CC Gran Alacant Local 9, 03130 Santa Pola, Spain*

[3]*ARPA Emilia-Romagna, Struttura Oceanografica Daphne, V le Vespucci 2, 47042 Cesenatico (FC), Italy*

9.1 Introduction

Eutrophication – the presence of abnormally high levels of nutrients in water bodies, often caused by agricultural fertilisers leaching into lakes and coastal waters – often leads to an increase of phytoplankton in the water and to the development of nuisance phytoplanktonic blooms. This often results in discoloration and reduced transparency of the water column, and in severe cases may significantly impact the benthos. Eutrophication is a global problem and has increased considerably in the last few decades in coastal and shelf waters (Fletcher, 1996). In the most serious manifestations, microalgal blooms are accompanied by a massive growth of submersed and floating macrophytes (Vollenweider, 1968, 1981). Other undesirable effects may follow, such as reduced biodiversity and formation of toxic nutrients (H_2S, CH_4, CO_2 and NH_3). These phenomena have been observed and documented with increasing frequency, raising concerns over whether the monitoring of such events is adequate and whether current practice affords a sufficient sampling frequency to manage and protect coastal function effectively.

Statistical Methods for Trend Detection and Analysis in the Environmental Sciences, First Edition.
Richard E. Chandler and E. Marian Scott.
© 2011 John Wiley & Sons, Ltd. Published 2011 by John Wiley & Sons, Ltd.

The impact of eutrophication (e.g. algal blooms, death of benthic fauna, shellfish poisoning and oxygen depletion) can show considerable spatial and temporal differences depending on where, when and in what context the bloom occurs. This high variability is often related to the presence of large rivers that have a dominating influence over the eutrophication level (trophic state) of a region. This is true for the Northern Adriatic Sea and the coastal stretch in the Emilia-Romagna region of Italy (see Figure 9.1), which is

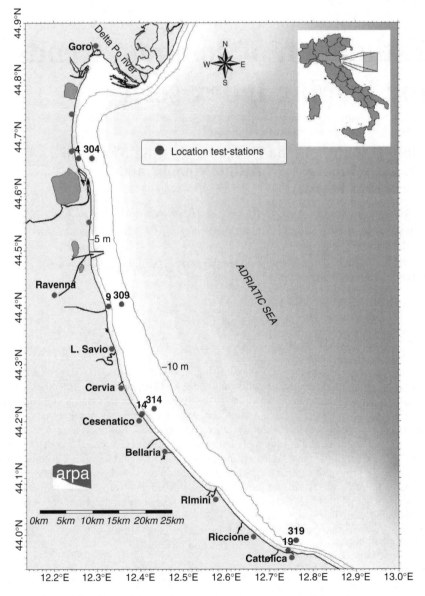

Figure 9.1   Map of the study area. Note the river Po at the northern end.

under the direct influence of the inputs of the river Po (Rinaldi and Montanari, 1988; Vollenweider, Rinaldi and Montanari, 1992) and is the area of interest in the present study. When setting management objectives, it is important to know whether the water column trophic state is largely determined by riverine water quality or by the extent and frequency of the microalgal blooms (Pompei *et al.*, 1995).

The areas facing the Po delta and the coast of the Emilia-Romagna region are directly influenced by the river inputs of the hydrographic basin of the Po valley, conveyed by the River Po itself, as well as the river inputs of the coastal basins. The influence of these river inputs is evident from the salinity of the sea water, which falls noticeably along the coastal area by comparison with the open sea. The great mass of freshwater transported by the river Po (about $1500 \, \mathrm{m}^3 \, \mathrm{s}^{-1}$) has a profound influence on the northwestern basin of the Adriatic Sea and determines many of the ecological processes of the coastal system. The unique hydrography and oceanography of the area means that the region is heavily impacted by eutrophication: as such, it is one of the most problematic regions of the whole Mediterranean Sea.

One approach to the study of eutrophication is to analyse a wide range of possible ecosystem responses involving many variables such as nutrient flux, chlorophyll-a and oxygen. An alternative is to define a synthetic index function, allowing the investigator to reduce a complex biological system to a single site-specific numeric score (or category) that can be interpreted by a nonspecialist within a 'good' versus 'bad' continuum, often to meet a minimum legislative requirement (for review, see Diaz, Solan and Valente, 2004). This methodology is popular with resource managers and endorsed by many applied environmental scientists because of the ease in which assessments can be made, interpreted and explained to policy makers and the public (Barbour, Stribling and Karr, 1995; Fausch *et al.*, 1990).

In this chapter, we use the multimetric TRophic IndeX (TRIX) (Giovanardi and Vollenweider, 2004; Vollenweider, Montanari and Rinaldi, 1998), defined as

$$\mathrm{TRIX} = \left[\log_{10} \left(\mathrm{Cha} \times \mathrm{aD\%} \times \mathrm{Oxygen} \times \mathrm{N} \times \mathrm{P} \right) + 1.5 \right] / 1.2,$$

where:

- Cha denotes chlorophyll-a concentration ($\mu\mathrm{g} \, \mathrm{L}^{-1}$);

- aD% denotes oxygen (absolute % derivate for saturation);

- N denotes mineral nitrogen ($\mathrm{N-NO_3} + \mathrm{N-NO_2} + \mathrm{N-NH_4}$; as $\mu\mathrm{g} \, \mathrm{L}^{-1}$); and

- P denotes total phosphorus ($\mu\mathrm{g} \, \mathrm{L}^{-1}$).

The index is characterised by factors that are directly related to primary productivity, including chlorophyll-a, oxygen saturation and nutritional factors (mineral, total nitrogen and phosphorus); it is therefore assumed to be a good indicator of eutrophication levels (trophic states). Numerically, the index is scaled from 0 to 10, covering a wide range of eutrophication levels from oligotrophic (nutrient-poor) to eutrophic. Since the use of TRIX is well developed for monitoring coastal and oceanic waters, it has become entrenched in Italian environmental legislation and must be applied when monitoring all Italian coastal seawaters (Leg. Dec. 258/00).

We use data from eight stations in the study area (Figure 9.1). They represent a north–south and inshore–offshore trophic gradient. The northern stations (4, 304, 9, 309) are influenced by inputs from the rivers Po and Reno, and hence are subject to a higher risk of eutrophication. The southern stations (14, 314 19, 319) have lower nutrient concentrations and productivity. TRIX time series are available from each station from January 1982 to December 2005; however, due to several missing values at some stations in the first couple of years, only data from 1984 onwards are used in the analyses below. Samples were taken at various times per month but we use average monthly values to obtain regularly spaced series.

At each location, salinity was measured: this may be an important covariate for TRIX since it indicates the extent to which the location is influenced by freshwater inputs from rivers and coastal basins. Another possible covariate is the month of sampling as it represents the season.

The objective of the study is to describe the effects of pollution (as quantified by the TRIX index) over time, in order to inform management strategies such as clean-up operations and to set new regulations. It is therefore useful to know whether all eight TRIX time series follow the same trend over time or whether there are different trends. To answer this question, we have to adopt a data analysis strategy that takes care of seasonality and the effects of salinity. The reason for this is as follows (see also the discussion in Section 3.2). If all TRIX time series follow a seasonal (i.e. monthly) pattern (and indeed they do), then the TRIX series are highly correlated but only due to seasonality. However, the seasonality is not the common pattern that we are interested in. The same holds for salinity; if all TRIX series are driven by salinity we need to know the common residual patterns. This means that we either have to remove the seasonal and salinity patterns from TRIX or include the seasonality component into the statistical models and simultaneously estimate the seasonal components, the salinity effects and the common trends. An extra problem may be that salinity itself exhibits a seasonal pattern.

We present two different statistical approaches. In the first we use additive models (Wood, 2006; see also Section 4.2) incorporating temporal autocorrelation along with a spatial residual correlation structures. Additive models represent the relationships between TRIX, salinity and the temporal trends simultaneously. The disadvantage is that it is difficult to build up a simultaneous model for time series at multiple stations, although not impossible, as we will show in this chapter (for an alternative line of attack in such situations, see Chapter 10). In the second approach, dynamic factor analysis (DFA) is used (Harvey 1989; Zuur et al., 2003a,b). This multivariate time series method extracts common trends in a similar way as principal component analysis extracts axes (see Section 6.2). However, DFA is especially designed for multiple time series and can incorporate (i) effects of covariates (e.g. month and salinity), (ii) common trends and (iii) residual correlation. A detailed mathematical description of DFA can be found in Zuur, Tuck and Bailey (2003) and Zuur et al. (2003), and various ecological examples can be found in Erzini (2005), Erzini, Inejih and Stobberup (2005), Chen et al. (2006), Laine et al. (2006), Zuur and Pierce (2004), Zuur, Ieno and Smith (2007) and others.

Rather than just presenting our final models, we will show the process of how they were obtained as well as discussing the relative merits of the intermediate models. As such, this chapter can be regarded as a guide to the model-building process for practitioners. In our view the road to the optimal models is as informative as the final model itself, if not more so. The dynamic factor analysis and data exploration were carried out in the software package **Brodgar**[1] version 2.6.4, and the additive modelling was done in R using the mgcv library (Wood, 2006).

---

[1] www.brodgar.com.

9.2   Data exploration

We start with an exploratory analysis of the data, following the protocol given in Zuur, Ieno and Elphick (2010) and using some of the techniques described in Chapter 2. As usual, the starting point is to plot each of the TRIX time series (Figure 9.2). Notice that the series from the offshore stations have missing values, especially in the first few years. There are no observations with extremely large or small TRIX values: this is confirmed by Cleveland dotplots (not shown here) and by boxplots. Thus, initially at least, we can work with the TRIX series as they stand without any need for data transformation. Similar graphs (not shown here) have been made for salinity, in this case to see if there was need for a data transformation to reduce the influential effects of outliers (see Section 3.1.5 for a discussion of the effects of outlying covariate values when fitting regression models). Once again, there are no obvious outliers so no transformation is required.

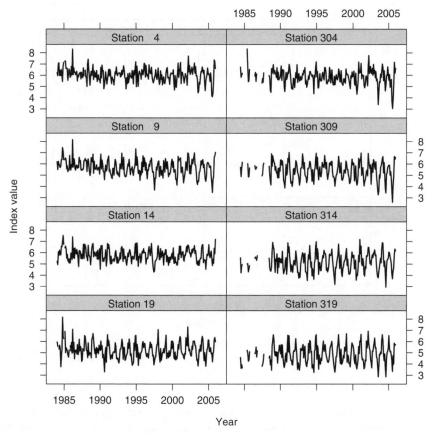

Figure 9.2   TRIX series from the eight stations. The plots are sorted from north (top) to south (bottom). The left column contains the inshore stations at 0.5 kilometres from the coast and the right column the offshore stations at 3 kilometres. Note the missing values at the start of the offshore series.

Next we examine the sample autocorrelation function (see Section 2.1.3) for each of the TRIX and salinity time series. The results in Figures 9.3 and 9.4 show that all series exhibit seasonal patterns. However, the seasonality is considerably stronger for TRIX than for salinity – thus any model that describes TRIX as a function of salinity may need to include additional covariates to capture fully the seasonal variation in the TRIX series – and is also stronger at the southern stations.

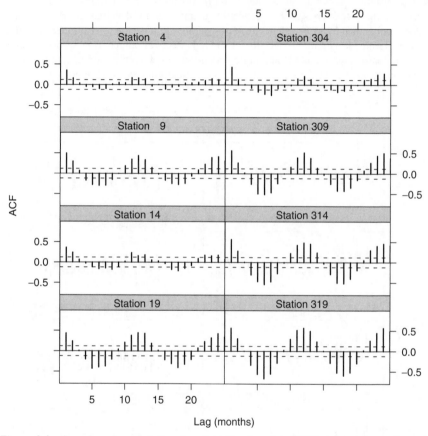

Figure 9.3   Sample autocorrelation functions for the eight TRIX time series at lags up to two years. Note that the southern stations (14, 314, 19, 319) have a stronger seasonal pattern.

A key aspect that should be investigated in a data exploration is collinearity: the correlation between covariates. As discussed in Section 3.2, a good tool to assess collinearity is the variance inflation factor (VIF); see Zuur, Ieno and Smith (2007) for various applications and Montgomery and Peck (1992) or Draper and Smith (1998) for a statistical discussion. To calculate the VIF for a particular covariate, it is regressed on all of the other covariates and the R^2 value (see Section 3.1) is calculated. A high R^2 indicates that most of the variation in the covariate of interest is explained by the other covariates,

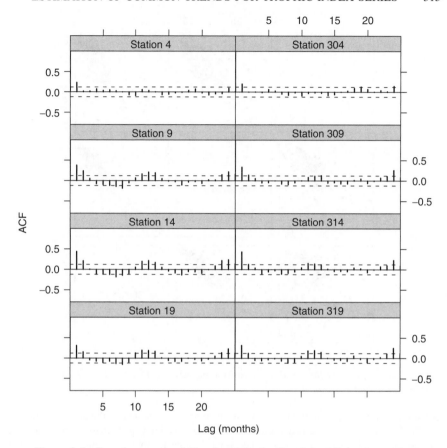

Figure 9.4   Sample autocorrelation functions for the eight salinity time series.

thereby indicating collinearity. The VIF is defined as

$$VIF = \frac{1}{1 - R^2}.$$

A VIF greater than 5 or 10 is generally considered as large (Montgomery and Peck, 1992; Quinn and Keough, 2002).

Here, we use VIFs to assess collinearity among the salinity time series. For all eight series, the VIFs exceed 5, thereby indicating severe collinearity. To find a subset of the series with low collinearity we can remove the series with the highest VIF and recalculate the VIFs for the remaining series, then repeat the process a couple of times until remaining VIFs are smaller than 5. After carrying out this process, the only salinity series remaining are those at stations 4, 9 and 19. Despite small VIFs, however, there is still a certain degree of collinearity between them, as illustrated by the scatterplots in Figure 9.5. In case the collinearity is due to shared seasonal patterns, we have also computed VIFs for salinity series in which seasonality was removed by subtracting the seasonal component obtained from an STL decomposition (see Section 4.3.1); the results were similar. Hence, the eight salinity time series behave in a very similar manner over time. This does not

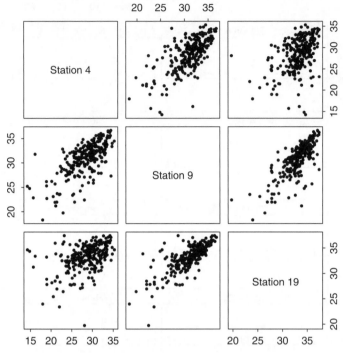

Figure 9.5   A pairplot for the three salinity time series with low VIF values. Units are permils (parts per thousand).

come as a surprise because the sampling distance between stations in the area is relatively small. However, from a statistical modelling perspective it suggests that if more than one salinity time series is used as a covariate for the same response variable, the individual effects of the different series will be difficult to disentangle from each other and hence will be estimated imprecisely.

As a last step in the data exploration part of our analysis, we investigate the relationship between TRIX and salinity. The scatterplots in Figure 9.6 indicate a clear negative relationship between the two variables at each station. There is also some suggestion that the strength of this relationship may increase moving from northwest (station 4) to southeast (station 319): this indicates that we should consider including interaction terms (see Section 3.2.2) in the models.

9.3   Common trends and additive modelling

In this section, we discuss a univariate smoothing method to estimate underlying patterns in multivariate time series. We use additive modelling (Faraway, 2005; Wood, 2006; Zuur, Ieno and Smith, 2007; Zuur *et al.*, 2009), taking a step-by-step approach to the development of an appropriate model for the TRIX series. Initially we fit separate models to the data from each station. The first such model has the form:

$$\text{TRIX}_t = \alpha + f_1(t) + f_2(\text{Salinity}_t) + \varepsilon_t \quad \text{where} \quad \varepsilon_t \sim N(0, \sigma^2). \quad (9.1)$$

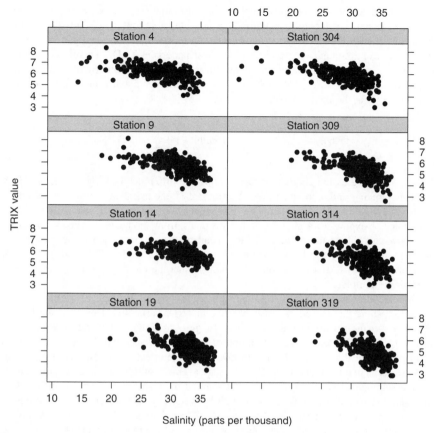

Figure 9.6   Scatterplots showing the TRIX–salinity relationship at all eight stations. All plots are on a common scale.

Here, 'TRIX$_t$' is the value of TRIX at time t (measured in years) for a particular station and 'Salinity$_t$' is the corresponding salinity; the notation $f_j(\cdot)$ refers to a smooth function (see Chapter 4, where such smooth functions were denoted by $m_j(\cdot)$), which is constrained to sum to zero so that the intercept α represents the overall mean TRIX value over time. The noise sequence (ε_t) is independently and normally distributed. Although Figure 9.6 indicated that the relationship between salinity and TRIX might be linear, the smooth components give the model a bit more flexibility if necessary. After all, if the relationship is really linear then a smoothing method should indicate this, and we can still revert to a linear regression model afterwards. To do the reverse (i.e. use linear regression and then inspect the residuals for any patterns) is feasible, although it is a bit more difficult and time consuming. The term $f_1(t)$ represents the long-term trend in the data and is of prime interest. To estimate the amount of smoothing for each smoothing curve, cross-validation (Wood, 2006) is used. Model (9.1) is fitted separately to data from each station to give estimated smoothing curves, intercepts and variances. Technical details of how to estimate smoothing functions are discussed in Chapter 4; see also Wood (2006) and Hastie and Tibshirani (1990).

The residuals of each model were plotted against month (not shown here) and showed a clear seasonal (i.e. monthly) pattern. This indicates a serious model misspecification. Hence, as already indicated by the autocorrelation function, the seasonal pattern in salinity is not strong enough to explain the seasonal pattern in TRIX. Therefore we add a categorical 'month' covariate to the model. In practice this is done by defining dummy variables $\{m_{jt} : j = 1, \ldots, 11\}$ such that m_{jt} takes the value 1 if time t corresponds to month j, so that the model becomes

$$\text{TRIX}_t = \alpha + f_1(t) + f_2\left(\text{Salinity}_t\right) + \sum_{j=1}^{11} \gamma_j m_{jt} + \varepsilon_t. \tag{9.2}$$

Here, γ_j is a regression coefficient representing the difference between the mean TRIX levels in month j and in month 12 (see Section 3.2.1); the intercept α now corresponds to the mean TRIX level in December. The error sequence (ε_t) is still independently normally distributed. Boxplots (not shown) of the residuals from model (9.2) versus month do not show any clear patterns in either the mean or the variance; hence the model now accounts successfully for the seasonality. A comparison of models (9.1) and (9.2), using an approximate F test (see Section 4.2), confirms that the monthly covariates are indeed needed by the model.

Model (9.2) assumes that the residual variance is constant. Biologically, it makes more sense to assume that there is a seasonal cycle in TRIX variability. However, the lack of seasonal structure in the residual variances suggests that any systematic differences are relatively small. Therefore the assumption of constant variance will be adequate for the main purpose of this study, which is to investigate the effect of trends. Another thing to look out for is that the monthly TRIX values are averages of different numbers of individual measurements (see Section 9.1) so that some of them are more precise than others. If this were problematic, it could be dealt with by weighting each observation according to its precision as in generalised least squares (GLS; see Section 3.3.3). Fortunately, differences in the precision of individual observations are not too much of a problem providing they are not related systematically to the covariates of interest (Section 3.1.1). In this case, there is no evidence from residual plots that the variances are related systematically to any of the covariates in the model.

Model (9.2) also assumes independence of the residuals. However, since the data are time series, the residuals may be autocorrelated. In fact, the sample autocorrelation functions (not shown) of the residuals at each station show that autocorrelation is indeed present. If we ignore this, there is a risk of erroneously concluding that covariates have a significant effect when in fact they do not (see Section 3.1.1). The authors of this chapter have been involved in various studies in which covariates that initially appeared to have a significant influence upon the response variable were subsequently found to be borderline or nonsignificant after accounting for autocorrelation. To improve the model, therefore, we will add an error autocorrelation structure.

9.3.1   Adding autocorrelation to the additive model

Pinheiro and Bates (2000) discuss various options for incorporating a temporal autocorrelation structure into additive or linear regression models. A model for the error structure needs to be chosen. Possible options are the AR(1), ARMA(p, q) and compound error structures (see Chapter 5). Ecological applications of different error models can be found in Zuur, Ieno and Smith (2007). For guidance on the choice of different correlation structures, see Section 5.1 or Pinheiro and Bates (2000, Section 5.3). Biological knowledge suggests that residual autocorrelation in TRIX values is unlikely to persist for many months; the residual correlograms from model (9.2), along with the corresponding partial autocorrelation functions (see Section 5.1.3), support this. Initially, therefore, we opt for the easiest model structure: a first-order autoregressive or AR(1) structure (see Section 3.3.3). Consequently, the correlation between two error components separated by a lag of h time units becomes

$$\text{Corr}\,(\varepsilon_t, \varepsilon_{t+h}) = \rho^{|h|}. \tag{9.3}$$

The parameter ρ, which must be estimated from the data, can be interpreted as the correlation between two successive months' residuals.

Figures 9.7 and 9.8 show, for each station, the estimated smoothing curves for salinity and year obtained from the additive model with an AR(1) autocorrelation structure. Note that salinity has a negative linear relationship for nearly all stations. For station 304, there seems to be a nonlinear salinity effect, but this is due to only a few observations with small salinity values. For the four northern stations (4, 304, 9, 309) the smoother for year indicates a general linear decrease of TRIX values over time, but for the other stations there is a slight increase after the mid 1990s.

The vertical axes in Figures 9.7 and 9.8 show the contribution of the smoother to the fitted TRIX values. Note that the range of the salinity contributions (the axis labels

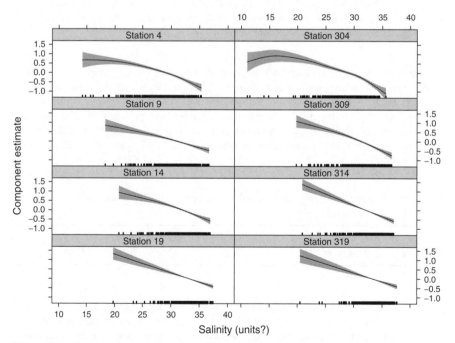

Figure 9.7   Estimated smoothing curves and pointwise 95% uncertainty bands for the salinity smoother for each station obtained by the additive model combining Equations (9.2) and (9.3).

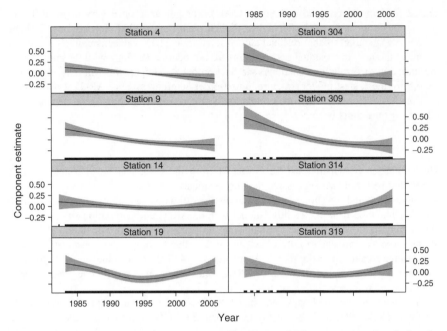

Figure 9.8   Estimated smoothing curves and pointwise 95 % uncertainty bands for the year smoother for each station obtained by the additive model combining Equations (9.2) and (9.3). This smoother represents the trend at each station.

run from -1 to 1.5) is considerably larger than that of the trend (from -0.25 to 0.5). This means that salinity has a much stronger effect on TRIX than the trend. According to approximate F tests, all salinity relationships are significantly different from zero at the 5 % level; for the trends, this is true for all stations except 4 and 14. The estimated autoregressive coefficients (ρ) at the eight stations are given in Table 9.1. These are all fairly similar except at the southernmost stations 19 and 319, where the autocorrelation seems weaker.

Table 9.1   Estimated autoregressive coefficients in the additive model fitted separately to data from each station.

Station	4	9	14	19	304	309	314	319
Coefficient	0.20	0.22	0.21	0.05	0.32	0.29	0.24	0.09

In case the estimated smooth functions are sensitive to the assumption of an AR(1) correlation structure, an additional set of models has been fitted at each station using an ARMA(1,1) correlation structure (see Section 5.1). Approximate likelihood ratio tests were used to compare the AR(1) and ARMA(1,1) models: at each station, the null hypothesis that the AR(1) structure is adequate was accepted at the 5 % level. Therefore we stick with the AR(1) structure for the subsequent modelling.

9.3.2 Combining the data from the eight stations

So far we have fitted additive models separately to the univariate time series from each station, resulting in eight models. The problem with this approach is that it ignores any structure that is shared between stations. To take advantage of any common structure that may be present, it may be beneficial to fit a single model to the data from all stations simultaneously. In this case, we will write TRIX_{st} to denote the TRIX at station s and time t.

Whenever models are fitted simultaneously to data from several stations, the residuals are likely to be correlated both in space and time. The implications of ignoring spatial (i.e. interstation) dependence are the same as those of ignoring temporal autocorrelation: effects can appear significant when in fact they are not. Thus it is important to account for possible interstation correlations. One way to do this is to specify a spatial correlation structure (for alternatives see Section 6.1, Chapter 8 of the present volume and Bowman, Giannitrapani and Scott, 2009). Figure 9.9 shows, for each pair of stations, the correlations between the residuals from the final additive models of the previous section, plotted as a function of interstation distance. Interstation correlation is clearly present, so we need to find a plausible correlation structure. One of the simplest such structures is the exponential:

$$\text{Corr}\left(\varepsilon_{ts_1}, \varepsilon_{ts_2}\right) = \exp\left(-d_{s_1,s_2}/\psi\right),$$

where d_{s_1,s_2} denotes the distance between stations s_1 and s_2. The dashed line in Figure 9.9 shows the fit of this model to the observed residual correlations. The fit is poor, over-estimating the correlations at small spatial lags and underestimating at large lags. This suggests that there may be a nugget effect (Webster and Oliver, 2001, Section 6.1), due

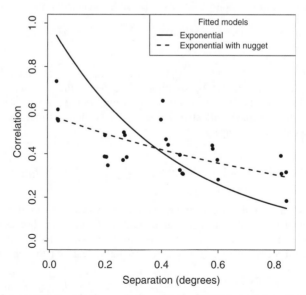

Figure 9.9   Interstation residual correlations for the additive model combining Equations (9.2) and (9.3), plotted against interstation distances (units are effectively degrees of latitude). Points represent the observed values; lines correspond to fitted correlation models.

either to measurement error or to fine-scale spatial TRIX variation. In this case, a better model for the spatial correlation structure would be

$$\text{Corr}\left(\varepsilon_{ts_1}, \varepsilon_{ts_2}\right) = \lambda \exp\left(-d_{s_1,s_2}/\psi\right) \tag{9.4}$$

for some $\lambda \in (0, 1)$. The solid line in Figure 9.9 shows the fit of this model, which seems reasonable. For the remainder of this section, therefore, we represent interstation dependence using an exponential correlation structure with a nugget effect.

The need to account for both spatial and temporal correlation causes a problem, because the mgcv library does not currently handle this (the correlation structures available are those provided by the nlme library of Pinheiro *et al.*, 2008). A solution is to account for the temporal correlation by including lagged TRIX values as additional covariates in the model. Note that this is directly equivalent to specifying an autoregressive correlation structure at each station, although now we need to be careful when interpreting the other model components (see Section 3.3.3).

To come up with a sensible model for TRIX at all stations simultaneously, it is helpful to try and separate model components that are likely to vary between stations from those that are likely to be the same. The exploratory analysis, together with the results from fitting models separately at each station, suggests that the salinity and trend effects may be station-dependent, along with the seasonal cycle. However, for the additive models with AR(1) correlation structures, all of the residual variance estimates were between 0.18 and 0.21, with the exception of station 309 for which the value is 0.16. Thus the residual variances appear approximately the same for all stations; this is consistent with the informal conclusions, reported above, from the analysis of residuals from model (9.2). Similarly, the AR(1) correlation parameters in Table 9.1 are fairly similar at all except the two southern stations.

Combining all of these considerations, a possible multistation model for TRIX is

$$\text{TRIX}_{st} = \alpha_s + f_{1,s}(t) + f_{2,s}\left(\text{Salinity}_{st}\right) + \sum_{j=1}^{11} \gamma_{sj} m_{jt} + \phi \text{TRIX}_{s(t-1)} + \varepsilon_{st}, \tag{9.5}$$

where $\varepsilon_{st} \sim N(0, \sigma^2)$ and $\text{Corr}\left(\varepsilon_{s_1t}, \varepsilon_{s_2t}\right) = \lambda \exp\left(-d_{s_1,s_2}/\psi\right)$. However, this model has 16 smoothing functions: it can be simplified (and hence the computational load can be reduced) by using the results from the previous section, which showed that the salinity–TRIX relationship is effectively linear at all stations so that the salinity effect can be represented parametrically:

$$\text{TRIX}_{st} = \alpha_s + f_{1,s}(t) + \left(\beta_s \times \text{Salinity}_{st}\right) + \sum_{j=1}^{11} \gamma_{sj} m_{jt} + \phi \text{TRIX}_{s(t-1)} + \varepsilon_{st}. \tag{9.6}$$

This model contains 'only' eight smoothing curves, which reduces the computational burden. Nonetheless, it still contains a large number of parameters because there are 11 monthly coefficients $\{\gamma_{sj} : j = 1, \ldots, 11\}$ at each of the eight stations. It may be more economical to use a circular smoother (see Section 4.2) to represent the seasonal structure:

$$\text{TRIX}_{st} = \alpha_s + f_{1,s}(t) + \left(\beta_s \times \text{Salinity}_{st}\right)$$
$$+ f_{3,s}\left(\text{Month}_t\right) + \phi \text{TRIX}_{s(t-1)} + \varepsilon_{st}, \tag{9.7}$$

where 'Month$_t$' is the month number (1 to 12) corresponding to time t.

A comparison of models (9.6) and (9.7), using an approximate likelihood ratio test, suggests that the seasonal effect can be represented as a smooth function with between 3 and 5 effective degrees of freedom at each station (see Section 4.1.4 for a discussion of effective degrees of freedom in smoothing models). This saves about 52 degrees of freedom in total. With the smoothing parameters chosen by cross-validation, model (9.7) contains 62.4 effective degrees of freedom and yields a log-likelihood of -1679.2.

At this point the advantage of fitting an additive model simultaneously to all data becomes apparent: we can look at the various terms in model (9.7) and see how they change from station to station. In particular, we can simplify the model by assuming, for example, that the salinity effect is the same at all stations. If we do this and retain the same error structure, the simplified models will be nested within (i.e. can be obtained as special cases of) model (9.7). Therefore likelihood ratio tests can be used to determine whether the simplifications are justified. When doing this, however, we should remember that currently available tests for comparing smoothing models are approximate at best (see Section 4.2). This means that we have to exercise an element of judgement when interpreting the test results.

The first simplification we try is to see whether the salinity effect is the same at all stations. This is done by fitting model (9.7) without any salinity–station interactions, i.e. in which the coefficients $\beta_1, \ldots, \beta_8$ are replaced by a single salinity coefficient β. The resulting model has 38.9 effective degrees of freedom and a log-likelihood of -1776.8. The likelihood ratio statistic (see Section 3.5.2) is therefore $2 \times (-1679.2 + 1776.8) = 195.3$. Compared with a chi-squared distribution on $62.4 - 38.9 = 23.5$ degrees of freedom, this yields a p-value well below 10^{-4}. Even allowing for the approximate nature of the test, this shows strong evidence that the TRIX–salinity relationship differs between stations (or at least for some stations). The mean of the estimated slopes $\{\beta_s\}$ is -0.084 and is significantly different from zero at the 0.1 % level. The individual slopes show that the TRIX–salinity relationship becomes stronger (more negative) as one moves from northwest to southeast: this is consistent with the earlier exploratory analysis.

Next we look to see whether station-specific seasonal effects are needed. Fitting a model with a common seasonal component yields a log-likelihood of -1733.2, with 34.4 effective degrees of freedom. Comparing this with model (9.7) yields a likelihood ratio statistic of 108.1 on 28 degrees of freedom: once again, this provides convincing evidence that the seasonal cycle varies between stations.

Interestingly, the degrees of freedom chosen using cross-validation for the trend smoothers $f_{1,1}(\cdot), \ldots, f_{1,8}$ are all equal to 1: the estimated trends are all linear. This suggests that we could switch to a parametric representation of the trends as well as the salinity. However, for the moment we stick with the nonparametric analysis. The last question we address here is whether we actually need the interaction between the trends and stations or whether one smoother will do the job. We will therefore compare model (9.7) with the following model:

$$\text{TRIX}_{st} = \alpha_s + f_1(t) + \left(\beta_s \times \text{Salinity}_{st}\right)$$
$$+ f_{3,s}(\text{Month}_t) + \phi \text{TRIX}_{s(t-1)} + \varepsilon_{st}. \tag{9.8}$$

Note that in model (9.8) there is only one trend whereas (9.7) has eight trends. The likelihood ratio statistic for comparing the models is 23.73 on 5.8 effective degrees of freedom, yielding an approximate p-value of 0.0005. This suggests that the trends do indeed vary between stations.

So, is model (9.7) our final model? With 62.4 effective degrees of freedom it is quite complex, even if it does describe the variation of TRIX at eight locations simultaneously. One possible simplification is to try and estimate common trends within 'similar' groups of stations. In the earlier station-specific analysis, the shape of the trends in Figure 9.8 indicated that there may be north–south and inshore–offshore differences. This could be due to dilution of nutrients from the river input due to currents. To investigate this we try fitting a model with four different smoothers for the trend so that the site pairs (4, 9), (304, 309), (14, 19), (314, 319) each share a common trend. This is again a special case of (9.7). The likelihood ratio statistic is 1.5 on 4.1 effective degrees of freedom, leading to an approximate p-value of 0.84. Therefore this simplification is justified, suggesting that the stations can be subdivided into four pairs as far as trends are concerned.

The estimated trends for the four pairs of stations are given in Figure 9.10. Note that all are linear, with the strongest trends in the north: the estimated trends for the southern stations do not differ significantly from zero. When interpreting these trends, it should be remembered that the model also includes lagged TRIX values so that the modelled trends do not correspond to the actual change in TRIX over time. However, since the modelled trends are linear, the corresponding changes in TRIX are also linear (see Section 3.3.3). Moreover, the temporal correlation is relatively weak (the regression coefficient for the lagged TRIX value is 0.11) so that the trends displayed in Figure 9.10 provide a reasonable representation of overall change.

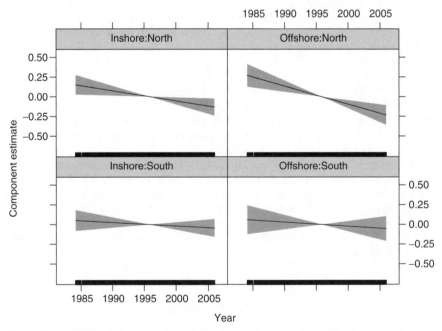

Figure 9.10   Estimated temporal trends for four pairs of stations. 'Northern' stations are numbers 4, 9, 304 and 309; 'inshore' stations are numbers 4, 9, 14 and 19.

9.3.3   Doing it all within a parametric model

Possible criticisms of the nonparametric approach of the previous section are firstly that graphs are needed to visualise the covariate effects, and secondly that model comparison is difficult. However, the results above suggest that both the trend and salinity effects are linear and that the seasonal effects can be modelled by a smooth function. Therefore a possible solution is to switch to a parametric framework, using generalised least squares (GLS) to account for correlation and potential changes in variance (which were not investigated in the previous section). To represent the smoothly varying seasonal cycle within a parametric framework, we can use a Fourier representation (see Section 3.2.1): since the seasonal smooths in the nonparametric model all had fewer than 6 effective degrees of freedom, it seems reasonable to consider sine and cosine components with cycle lengths of 12 months, 6 months and 4 months. We can also allow for possible nonlinear trends by including quadratic functions of time in the model. However, if we use t and t^2 as covariates then we will encounter collinearity: to avoid this it is advisable to centre the time index as $\tilde{t} = t - \bar{t}$, say, where $\bar{t}$ is the centre of the observation period, and then to work with covariates $\tilde{t}$ and $\tilde{t}^2$. This suggests the following parametric model (note that the coefficient notation differs from the previous section):

$$
\text{TRIX}_{st} = \alpha_s + \sum_{j=1}^{3} \left[\beta_{1js} \cos\left(\frac{2\pi j \times \text{Month}_t}{12} \right) + \beta_{2js} \sin\left(\frac{2\pi \times \text{Month}_t}{12} \right) \right]
$$

$$
+ \gamma_{1s}\tilde{t} + \gamma_{2s}\tilde{t}^2 + \left(\gamma_{3s} \times \text{Salinity}_{st} \right) + \phi\text{TRIX}_{s(t-1)} + \varepsilon_{st}. \tag{9.9}
$$

As before, we can allow for autocorrelation in the errors and investigate similarities and differences between stations by fitting simplified versions of model (9.9) and carrying out likelihood ratio tests. However, the computing time is considerably less than for the additive models of the previous section. This provides the opportunity to explore the model structure in more detail: in particular, we can look for an optimal model in terms both of error structure and of the covariates ($\tilde{t}$, $\tilde{t}^2$, salinity, seasonal components and interactions).

As far as the error structure is concerned, we have not yet formally addressed the question of whether the error variance is the same for all sites and months. While answering questions about the random components of a model, it is general practice to include as many covariates as possible: once an optimal model for the random components has been identified, the systematic part of the model (i.e. covariate structure) can be explored in more detail (Fitzmaurice, Laird and Ware, 2004). Likelihood based comparisons of different structures for the random component, but with the same covariates, require restricted maximum likelihood (REML) estimation – see Section 6.1. In contrast, comparisons of models with the same structure for the random components, but with different covariates, require maximum likelihood estimation (Pinheiro and Bates, 2000).

We first fit model (9.9) with the spatial autocorrelation structure (9.4) on the errors, and then a model in which we also allow different variances at each station. So instead of assuming that $\varepsilon_{st} \sim N(0, \sigma^2)$ we use $\varepsilon_{st} \sim N(0, \sigma_s^2)$ for $s = 1, \ldots, 8$. As these models are nested (the second is an extension of the first), a likelihood ratio test can be applied. Results indicate no evidence to reject the null hypothesis that residuals have the same spread at each station ($p = 0.39$). We next fit a model in which each month is allowed to have a different variance: the likelihood ratio test indicates a significant model improvement ($p = 0.012$). Hence, the residual variability differs by month, but not by station. A likelihood ratio test also confirms that the interstation correlation structure is required ($p < 0.001$). So, the optimal model in terms of error structure contains interstation correlations and different variances per month. We can now change the estimation method to maximum likelihood and determine the optimal set of fixed covariates. The results suggest that all of the covariates in (9.9) are significant (taking the seasonal sine and cosine terms as a single group) and that, with the exception of the nonlinear trend component $\bar{t}^2$, all effects vary between stations. This is essentially the same conclusion as for the nonparametric analysis: however, the parametric model has also revealed the seasonal changes in residual variation and suggests some nonlinearity in the trend that was not apparent from the simultaneous nonparametric model fit (although the nonparametric fits at individual stations did suggest some nonlinearity of this type).

9.4   Dynamic factor analysis to estimate common trends

The additive modelling in the previous section showed that the eight TRIX time series exhibit (a) mildly nonlinear long-term trends, (b) a linear relationship with salinity and (c) a seasonal pattern. The trends vary between stations and are strongest in the north. We now use dynamic factor analysis (DFA) as an alternative way to estimate common trends and effects of covariates.

9.4.1   The underlying model

DFA is closely related to the Kalman filter and smoother, in particular using random walks to represent trends (see Section 5.5). Mathematical details can be found in Zuur, Tuck and Bailey (2003) and Zuur *et al.* (2003); see also Shumway and Stoffer (1982, 2000) or Harvey (1989) for a full discussion of the model formulation. Other forms of DFA exist (Mendelsohn and Schwing, 2002; Molenaar, 1985) but are not discussed here. DFA models the eight TRIX time series as a function of M common trends and effects of covariates (salinity and month). In addition, interstation residual correlation is allowed.

In the additive modelling above we used station-specific salinity series as covariates for TRIX. However, although the DFA model in principle allows more than one continuous covariate, due to software restrictions we are only able to use one such covariate in this study. Therefore we use the average salinity value at all stations as a covariate in the DFA. This should not be too much of a problem, because the results from the data exploration and additive models show that most salinity series are highly correlated with each other and have similar effects on TRIX.

The mathematical notation used in this section is slightly different from the previous section. Let $\mathbf{y}_t$ be a vector containing the eight TRIX values in the tth month of the record and let $\mathbf{x}_t$ be a corresponding vector of covariates. The DFA model is given by the pair of equations

$$\mathbf{y}_t = \boldsymbol{\mu} + A\mathbf{z}_t + \boldsymbol{\beta}\mathbf{x}_t + \boldsymbol{\varepsilon}_t \qquad (9.10)$$

and

$$\mathbf{z}_t = \mathbf{z}_{t-1} + \boldsymbol{\eta}_t. \qquad (9.11)$$

The easiest way to explain this model is if only one common trend ($M = 1$) is used. In this case, $\mathbf{z}_t$ is a scalar and represents a random walk trend: $z_t = z_{t-1} + \eta_t$. The error term η_t is normally distributed with expectation zero and variance σ_η^2. For small values of this variance, the trend (z_t) evolves quite smoothly but larger values produce more wiggly curves. For the TRIX series with station-averaged salinity as a single covariate, if $M = 1$ then (9.10) becomes

$$
\begin{pmatrix} \text{TRIX}_{1t} \\ \text{TRIX}_{2t} \\ \vdots \\ \text{TRIX}_{8t} \end{pmatrix} = \begin{pmatrix} \mu_1 \\ \mu_2 \\ \vdots \\ \mu_8 \end{pmatrix} + \begin{pmatrix} a_1 \\ a_2 \\ \vdots \\ a_8 \end{pmatrix} z_t + \begin{pmatrix} \beta_1 \\ \beta_2 \\ \vdots \\ \beta_8 \end{pmatrix} \text{Salinity}_t + \begin{pmatrix} \varepsilon_1 \\ \varepsilon_2 \\ \vdots \\ \varepsilon_8 \end{pmatrix}. \qquad (9.12)
$$

Each TRIX time series (y_{st}) is modelled as an intercept μ_s (as in linear regression) along with contributions from the trend, salinity and noise. The trend affects each station differently according to the factor loadings $\{a_s\}$. Similarly, the effect of the univariate salinity variable is station-specific and is modelled as in linear regression. The noise vector ε_t is assumed to be normally distributed with expectation $\mathbf{0}$ and covariance matrix $\mathbf{H}$. The trend (z_t) represents the common information in the TRIX time series that is not already explained by salinity. A high value of the factor loading a_s indicates that the trend is important for the TRIX at station s, whereas if a_s is close to zero the trend is not important at this station. A negative factor loading means that the contribution of the trend to the fitted values is opposite to the trend pattern. To simplify the interpretation of the factor loadings it is useful to standardise the TRIX time series.

What happens if there are two common trends $(M = 2)$? The random walk model (9.11) becomes $\mathbf{z}_t = \mathbf{z}_{t-1} + \boldsymbol{\eta}_t$, where $\mathbf{z}_t = (z_{1t}, z_{2t})'$, and (9.10) is identical to (9.12) except that the second term on the right-hand side becomes

$$
\begin{pmatrix} a_{11} & a_{12} \\ a_{21} & a_{22} \\ \vdots & \vdots \\ a_{81} & a_{82} \end{pmatrix} \begin{pmatrix} z_{1t} \\ z_{2t} \end{pmatrix} = \mathbf{A}\mathbf{z}_t.
$$

The 8×2 matrix $\mathbf{A}$ contains the factor loadings, the magnitudes of which indicate whether each individual TRIX series is related to either, both or neither of the two trends. To choose between models with different numbers of trends or covariates, Akaike's Information Criterion (AIC) can be used (see Section 3.2.3). It is a combined measure of goodness of fit and model complexity, and its definition for the DFA model is given in Zuur, Tuck and Bailey (2003) and Zuur et al. (2003).

Finally, we discuss the error covariance matrix $\mathbf{H}$. Two options used in Zuur, Tuck and Bailey (2003) and Zuur et al. (2003) are a diagonal matrix with different variances and an unstructured covariance matrix with nonzero off-diagonal elements. Under the first option, the residuals at different stations are assumed to be uncorrelated with each other but the nondiagonal option allows dependence between stations. The price to be paid for this is an increase in the number of parameters to be estimated, although the structure could in principle be simplified (and the number of parameters reduced) by using a spatial correlation model such as (9.4).

Just as in linear regression models, seasonal patterns can be included using a categorical 'month' variable. In this case the model becomes

$$
\mathbf{y}_t = \boldsymbol{\mu} + \mathbf{A}\mathbf{z}_t + \boldsymbol{\beta}\mathbf{x}_t + \sum_{j=1}^{11} \boldsymbol{\gamma}_j m_{jt} + \boldsymbol{\varepsilon}_t,
$$

where the dummy variable m_{jt} is defined as in Section 9.3 and γ_j is a regression parameter vector of dimension 8×1, allowing for a different month effect at each station.

Various DFA models with different sets of covariates have been fitted. Table 9.2 shows a selection of these, with their AIC values. In addition to the models shown, we fitted models with a nondiagonal covariance matrix H and only salinity as covariate. However, numerical problems were encountered in the estimation of these models, probably because the shared seasonality in the TRIX series was interpreted as very high interstation correlation, which led to a singular covariance matrix H.

In the models with no covariates and one common trend ($M = 1$) the 'trend' estimates correspond to the seasonal cycle: all stations are related to this because all TRIX series show strong seasonality. The AICs in Table 9.2 clearly show that the best models are those with (a) month and salinity as covariates, (b) a nondiagonal error covariance matrix and (c) $M = 1$ or $M = 2$ common trends. The difference in the AIC between these two models is rather small (3342.5 versus 3341.6) and the choice as to which one should be presented should be based on (a) biological interpretation and (b) a model validation. In the model with $M = 2$, the factor loadings for the second common trend were considerably smaller than those for the first trend. This indicates that the second trend is considerably less important. Moreover, the model validation (plots of residuals versus time, histograms, quantile–quantile plots and other model validation tools) gave no reason to prefer the second model ($M = 2$) above that of the first ($M = 1$). Therefore we present the model with one common trend below.

Table 9.2   DFA models fitted to the TRIX series. M stands for the number of common trends and H is the error covariance matrix. The lower the AIC, the better the model.

H diagonal			H nondiagonal		
M	Covariates	AIC	M	Covariates	AIC
1	–	4035.5	1	–	3691.4
2	–	3814.3	2	–	3685.3
3	–	3817.8	3	–	3682.3
			4	–	3692.9
1	Month	3829.1	1	Month	3490.0
2	Month	3708.8	2	Month	3492.3
3	Month	3709.8			
1	Month and Salinity	3765.7	1	Month and Salinity	**3342.5**
2	Month and Salinity	3653.6	2	Month and Salinity	**3341.6**
3	Month and Salinity	3657.0	3	Month and Salinity	3354.2

The estimated trend is presented in Figure 9.11(a). The first year of the trend should be interpreted with care, as there were many missing values. Nevertheless, it is clear that there is a general decrease during the 1980s and early 1990s, followed by a stable

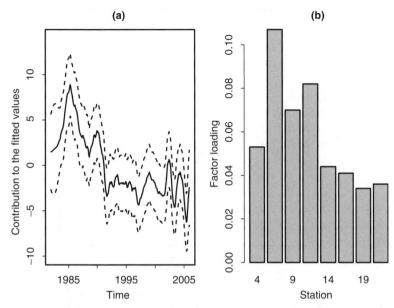

Figure 9.11   Results from DFA using only one common trend: (a) estimated trend (solid line) and 95% confidence bands, (b) factor loadings for each station corresponding to the trend in (a).

period of minor changes. However, for which station(s) is this trend important? The factor loadings in Figure 9.11(b) show that it mainly affects the northern four stations (4, 304, 9 and 309).

The error covariance matrix *H* can either be presented in a table or visualised by converting to a correlation matrix and then using metric multidimensional scaling (MDS) to show which stations have similar residual patterns. MDS (Everitt and Dunn, 2001; Zuur, Ieno and Smith, 2007) is a multivariate technique that takes a set of dissimilarities and returns a set of points such that the distances between the points are approximately equal to the dissimilarities. In our case, the dissimilarities are calculated by subtracting the interstation correlations from 1: thus two stations with perfect correlation will have zero dissimilarity. Figure 9.12 shows the MDS ordination. The first axis represents 38% of the variation and is twice as important as the second axis. Note that there is a clear grouping of stations along the first axis: stations 4, 304, 14 and 314 versus stations 9, 309, 19 and 319. Hence, there is a certain degree of structure in the unexplained variation.

The regression parameters for salinity are presented in Table 9.3. Note that all slopes are negative and significantly different from zero at the 5% level. This is consistent with the results from the additive modelling exercise previously; however, the DFA results suggest, if anything, that the salinity–TRIX relationships are weaker at the southern stations whereas before they appeared stronger there.

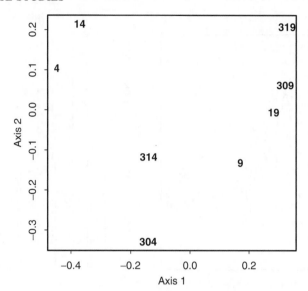

*Figure 9.12   MDS ordination of the error covariance matrix **H**, after conversion to a correlation matrix. The numbers in the graph refer to the stations. Stations with similar residual patterns appear close to each other on this plot.*

Table 9.3   Estimated regression coefficients $\{\beta_s\}$ for salinity in model (9.12) fitted to the TRIX series.

Station	Estimated slope	Standard error	t-value
4	−0.20	0.02	−12.14
304	−0.20	0.02	−11.75
9	−0.14	0.01	−9.74
309	−0.17	0.01	−12.96
14	−0.16	0.02	−9.23
314	−0.15	0.01	−10.14
19	−0.12	0.01	−8.08
319	−0.11	0.01	−7.63

9.5   Discussion

In this chapter we have used two rather different modelling approaches to investigate trends in eutrophication, as measured using TRIX, along the coast of the Emilia-Romagna region. Additive modelling indicates an overall seasonal pattern, a salinity effect and, for the four northern stations, a decreasing temporal trend. The decrease from the early 1990s is consistent with a nutrient reduction policy that comprised mainly the elimination of phosphorus from detergents and, where possible, the introduction of fertirrigation, lagoonage and wastewater treatment by phytodepuration (for an overview of some of the measures that have been taken, see De Wit and and Bendoricchio, 2001, and references therein). Further improvements within the additive model allow for different variances in each month (as suggested by the GLS), although this conclusion is not immediately clear from the additive model residual plots.

The DFA also shows a salinity and month effect, along with a single common trend that once again indicates a decrease in TRIX values everywhere except at the southernmost stations 19 and 319. The DFA trend estimate is less smooth than that from the additive model, indicating a decrease between the mid 1980s and early 1990s and then a period of stability.

In the literature, DFA has been applied to various ecological data sets. In some of these publications, the length of the time series were rather short. For example, Chen *et al.* (2006) used annual squid time series of length 12–14 years. They found that slightly different results were obtained if the DFA model was refitted with different starting values for the parameters, and compromised by presenting average AIC values from five DFA runs. Zuur and Pierce (2004) used 24 time series of 20–25 years and found that the numerical algorithm for parameter estimation converged rather slowly due to the large number of parameters in the model. Hence, care is required if DFA is applied to time series with relatively few observations. In the present example, however, there were relatively few time series with few missing values and the series were all considerably longer than in the cited references. The AIC values, regression parameters, error covariances and other numerical parameters were identical up to the third digit for multiple DFA runs with different starting values.

The main advantage of the additive modelling approach is that the user gains a much greater insight into the structure of the data than is possible with DFA. However, more time and expertise is required to interpret the additional numerical output and, moreover, the DFA is more capable of detecting rapid changes due to the random walk representation of the trend. This fundamental difference in trend representations – the one as a smooth function of time with autocorrelated errors and the other as a random walk – should be kept in mind when considering the differences between the modelling results above. In qualitative terms, however, the two sets of conclusions are similar.

Building on analyses such as those reported above, an additional possibility is to use the modelling results to improve the sampling effort. Zuur, Ieno and Smith (2007, Chapter 20) showed how an additive model can be used to obtain some insight into required sample sizes, and similar ideas could also be adopted within both of the frameworks presented in this chapter.

Acknowledgement

The authors would like to thank Martin Solan for useful comments on an earlier draft.

References

Barbour, M. T., Stribling, J. B. and Karr, J. (1995) Multimetric approach for establishing biocriteria and measuring biological conditions. In *Biological Assessment and Criteria – Tools for Water Resources Planning and Decision Making* (eds. W. S. Davis and T. P. Simon). Lewis Publishers, Boca Raton, Florida. pp. 63–77.

Bowman, A., Giannitrapani, M. and Scott, E. M. (2009) Spatiotemporal smoothing and sulphur dioxide trends over Europe. *Journal of the Royal Statistical Society, Series C*, **58**, 737–752.

Chen, C. S., Pierce, G. J., Wang, J., Robin, J. P., Poulard, J. C., Pereira, J., Zuur, A. F., Boyle, P. R., Bailey, N., Beare, D. J., Jereb, P., Ragonese, S., Mannini, A. and Orsi-Relini, L. (2006) The apparent disappearance of Loligo forbesi from the south of its range in the 1990s: trends in Loligo spp. abundance in the northeast Atlantic and possible environmental influences. *Fisheries Research*, **78**, 44–54.

De Wit, M. and Bendoricchio, G. (2001) Nutrient fluxes in the Po basin. *Science of the Total Environment*, **273**, 147–161.

Diaz, R., Solan, M. and Valente, R. (2004) A review of approaches for classifying benthic habitats and evaluating habitat quality. *Journal of Environmental Management*, **73**, 165–181.

Draper, N. R. and Smith, H. (1998) *Applied Regression Analysis*, 3rd edition. John Wiley and Sons, Inc., New York.

Erzini, K. (2005) Trends in NE Atlantic landings (southern Portugal): identifying the relative importance of fisheries and environmental variables. *Fishing and Oceanography*, **14**, 195–209.

Erzini, K., Inejih, C. and Stobberup, K. A. (2005) An application of two techniques for the analysis of short, multivariate non-stationary time series of Mauritanian trawl survey data. *ICES Journal of Marine Science*, **62**, 353–359.

Everitt, B. S. and Dunn, G. (2001) *Applied Multivariate Data Analysis*, 2nd edition. Arnold, London.

Faraway, J. J. (2005) *Extending the Linear Model with R: Generalized Linear, Mixed Effects and Nonparametric Regression Models*. Chapman & Hall/CRC, Boca Raton, Florida.

Fausch, K. D., Lyons, J., Karr, J. R. and Angermeier, P. L. (1990) Fish communities as indicators of environmental degradation. *American Society Symposium*, **8**, 123–144.

Fitzmaurice, G. M., Laird, N. M. and Ware, J. H. (2004) *Applied Longitudinal Analysis*. John Wiley & Sons, Inc., Hoboken, New Jersey.

Fletcher, R. L. (1996) The occurrence of green tides – a review. In *Marine Benthic Vegetation: Recent Changes and the Effects of Eutrophication* (eds. W. Schramm and P. H. Neinhuis). Springer, New York. pp. 7–44.

Giovanardi, F. and Vollenweider, R. A. (2004) Trophic conditions of marine coastal waters: experience in applying the trophic index TRIX to two areas of the Adriatic and Tyrrhenian seas. *Journal of Limnology*, **63**(2), 199–218.

Harvey, A. C. (1989) *Forecasting, Structural Time Series Models and the Kalman Filter*. Cambridge University Press, Cambridge.

Hastie, T. and Tibshirani, R. (1990) *Generalized Additive Models*. Chapman & Hall, London.

Laine, A. O., Andersin, A. B., Leiniö, S. and Zuur, A. F. (2006) Stratification induced hypoxia as a structuring factor of macrozoobenthos in the open Gulf of Finland (Baltic Sea). *Journal of Sea Research*.

Mendelsohn, R. and Schwing, F. B. (2002) Common and uncommon trends in SST and wind stress in the California and Peru–Chile current systems. *Progress in Oceanography*, **53**(2–4), 141–162.

Molenaar, P. C. M. (1985) A dynamic factor model for the analysis of multivariate time series. *Psychometrika*, **50**, 181–202.

Montgomery, D. C. and Peck, E. (1992) *Introduction to Linear Regression Analysis*. John Wiley & Sons, Inc., New York.

Pinheiro, J. C. and Bates, D. M. (2000) *Mixed-Effects Models in S and S-PLUS*. Springer-Verlag, New York.

Pinheiro, J. C., Bates, D. M., DebRoy, S. and Sarkar, D. (2008) nlme: *Linear and Nonlinear Mixed Effects Models*. R package version 3.1-89.

Pompei, M., Ghetti, A., Milandri, A. and Mazziotti, C. (1995) Fioriture microalgali ed evoluzione dei principali popolamenti fitoplanctonici nelle acque costiere Emiliano-Romagnole dal 1982 al 1994. In *Proceedings of Conference 'Evoluzione dello stato trofico in Adriatico analisi di interventi attuati e future linee di intervento'*, pp. 51–60, Marina di Ravenna, Italy, September.

Quinn, G. P. and Keough, M. J. (2002) *Experimental Design and Data Analysis for Biologists*. Cambridge University Press, Cambridge.

Rinaldi, A. and Montanari, G. (1988) Eutrophication in Emilia-Romagna coastal waters in 1984–1985. *Annals of New York Academy of Science*, **534**, 959–977.

Shumway, R. H. and Stoffer, D. S. (1982) An approach to time series smoothing and forecasting using the EM algorithm. *Journal of Time Series Analysis*, **3**, 253–264.

Shumway, R. H. and Stoffer, D. S. (2000) *Time Series Analysis and Its Applications*. Springer-Verlag, New York.

Vollenweider, R. A. (1968) Scientific fundamentals of eutrophication of lakes and flowing waters, with particular reference to nitrogen and phosphorus as factors in eutrophication. Technical Report, DAS/CSI/68.27., OECD, Paris.

Vollenweider, R. A. (1981) Eutrophication a global problem. *WHO Water Quality Bulletin*, **6**(3), 59–62.

Vollenweider, R. A., Montanari, F. G. G. and Rinaldi, A. (1998) Characterization of the trophic conditions of marine coastal waters, with special reference to the NW Adriatic Sea: proposal for a trophic scale, turbidity and generalized water quality index. *Environmetrics*, **9**, 329–357.

Vollenweider, R. A., Rinaldi, A. and Montanari, G. (1992) Eutrophication, structure and dynamics of a marine coastal system: results of 10-year monitoring along the Emilia-Romagna coast (northwest Adriatic Sea). In *Science of the Total Environment, Supplement 1992 – Marine Coastal Eutrophication* (eds. R. A. Vollenweider, R. Marchetti and R. Viviani), pp. 63–106.

Webster, R. and Oliver, M. A. (2001) *Geostatistics for Environmental Scientists*. John Wiley & Sons, Ltd, Chichester.

Wood, S. N. (2006) *Generalized Additive Models: An Introduction with* R. Chapman & Hall/CRC, Boca Raton, Florida.

Zuur, A. F., Ieno, E. N. and Elphick, C. S. (2010) A protocol for data exploration to avoid common statistical problems. *Methods in Ecology and Evolution*, **1**(1), 3–14.

Zuur, A. F., Ieno, E. N. and Smith, G. M. (2007) *The Analysis of Ecological Data*. Springer-Verlag, New York. 700 pp.

Zuur, A. F. and Pierce, G. J. (2004) Common trends in northeast Atlantic squid time series. *Journal of Sea Research*, **52**, 57–72.

Zuur, A. F., Tuck, I. D. and Bailey, N. (2003) Dynamic factor analysis to estimate common trends in fisheries time series. *Canadian Journal of Fisheries and Aquatic Sciences*, **60**, 542–552.

Zuur, A. F., Fryer, R. J., Jolliffe, I. T., Dekker, R. and Beukema, J. J. (2003) Estimating common trends in multivariate time series using dynamic factor analysis. *Environmetrics*, **14**(7), 665–685.

Zuur, A. F., Ieno, E. N., Walker, N., Saveliev, A. A. and Smith, G. M. (2009) *Mixed Effects Models and Extensions in Ecology with* R. Springer-Verlag, New York. xxii + 574 pp.

10

A space–time study on forest health

Thomas Kneib[1] and Ludwig Fahrmeir[2]

[1]*Department of Mathematics, Carl von Ossietzky University Oldenburg, D-26111 Oldenburg, Germany*
[2]*Department of Statistics, Ludwig-Maximilians-University Munich, Ludwigstraße 33, D-80539 München, Germany*

10.1 Forest health: survey and data

Vital forests play an important role in the ecosystem due to their regulatory impact on both the climate and the water cycle. However, a prerequisite for these balancing functions is a sufficiently healthy forest. In this case study, we will analyse longitudinal data on forest health to identify potential factors influencing the health status of trees. In addition to covariates characterising a tree and its stand, the exact locations of the trees are known, and we will make use of this spatial information. Therefore this case study will focus on detecting temporal and spatial trends while accounting for further covariate effects in a flexible manner.

Our data have been collected in annual visual forest health inventories carried out between 1983 and 2004 in a northern Bavarian forest district. The average health status at 83 observation plots with beeches (*Fagus sylvatica*) within an observation area extending 10 km from south to north and 15 km from west to east was assessed on an ordinal scale, where the nine possible categories denote different degrees of defoliation. The domain is divided in 12.5 % steps, ranging from healthy trees (0 % defoliation) to trees with 100 % defoliation. Figure 10.1 shows a histogram of the nine defoliation classes

Statistical Methods for Trend Detection and Analysis in the Environmental Sciences, First Edition.
Richard E. Chandler and E. Marian Scott.
© 2011 John Wiley & Sons, Ltd. Published 2011 by John Wiley & Sons, Ltd.

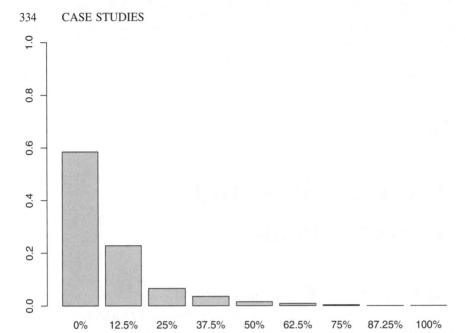

Figure 10.1   Forest health data: histogram of the nine defoliation classes for all obser-vation plots and all time points.

indicating that severe defoliation is relatively rare. Therefore we aggregated the data into three categories corresponding to healthy trees (0 % defoliation, category 1), slightly defoliated trees (between 12.5 % and 37.5 % defoliation, categories 2–4) and severely defoliated trees (more than 37.5 % defoliation, categories 5–9).

Since the data set has a longitudinal structure consisting of 83 time series of defo-liation states for the 83 observation plots, temporal correlations have to be considered appropriately. In addition, the trees are located within a relatively small observation area and spatial correlations due to comparable climate and site conditions are also likely to be encountered. Figure 10.2 shows the temporal development of the frequency of the three different defoliation states. Obviously, only a small percentage of the observation plots falls into the class with the highest defoliation degree. However, there seems to be a slightly increasing trend for this category. The lowest percentage of healthy trees is observed at the end of the 1980s. Afterwards, the population seems to recover, stabilis-ing at a percentage of approximately 60 % of healthy trees in the 1990s. All trends are relatively rough due to random variation in the data. Using a (nonparametric) regression model to estimate the trends allows the observation noise to be reduced, leading to more reliable estimates for the trends; it also controls for the effects of further covariates.

Figure 10.3 shows the distribution of the plots across the observation area. Each box corresponds to a specific observation plot and is shaded according to the percentage of years for which the trees were classified to show slight (light boxes) or severe (dark boxes) defoliation. Around Rothenbuch (the hole in the easterly part of the observation area) there seems to be an increased amount of defoliated trees. Employing spatial smoothing techniques will allow us to obtain clearer spatial patterns of healthy and defoliated trees and to estimate the spatial trend jointly with further covariate effects.

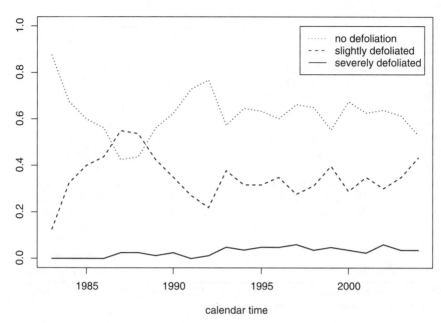

Figure 10.2   Forest health data: temporal development of the frequency of the three different defoliation states.

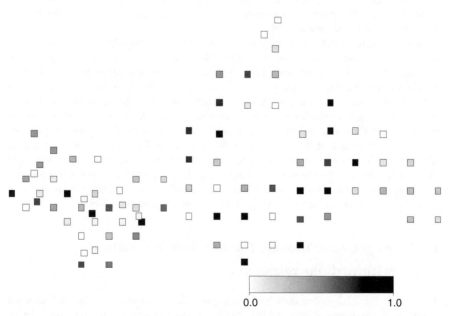

Figure 10.3   Forest health data: percentage of years for which an observation plot shows defoliation, averaged over the entire observation period.

Table 10.1   Forest health data: description of covariates.

Covariate	Description	Type	Range/coding
Age	Age of stand (years)	Con	[7, 234]
Elevation	Elevation (metres above sea level)	Con	[250, 480]
Inclination	Slope inclination (percent)	Con	[0, 46]
Soil	Soil layer depth (cm)	Con	[9, 51]
pH	Soil pH at 0-2cm depth	Con	[3.28, 6.05]
Canopy	Forest canopy density (%)	Con	[0, 1]
Stand	Stand type	Ca(2)	1 = 'deciduous', −1 = 'mixed'
Fertilised	Fertilisation applied?	Ca(2)	1 = 'yes', −1 = 'no'
Humus	Humus layer thickness	Or(5)	1 = 'low' to 5 = 'high'
Moisture	Soil moisture level	Or(3)	1 = 'moderately dry', 2 = 'moderately moist', 3 = 'moist or temporarily wet'
Saturation	Base saturation	Or(4)	1 = 'low' to 4 = 'high'

Key to 'Type' column: Con = continuous, Ca(n) = categorical with n categories, Or(n) = ordinal with n categories.

In addition to temporal and spatial information, numerous other covariates characterising the stand and the site are available. Table 10.1 contains covariates that either turned out to be significant in preliminary analyses or are of interest for substantive research questions. The set of covariates comprises both categorical and continuous covariates that have to be considered appropriately.

For example, Figure 10.4 displays the temporal development of the proportion of observation plots with defoliation for young trees (age≤75), middle-aged trees (75<age≤150) and old trees (age>150). Obviously, the three resulting time trends share some common features, e.g. the peak at the end of the 1980s, but there is also a clear evidence of a time-varying difference between the trends. To account for this, we have to include appropriate interaction terms between age and calendar time, e.g. by categorising the trees into three age groups and estimating time trends for each group separately. However, a drawback of this approach is the arbitrary definition of the age groups. In addition, employing a large number of age classes to obtain a realistic model may lead to unstable estimates for the trends. Therefore, a more sophisticated idea is to introduce a flexible interaction surface between age and calendar time. Similarly, nonparametric main effects of other covariates may be needed to obtain a realistic model.

10.2   Regression models for longitudinal data with ordinal responses

This section briefly describes three basic concepts that allow extensions of ordinal response models for cross-sectional data to a longitudinal setting. These are marginal models based on estimating equations, autoregressive-type transition models and random effects models. For notational simplicity, and in accordance with the data structure of the forest health

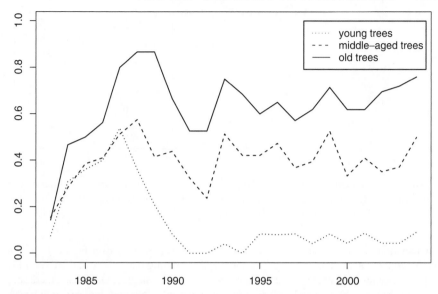

Figure 10.4   Forest health data: temporal development of the proportion of observation plots with defoliation for young trees (age≤75), middle-aged trees (75<age≤150) and old trees (age>150).

study, we assume that ordinal responses Y_{it} are observed for $i = 1, \ldots, n$ objects for the same time points or periods $t = 1, \ldots, T$, together with covariates. However, extensions to more general longitudinal data structures are of course possible.

Regression models for ordinal response variables are most easily defined based on continuous latent variables. In our specific application, a latent variable can be thought of as representing the true but unknown defoliation degree D_{it} (or more generally health state) of tree i at time t measured on a continuous scale. As in usual linear models, D_{it} can then be related to covariates x_{it} to form a linear predictor $x'_{it}\beta$ through

$$D_{it} = x'_{it}\beta + \varepsilon_{it}. \tag{10.1}$$

Keeping t fixed for the moment, ε_{it} are i.i.d. random error terms with a cumulative distribution function F. In a second step, the latent variable is linked to the categorical response variable based on a set of ordered thresholds

$$-\infty = \theta^{(0)} < \theta^{(1)} < \cdots < \theta^{(k)} = \infty.$$

These thresholds divide the real axis into k slices, and an observation is assigned to the class corresponding to its latent defoliation degree D_{it} via the threshold mechanism

$$Y_{it} = r \quad \Leftrightarrow \quad \theta^{(r-1)} < D_{it} \leq \theta^{(r)}. \tag{10.2}$$

Figure 10.5 gives an intuitive interpretation of the latent variable representation of cumulative regression models. Suppose that $f = F'$ is the density of ε_{it}. Then the distribution of D_{it} follows the same density shifted by $x'_{it}\beta$ and the regression coefficients β determine the direction and strength of the shift caused by a specific covariate

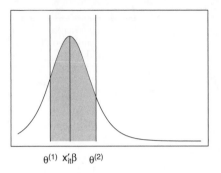 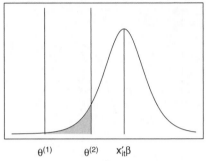

$\theta^{(1)}$ $x'_{it}\beta$ $\theta^{(2)}$ $\theta^{(1)}$ $\theta^{(2)}$ $x'_{it}\beta$

Figure 10.5   Interpretation of cumulative regression models: the linear predictor $x'_{it}\beta$ shifts the density of D_{it}. The shaded region denotes the probability $P(Y_{it} = 2)$ for an observation plot to be classified as being slightly defoliated.

combination x_{it}. This is illustrated for two different values of $x'_{it}\beta$ in Figure 10.5. The probability for a specific category r is obtained by splitting the density at the thresholds. As an example, the probability for category $r = 2$ is shaded in grey. Obviously, increasing the shift reduces the probability for a low category and conversely increases the probability for higher categories. Hence, positive regression coefficients are associated with a negative influence of the corresponding covariate on forest health in our example.

From (10.2), cumulative probabilities of Y_{it} can be easily obtained as

$$P(Y_{it} \leq r) = P(\varepsilon_{it} \leq \theta^{(r)} - x'_{it}\beta) = F(\theta^{(r)} - x'_{it}\beta), \quad r = 1, \ldots, k. \tag{10.3}$$

Hence the model defines cumulative probabilities, and is therefore usually called a cumulative regression model. Choosing a specific distribution for ε_{it} completes the model formulation. Popular variants are the logistic distribution or the standard normal distribution resulting in the cumulative logit and the cumulative probit model respectively. In the following, we will employ cumulative probit models but, as for binary responses, results obtained from logit and probit models are usually quite close to each other; see Fahrmeir and Tutz (2001, Chapters 2 and 3).

Three concepts may be distinguished for extending the basic cross-sectional model (10.3) to longitudinal data. The choice between them depends on the data structure and on the objectives of the study.

Marginal models are mostly used when longitudinal data consist of many short individual time series, i.e. for large n and very small T, and when the marginal effect of covariates on the response is of primary interest, whereas temporal correlation and time trends are of secondary interest (or even considered as nuisance components). Inference is usually based on a generalised estimating equations (GEE) approach, introduced by Liang and Zeger (1986) and Zeger and Liang (1986) for univariate, for example binary, responses. GEEs associate the seemingly separate marginal models (10.3) for $t = 1, \ldots, T$, through working correlations. Marginal models have been extended in various ways, in particular to categorical responses and to likelihood based models; see the book by Diggle *et al.* (2002). Because T is comparably large in our study and we aim at modelling and exploring temporal and spatial trends in the forest health data, marginal models are less appropriate. Also, our inferential concept is based on full likelihoods, whereas GEEs usually specify only first and second moments.

Conditional models introduce temporal correlation by including lagged values of the response variables in the predictor, i.e. by considering autoregressive models of the form

$$D_{it} = x_{it}'\beta + \alpha Y_{i,t-1} + \varepsilon_{it}. \tag{10.4}$$

In addition, interactions between the lagged variable $Y_{i,t-1}$ and other covariates may be considered or lags of higher order could be included. Conditional models of the form (10.4) induce Markov-type transition probabilities for the transition between $Y_{i,t-1}$ and Y_{it}. In contrast to simple Markov models, the transition probabilities additionally depend on covariates. Although conditional models can be easily fitted using standard software, they have the innate problem that covariate effects may be hidden by the autoregressive part of the model. Since $Y_{i,t-1}$ depends on essentially the same covariate information as Y_{it}, covariates may turn out to be insignificant in a conditional model although they are important to describe marginal properties of Y_{it}. In addition, the marginal expectation of D_{it} defined implicitly in Equation (10.4) is given by

$$E(D_{it}) = \left(\sum_{s=0}^{t-1} \alpha^s x_{is}'\right)\beta,$$

which is difficult to interpret (see Section 3.3.3 for further discussion of this issue). Therefore conditional models are well suited if prediction is of primary interest, but not the marginal expectation. On the other hand, they can also be applied if n is small and T is large. For the objectives of the forest health study, they are less useful because we are interested in detection of covariate effects and trends but not in prediction.

Mixed models induce temporal correlation by additional inclusion of random effects into the predictor of the linear model (10.1) and, as a consequence, into the predictor of the basic model (10.3). Generally, the standard form of a linear mixed model for (latent) longitudinal data is

$$D_{it} = x_{it}'\beta + z_{it}'b_i + \varepsilon_{it}, \tag{10.5}$$

where z_{it} is a design vector constructed from covariates, for example a subvector of x_{it}. The individual (plot-) specific random vectors b_i are i.i.d. Gaussian, $b_i \sim N(0, Q)$. In the simplest case $z_{it} = 1$, the model contains a random intercept only. Rewriting the linear mixed model as

$$D_{it} = x_{it}'\beta + \delta_{it}, \qquad \delta_{it} = z_{it}'b_i + \varepsilon_{it},$$

it is seen that the marginal expectation (i.e. integrating out the random effects b_i) is still of simple linear parametric form.

It is obviously difficult to model nonlinear time trends as in Figure 10.2 using simple parametric functions. Moreover, some effects of continuous covariates may be nonlinear too. To deal with this issue, all three basic types of longitudinal data models have been extended to incorporate additive nonparametric functions in a generalised additive model fashion; see, for example, Wild and Yee (1996) for additive extensions of GEE-based models, Lin and Zhang (1999) for additive mixed models and Sections 6.2 and 7.6 in Fahrmeir and Tutz (2001).

Mixed models are particularly attractive for flexible semiparametric approaches as needed for our health study. Temporal and spatial trends as well as nonparametric functions of continuous covariates can be incorporated through additional random effects, which are serially or spatially correlated, and inference can be performed within a unified framework. The next section suggests a broad class of spatiotemporal models through such an approach.

10.3   Spatiotemporal models

Random effects models allow for some specific types of correlation of the latent variables, but were mainly developed to account for the dependencies arising from unobserved heterogeneity in the data. In our example, the dependence structure is more complicated, since temporal and spatial correlations induced by a combination of unobserved covariates have to be considered. Furthermore, the assumption of purely parametric covariate effects may be too restrictive, and more general models allowing for flexible, nonlinear effects may be able to account for this. In the following, we will generalise the parametric models for ordinal responses to a general class of spatiotemporal or geoadditive regression models that allows for spatial and temporal correlations as well as nonlinear effects of covariates and unobserved heterogeneity.

Conceptually, the basic form of such a model has the form

$$D_{it} = f_1(v_{it1}) + \cdots + f_p(v_{itp}) + f_{\text{time}}(t) + f_{\text{spat}}(s_i) + x'_{it}\beta + \varepsilon_{it}, \tag{10.6}$$

where $f_1(v_{it1}), \ldots, f_p(v_{itp})$ are nonlinear effects of continuous covariates, $f_{\text{time}}(t)$ is a flexible time trend accounting for temporal correlations, $f_{\text{spat}}(s_i)$ is a spatial effect accounting for spatial correlations and $x'_{it}\beta$ comprises all further covariates with parametric effects in appropriate coding. Extended versions may be obtained by including tree-specific random intercepts (or slopes)

$$D_{it} = \cdots + b_i + \cdots$$

as in the linear mixed model (10.5) or interactions between some of the covariates

$$D_{it} = \cdots + f_{jk}(v_{itj}, v_{itk}) + \cdots$$

Tree-specific random intercepts can also be interpreted as an unstructured spatial effect that accounts for unobserved, locally varying covariates. In the geostatistical terminology, such a random effect would be called a nugget effect.

Both temporal and nonparametric effects can be estimated based on penalised splines; see Section 10.3.1 for details. Two-dimensional extensions of penalised splines can be employed both for the estimation of interaction surfaces and spatial effects; see Section 10.3.2. Competing methods for the latter are Markov random fields, and stationary Gaussian random fields, as described in Section 10.3.3.

In our specific application, a suitable model is given by

$$D_{it} = f_1(\text{age}_{it}) + f_2(\text{inclination}_i) + f_3(\text{canopy}_{it}) + f_{\text{time}}(t)$$

$$+ f_4(t, \text{age}_{it}) + f_{\text{spat}}(s_i) + b_i + x'_{it}\beta + \varepsilon_{it}. \tag{10.7}$$

Both temporal and spatial correlations are included as well as nonlinear effects of age, inclination of slope and the canopy density. These continuous covariates were identified in preliminary analysis as covariates with potentially nonlinear influence on the response.

In addition, the age effect is allowed to be time-varying, i.e. an interaction between the calendar time t and age is included in the model. The random intercept b_i accounts for tree-or location-specific heterogeneity.

10.3.1 Penalised splines

An approach to the estimation of nonparametric effects of continuous covariates and time trends that is both parsimonious and flexible was proposed by Eilers and Marx (1996). Their idea is based on a basis function representation of functions $f(v)$; i.e. $f(v)$ is written as a weighted sum of basis functions $B_m(v)$:

$$f(v) = \sum_{m=1}^{d} \xi_m B_m(v),$$

where $\xi = (\xi_1, \ldots, \xi_d)'$ corresponds to a vector of unknown regression coefficients. The specific basis functions chosen by Eilers and Marx (1996) are B-splines of degree l. Each of the basis functions is constructed as a combination of piecewise polynomials of degree l, which are connected in a way that ensures that $B_m(v)$ is $(l-1)$ times continuously differentiable (see Dierckx, 1993, for a formal definition of B-splines and rigorous developments of their properties). B-splines form a numerically stable basis with local support, resulting in mathematically favourable properties. As in Section 4.1.3, the points at which the splines are connected are called *knots*.

If the number and the locations of the basis functions are fixed, the regression coefficients ξ can be obtained by standard maximum likelihood techniques as in parametric regression models. Figure 10.6 illustrates such a B-spline fit for a Gaussian regression model with one nonparametric effect of sinusoidal form. Figure 10.6(a) shows a full set of B-spline basis functions of degree $l = 3$. In Figure 10.6(b) the basis functions are scaled according to the corresponding (estimated) regression coefficients $\hat{\xi}_m$. Obviously, basis functions in regions with many negative (positive) observations are assigned a negative (positive) weight. Summing the weighted basis functions finally yields the fit presented in Figure 10.6(c). The true function (a sine curve) is included as a dashed line for comparison.

When using such procedures, the choice of knot locations can have a substantial effect on the results. With a small number of knots, the estimated curve will be very smooth and therefore important details may be lost. On the other hand, a large number will lead to a wiggly estimate that is close to an interpolation of the data. To avoid the difficult task of choosing the number and position of the knots in an optimal way, Eilers and Marx (1996) propose to use a large number of basis functions but to penalise differences between adjacent parameters ξ_m to guarantee sufficient smoothness. This approach has also been employed in the illustrative fit in Figure 10.6. The coefficients ξ_m vary smoothly over the domain of x, resulting in a reasonable amount of smoothing.

Penalised spline smoothing for nonlinear effects is closely related to the spline smoothing approach discussed in Section 4.1.3 based on natural cubic smoothing splines: the basis is slightly different, but the difference penalty considered for penalised splines is a simple approximation to penalties based on integrated squared derivatives. Although penalised splines cannot be motivated as a solution of an optimisation criterion, they still provide a versatile alternative smoothing approach. Moreover, optimality of the smoothing spline in Section 4.1.3 only holds when all distinct observation points are used as knots, which is only rarely done in practice.

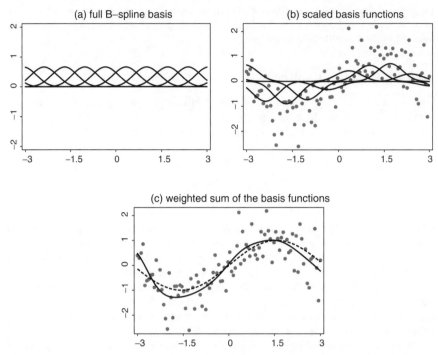

Figure 10.6 Spatiotemporal models: fitting nonparametric functions with B-splines. In (c) the dashed line represents the true curve and the solid line the corresponding B-spline estimate.

From a Bayesian perspective, penalisation of differences of adjacent parameters corresponds to the assumption of random walk priors for the coefficients. First- and second-order random walks are defined by

$$\xi_m = \xi_{m-1} + u_m \qquad \text{and} \qquad \xi_m = 2\xi_{m-1} - \xi_{m-2} + u_m, \qquad (10.8)$$

where the u_m are Gaussian error terms $u_m \sim N(0, \tau^2)$ and diffuse priors are assigned to the initial values. A first-order random walk penalises deviations of the function $f(v)$ from a constant value, whereas a second-order random walk penalises deviations from a function linear in v. The strength of the penalisation is controlled by the variance τ^2 of the error terms. If τ^2 is small, the regression coefficients will be almost deterministic, resulting in a constant or a linear effect depending on the specific random walk. If, in contrast, the variance is large, ample deviations from the deterministic trend are allowed and highly fluctuating functions are obtained. In contrast to unpenalised B-splines, the compromise between flexibility and fidelity to the data is controlled by one single parameter, the smoothing parameter τ^2. Hence, the complex problem of choosing optimal knots is replaced by the much simpler problem of finding an optimal smoothing parameter. We will discuss two different approaches to this problem in Section 10.3.4.

10.3.2   Interaction surfaces

If bivariate interaction surfaces in the spirit of univariate penalised splines need to be derived, we first have to define appropriate two-dimensional basis functions. A general way to construct bivariate interactions between two sets of univariate functions is to consider all pairwise products, i.e. the *tensor product* of the univariate basis functions. To be more specific, if we want to model an interaction term $f(x, z)$, a set of bivariate basis functions can be obtained as

$$B_{jk}(v_1, v_2) = \bar{B}_j(v_1)\tilde{B}_k(v_2),$$

where $\bar{B}_j$ and $\tilde{B}_k$ are univariate basis functions defined in the v_1 and v_2 domain respectively. Figure 10.7 shows two such basis functions, when the underlying univariate bases are B-spline bases of degree $l = 1$ and $l = 3$. Obviously, tensor product B-splines form a nonradial basis but the amount of nonradiality decreases rapidly with increasing degree. In the limit, appropriately normalised tensor product B-splines converge to the density of a standard normal distribution.

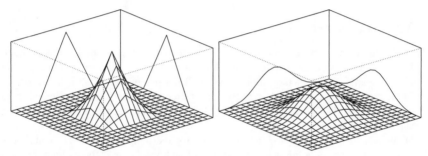

Figure 10.7   Spatiotemporal models: surface plots of tensor product B-splines of degree $l = 1$ and $l = 3$.

Based on the bivariate basis functions, nonparametric surface fits can be estimated in complete analogy to the univariate case described in the previous section. However, if a smoothness penalty is to be included, bivariate random walks have to be introduced for the regression parameters. Since the tensor product basis functions are aligned on a regular grid, it seems natural to consider a set of k nearest locations on the grid as neighbours of the basis function at hand. Two such neighbourhood definitions are illustrated in Figure 10.8, where the neighbours of basis function B_{jk} are represented as grey

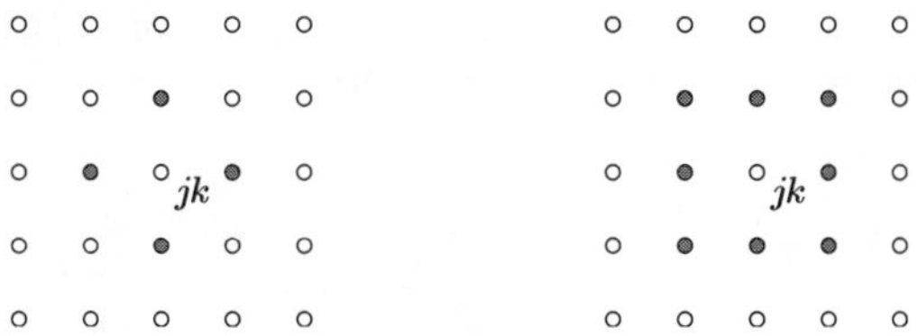

Figure 10.8   Spatiotemporal models: neighbourhoods on a regular lattice. The four (eight) next neighbours of location jk are indicated as grey points.

dots. Based on these neighbourhoods, a first-order random walk is defined by assuming that the expectation of the regression parameter at location jk is given by the average of the regression parameters at adjacent sites and its variance is inversely proportional to the number of neighbours. We will discuss this in more detail in the following section, where more general first-order random walks on irregular lattices will be considered for the construction of spatial smoothness priors. Random walks of higher order on regular lattices are treated extensively in Rue and Held (2005, Chapter 3).

Of course, interaction surfaces can also be employed to estimate spatial trends, if longitude and latitude are identified with the interacting covariates. As we will see in the next subsection, there is also a close correspondence between surface smoothers and more commonly applied spatial smoothing approaches.

Assuming a bivariate random walk prior for the coefficients of the tensor product basis functions induces a smoothness prior similar to the univariate case. Again the variance of the random walk plays a crucial role when determining the interaction effect since it acts as a smoothing parameter. In our definition, the smoothing parameter is the same for both the v_1 and the v_2 directions, but extensions with direction-specific smoothness penalties can also be defined (see, for example, Eilers and Marx, 2003).

10.3.3   Spatial trends

Although interaction surfaces can in principle be used to estimate spatial effects based on their x and y coordinates, more specialised approaches have been introduced that exploit the spatial nature of the problem. We will discuss two of these in more detail. While the first can be considered an extension of random walks from one dimension to two, the second explicitly models dependence between the spatial locations in terms of a correlation function.

In models for trend analysis, random walks for the temporal effect are commonly applied. If $f(t)$ represents the value of the trend function at time t, the assumption of a first-order random walk for $f(t)$ is equivalent to assuming

$$f(t)|f(t-1), f(t+1) \sim N\left(\frac{1}{2}[f(t-1) + f(t+1)], \frac{\tau^2}{2}\right); \qquad (10.9)$$

i.e. the expectation of $f(t)$ given the neighbours is simply the average of the two adjacent values $f(t-1)$ and $f(t+1)$. To generalise this to the bivariate setting, we first have to define which points s' shall be considered as spatial neighbours of a point s. While natural definitions exist for data on regular grids (as we have seen in the previous section), the definition of neighbours is less clear-cut for irregularly spaced observations as in our example. A simple rule to obtain a neighbourhood is to define two points as neighbours if their distance drops below a certain value. If ∂_s is the set of neighbours of location s we can generalise the temporal random walk (10.9) to

$$f(s)|f(s'), s' \neq s, \tau^2 \sim N\left(\frac{1}{N_s}\sum_{s' \in \partial_s} f(s'), \frac{\tau^2}{N_s}\right), \qquad (10.10)$$

where $N_s = |\partial_s|$. The variance τ^2 again plays the role of a smoothing parameter controlling the flexibility of the spatial function estimate. Implicitly, (10.10) defines correlations between locations which are close to each other since it forces adjacent function evaluations to be of similar magnitude.

An approach that explicitly considers correlations between locations, commonly known as *kriging* in the geostatistics literature, is based on stationary Gaussian random fields. In this case, $f(s)$ is assumed to follow a zero-mean Gaussian stochastic process with an intrinsic correlation function $\rho(u)$, where $u = ||s - s'||$ is the distance between two points s and s'. This means that correlations only depend on the (Euclidean) distance of two points and not on their specific location and direction. Choosing an appropriate correlation function $\rho(u)$ is important since the resulting estimates for the spatial effect inherit properties such as continuity and differentiability from the correlation function (Stein, 1999, Section 2.4). Several proposals have been made in the geostatistics literature, among which the Matérn family $\rho(u; \nu, \alpha)$ is highly recommended due to its flexibility (Stein, 1999, page 31).

While evaluation of $\rho(u; \alpha, \nu)$ in general requires the numerical evaluation of modified Bessel functions, simple forms $\rho(u; \alpha)$ are obtained for predetermined values $\nu = m + 0.5$, $m = 0, 1, 2, \ldots$, of the smoothness parameter ν, for example

$$\rho(u; \alpha, \nu = 0.5) = \exp(-|u/\alpha|),$$

$$\rho(u; \alpha, \nu = 1.5) = \exp(-|u/\alpha|)(1 + |u/\alpha|),$$

$$\rho(u; \alpha, \nu = 2.5) = \exp(-|u/\alpha|)(1 + |u/\alpha| + \tfrac{1}{3}|u/\alpha|^2).$$

Figure 10.9 displays the first two Matérn correlation functions with $\nu = 0.5$ and $\nu = 1.5$.

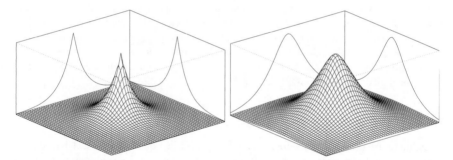

Figure 10.9   Spatiotemporal models: surface plots of Matérn correlation functions with smoothness parameters $\nu = 0.5$ and $\nu = 1.5$.

The scale parameter α controls how fast correlations die out with increasing distance u. To simplify the estimation procedure, α may be determined in a preprocessing step based on the rule

$$\hat{\alpha} = \max_{i,j} ||s_i - s_j||/c. \tag{10.11}$$

The constant $c > 0$ is chosen such that $\rho(c, \alpha = 1)$ is small (for example, 0.001); i.e. c specifies the desired effective range of the correlation function. Therefore, the different values of $||s_i - s_j||/\hat{\alpha}$ are spread out over the u axis of the correlation function and scale invariance of the estimation procedure is ensured.

The assumption that $f(s)$ follows a Gaussian stochastic process can also be interpreted as a smoothness prior for the spatial effect. In this case, the inverse of the correlation matrix of f_s, evaluated at the observation locations s_i, $i = 1, \ldots, n$, acts as a penalty matrix, while the variance of the process acts as the smoothing parameter that controls the influence of the penalty. In simple terms, two function evaluations $f(s_i)$ and $f(s_j)$ are forced to be close to each other if their correlation is large.

If we compare the tensor product B-spline bases depicted in Figure 10.7 and the Matérn correlation functions displayed in Figure 10.9, both look very similar. This is not a coincidence: kriging estimates can be interpreted as surface smoothers based on radial bases derived from the correlation functions (see Nychka, 2000, for a thorough discussion). Each of these basis functions is located at an observation point, which can therefore be considered as one of the knots of the surface smoother. This interpretation in terms of basis functions also provides the motivation for a dimension reduction technique that can be applied to reduce the numerical burden associated with estimating Gaussian fields. If we use only a subset of the original observation locations as knots associated with basis functions, this leads to low-rank kriging as considered by Kammann and Wand (2003).

Since spatial effects are usually introduced into regression models to account for spatial correlations induced by unobserved, spatially varying covariates, it can be useful to split the spatial effect into a spatially correlated (structured) part f_{str} and a spatially uncorrelated (unstructured) part f_{unstr}:

$$f_{spat}(s) = f_{str}(s) + f_{unstr}(s).$$

A rationale is that a spatial effect is usually a surrogate of many unobserved influential factors, some of which may obey a strong spatial structure while others may be present only locally. By estimating a structured and an unstructured spatial component we aim to distinguish between the two kinds of influential factors. The uncorrelated part f_{unstr} may be estimated based on location-specific i.i.d. Gaussian random effects, while Markov random fields, stationary Gaussian random fields or interaction surfaces can be used for the smooth spatial effect f_{str}. In our basic model (10.7) an unstructured spatial effect is already included via the random intercept b_i.

10.3.4   Inference in spatiotemporal models

To estimate the general spatiotemporal model defined in Equation (10.6), it is useful to express the model in matrix notation. It turns out that each of the predictor components can be expressed as the product of a design matrix V_j and a possibly high-dimensional vector of regression coefficients. For example, for penalised splines the design matrix consists of the basis functions evaluated at the observed values of the covariate and the regression coefficients are given by the amplitudes of the basis functions. Analogously, interaction surfaces and spatial effects can be expressed in a similar way. This yields the expression

$$D = V_1\xi_1 + \cdots + V_p\xi_p + V_{time}\xi_{time} + V_{spat}\xi_{spat} + X\beta + \varepsilon, \qquad (10.12)$$

where D and ε are stacked vectors of the individual observations and X corresponds to the usual design matrix of fixed effects.

From a Bayesian viewpoint, model formulation is completed by defining appropriate priors for the vectors of regression coefficients. While diffuse priors can be assigned to the thresholds $\theta^{(r)}$ (subject to the ordering constraint) and the fixed effects β, informative smoothness priors are employed for the remaining effects. It turns out that all the smoothness priors discussed in Sections 10.3.1 to 10.3.3 can be expressed as multivariate (partially improper) Gaussian priors of the form

$$p(\xi_j|\tau_j^2) \propto \exp\left(-\frac{1}{2\tau_j^2}\xi_j' K_j \xi_j\right), \tag{10.13}$$

where the precision matrix K_j acts as a penalty matrix that shrinks parameters towards zero or penalises abrupt jumps between neighbouring parameters. For example, for penalised splines the precision matrix is given by $K_j = D_j' D_j$, where D_j is a difference matrix of appropriate order. Hence, the precision matrix leads to the penalisation of differences of parameters corresponding to adjacent basis functions. For spatial effects, the precision matrix is either given by an adjacency matrix (Markov random fields) or as the inverse correlation matrix (kriging).

Having defined priors for the regression coefficients, we still have to specify assumptions on the variance parameters τ^2. In an empirical Bayes approach, the variances are treated as unknown constants, which have to be estimated from the corresponding marginal posterior. In contrast, fully Bayesian inference assigns additional hyperpriors to the variances and estimates them based on the joint posterior of all parameters. We discuss empirical Bayesian inference first and describe how mixed model methodology can be utilised for estimation. Afterwards, we give a brief description of fully Bayesian inference based on Markov chain Monte Carlo (MCMC) simulation techniques. Both approaches are implemented in the public domain software package **BayesX** (Belitz et al. 2009).

10.3.4.1 Mixed model based inference

The foundation for empirical Bayes inference is a different viewpoint on the model formulated in Equations (10.12) and (10.13). If we ignore the fact that the predictor in (10.12) in fact describes a semiparametric spatiotemporal regression model, it has a structure completely equivalent to that of a model with random effects ξ_j, where prior (10.13) defines a correlated random effects distribution. Therefore it is tempting to interpret the spatiotemporal model as a mixed model and to employ mixed model methodology to estimate it, in particular using penalised likelihood to estimate the regression coefficients and restricted maximum likelihood for the variance parameters. However, care has to be taken since the precision matrices of penalised spline and Markov random field priors are rank-deficient and therefore the corresponding prior distributions are partially improper.

Standard mixed model technology assumes that random effects are uncorrelated as in the basic model (10.5) for longitudinal data. If the multivariate prior (10.13) is proper, that is if the precision matrix K_j has full rank, this can be easily achieved by reparametrising ξ_j to $b_j = K_j^{0/5}\xi_j$, so that $b_j \sim N(0, \tau_j^2 I)$. This is the case if the spatial component is modelled through a GRF kriging prior. More care has to be taken for penalised spline and Markov random field priors, where the precision matrix is rank deficient and the corresponding prior (10.13) is partially improper. In this case it is possible to decompose ξ_j in reparametrised form into a 'fixed' parameter β_j with flat prior and a 'random' parameter b_j with proper Gaussian prior $N(0, \tau_j^2 I)$ of dimension rank(K_j). Details of this reparametrisation are given in Fahrmeir, Kneib and Lang (2004) for univariate response models and Kneib and Fahrmeir (2006) for categorical extensions. An extensive treatment of mixed model based inference in extended spatiotemporal regression models for several types of responses including survival times is presented in Kneib (2006).

In classical mixed model terminology, the reparameterised model is a variance components model with one variance component per nonparametric or spatial term. Mixed model methodology, adapted to our categorical setting, allows the estimation of both the regression coefficients based on penalised likelihood and restricted maximum likelihood (REML) for the variance parameters. As Harville (1974) showed, REML estimation is equivalent to the maximisation of the marginal posterior of the variance parameters leading to an interpretation of the derived estimates as empirical Bayes estimates (see Kneib and Fahrmeir, 2006, for details of the inferential scheme).

10.3.4.2   Inference based on Markov chain Monte Carlo simulation techniques

In a fully Bayesian approach, the variance parameters τ_j^2 can be estimated simultaneously with the regression coefficients ξ_j by assigning additional hyperpriors to them. The most common assumption is that the τ_j^2 are independently inverse gamma distributed, i.e. $\tau_j^2 \sim IG(a_j, b_j)$, with hyperparameters a_j and b_j specified a priori. A standard choice is to use $a_j = b_j = 0.001$. In some data situations (for example for small sample sizes), the estimated nonlinear functions f_j may depend considerably on the particular choice of hyperparameters. It is therefore good practice to estimate all models under consideration using a (small) number of different choices for a_j and b_j to assess the dependence of results on minor changes in the prior assumptions.

Efficient sampling schemes for cumulative probit models can be developed through incorporation of the Gaussian mixed model for the latent continuous defoliation degree D_{it} in combination with the threshold mechanism relating D_{it} to the observed categorical response Y_{it}. The model and the posterior are augmented with the latent variables D_{it}, which are generated as part of the MCMC sampling scheme. Sampling the D_{it} is relatively easy and fast because their full conditionals are truncated normal distributions. The full conditionals for the fixed effects β and the random effects ξ_j, given the latent variables and the other parameters, are Gaussian, and the full conditionals for the variances τ_j^2 are again inverse gamma distributions with updated parameters. Therefore, the MCMC algorithm can be implemented through Gibbs sampling, without a need for Metropolis–Hastings steps. Details are given in Fahrmeir and Lang (2001) and Brezger and Lang (2006).

10.4   Spatiotemporal modelling and analysis of forest health data

Apart from analysing the effects of the covariates given in Table 10.1, we aim to illustrate the possibilities and the flexibility of our spatiotemporal models for exploring temporal and spatial trends. In exploratory analysis we found age, inclination of slope and the density of the forest canopy to be of possibly nonlinear influence on the defoliation degree, while the remaining continuous covariate effects appeared approximately linear. Our basic spatiotemporal model is therefore specified by $D_{it} = \eta_{it} + \varepsilon_{it}$ with the spatiotemporal predictor

$$\eta_{it} = f_1(\text{age}_{it}) + f_2(\text{inclination}_i) + f_3(\text{canopy}_{it}) + f_{\text{time}}(t)$$

$$+ f_4(t, \text{age}_{it}) + f_{\text{spat}}(s_i) + x_{it}'\beta. \tag{10.14}$$

Both temporal and spatial trends are included as well as nonlinear main effects of age, inclination of slope and the canopy density. In addition, motivated by Figure 10.4, an interaction term between calendar time t and age is included in the predictor. The resulting probit model for the categories 'no defoliation' ($r = 1$) and 'slight defoliation' ($r = 2$) of the observed response Y_{it} is given by

$$P(Y_{it} \leq r | \eta_{it}) = \Phi(\theta^{(r)} - \eta_{it}), \quad r = 1, 2,$$

and the probability for 'severe defoliation' ($r = 3$) is obtained as

$$P(Y_{it} = 3 | \eta_{it}) = 1 - \Phi(\theta^{(2)} - \eta_{it}) = \Phi(-\theta^{(2)} + \eta_{it}).$$

We compared models with a Markov random field prior, a kriging prior and a bivariate P-spline prior for the spatial effect. For the Markov random field prior, two observation plots were considered as neighbours if their distance is less than 1.2 km. For the correlation function of the stationary Gaussian random field in the kriging approaches, we chose a Matérn correlation function with $\nu = 1.5$. The two-dimensional P-splines were defined as cubic tensor product B-splines with 12 inner knots for each direction and a bivariate first-order random walk penalty. Similarly, the interaction effect between age and calendar time was modelled using a bivariate cubic P-spline with a first-order random walk penalty but 20 inner knots for each direction. For the univariate nonparametric effects we employed cubic penalised splines with a second-order random walk penalty and 20 inner knots.

As an extension, we also considered models with linear predictor $\tilde{\eta}_{it} = \eta_{it} + b_i$, where the $\{b_i\}$ are uncorrelated tree- or location-specific random effects. For comparison we also fitted a model without spatial effects and a purely linear random effects model as in (10.5), where the nonlinear effects and the interaction are replaced by polynomials up to order three and all pairwise interactions between the polynomials of age and time. Thus, we considered the following models:

(M0) A purely parametric GLMM with third-order polynomials for the nonlinear effects and all pairwise interactions between the polynomials of age and time;

(M1) Model (10.14) with neither a spatial effect nor a random effect;

(M2) Model (10.14) with uncorrelated random effects as the spatial effect;

(M3) Model (10.14) with a Markov random field prior for the spatial effect;

(M4) Model (10.14) with a Kriging prior for the spatial effect;

(M5) Model (10.14) with a bivariate P-spline prior for the spatial effect;

(M6) Model (10.14) with a Markov random field prior plus uncorrelated random effects;

(M7) Model (10.14) with a Kriging prior plus uncorrelated random effects;

(M8) Model (10.14) with a bivariate P-spline prior plus uncorrelated random effects.

Estimation was carried out using both the empirical Bayes approach based on mixed model methodology and the fully Bayesian MCMC approach. For the latter, models including Kriging terms were excluded from the comparison since they would require the inversion of a full 83×83 matrix in each iteration, leading to quite long execution times.

Table 10.2 summarises several characteristics of the model fit for the various models. For the empirical Bayes approach, this includes minus twice the log-likelihood, the

Table 10.2   Spatiotemporal analysis: characteristics of the model fit for models (M0) to (M8).

Model	Empirical Bayes				Full Bayes		
	$-2l$	df	AIC	GCV	dev.	p_D	DIC
(M0)	1280.6	90.2	1461.1	0.64	1280.8	86.0	1452.7
(M1)	1390.7	96.4	1583.4	0.69	1394.4	89.1	1572.6
(M2)	1108.8	123.3	1355.4	0.54	1080.5	119.5	1319.6
(M3)	1119.5	119.4	1358.3	0.54	1089.3	116.8	1323.0
(M4)	1127.1	119.0	1365.1	0.55			
(M5)	1140.9	117.2	1375.3	0.56	1117.8	114.2	1346.3
(M6)	1109.2	123.2	1355.6	0.54	1080.8	119.3	1319.4
(M7)	1108.3	123.4	1355.2	0.54			
(M8)	1107.5	123.6	1354.8	0.54	1081.3	120.0	1321.3

effective number of parameters in the model (also called the degrees of freedom, see Section 4.1.4) df, Akaike's information criterion AIC and the generalised cross-validation statistic GCV. For the fully Bayesian approach, Table 10.2 includes the posterior deviance, the effective number of parameters p_D and the deviance information criterion (Spiegelhalter *et al.*, 2002). Obviously, including a spatial effect leads to an improved fit, regardless of the chosen parametrisation. Differences between the spatial models are smaller, but there seems to be a preference for models including an unstructured spatial effect or consisting only of an uncorrelated random effect. Differences between these four models are very small both in terms of AIC, GCV and DIC. There is also clear evidence that inclusion of nonparametric effects instead of simple polynomials results in a significantly better model fit.

Taking a closer look at the estimation results for the spatial effects reveals that in models with only a structured spatial effect the estimates are quite close to those obtained with only an uncorrelated plot-specific random effect. Probably this is due to the fact that our data set already contains several covariates describing forest site conditions that vary smoothly over the spatial domain. Therefore the remaining spatial effect mostly captures tree-specific, highly fluctuating unobserved covariates. In accordance with this hypothesis, the estimated structured spatial effect is generally quite small in models (M6), (M7) and (M8) where both an unstructured and a structured effect are included. For the Kriging and the bivariate P-spline prior in (M7) and (M8) the structured spatial effect is almost zero because these priors are in favour of smooth spatial effects or surfaces. In contrast, model (M6) reveals both an unstructured and a structured pattern for the spatial effect, although the structured effect is still small compared to the unstructured part. Therefore we will mostly concentrate on model (M6) in the following presentations, although it is not the best model according to AIC and DIC.

Figure 10.10 displays the empirical Bayes estimates for the spatial effects obtained in models (M2), (M3), (M5) and (M6). Obviously, results obtained with only a single term for the spatial effect are very close to each other, although no spatial correlation is assumed in model (M2). In model (M6), we obtain a clearer discrimination between two types of spatially varying influences. While observation plots in the westerly and the northern part of the observation area have a slightly negative structured spatial effect, positive effects

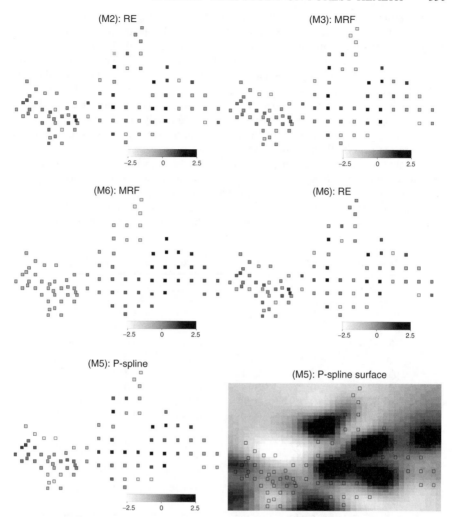

Figure 10.10   Posterior mode estimates of the spatial effects.

are observed in the easterly and the southern part. This corresponds to more healthy circumstances in the northwestern part of the map. The unstructured, location-specific spatial effect in model (M6) gives hints on locally varying influences on the health status of beeches. For example, adding up both effects, we recover the less favourable health status around Rothenbuch found in the descriptive analyses.

In contrast to models based on MRFs and random effects where the spatial effect is only defined at the observation locations, bivariate penalised splines and GRFs allow for the estimation of spatial surfaces. This is illustrated in Figure 10.10 for model (M5) with the spatial effect estimated by a bivariate P-spline. Although care has to be taken when interpreting the spatial effect in regions without any observations, the estimated surface still allows the progression of the spatial effect between two locations to be judged.

The estimated linear effects of the continuous and categorical covariates in the vector x_{it} were rather similar for the various models. Table 10.3 contains the estimation results from model (M6) (for a definition of the covariates compare Table 10.1). The defoliation degree is not affected by elevation above sea level, depth of soil layer or soil pH. This may be particularly surprising for the latter, but the reason is that the pH values do not differ too much and are all below a value of 7, implying acidic soil for the whole observation area. The effects of stand type and fertilisation give some evidence for their expected impact on health status, despite not being formally significant. The effects of the thickness of the humus layer and the level of soil moisture are significant and reveal an interesting pattern. From humus levels 1 to 3 there is an increased tendency towards defoliation; the effect is then roughly constant for the higher levels. The thickness of the humus layer tends to increase from pure broadleaved forests over mixed forests to coniferous forests. Thus the thickness of the humus layer can be regarded as a proxy for both stand and soil conditions: this leads to the interpretation that in pure beech stands the risk for defoliation is lower than in forests with a higher proportion of other tree species.

Table 10.3   Posterior mode estimates: Fixed effects in model (M6).

Variable	$\hat{\beta}_j$	std. dev.	p-value	95 % ci	
$\theta^{(1)}$	0.046	1.568	0.977	−3.027	3.120
$\theta^{(2)}$	3.603	1.577	0.022	0.511	6.695
elevation	0.001	0.003	0.741	−0.005	0.007
soil	−0.020	0.015	0.172	−0.049	0.009
ph	−0.037	0.212	0.860	−0.452	0.377
stand	0.253	0.147	0.086	−0.036	0.541
fertilised	−0.429	0.268	0.108	−0.954	0.095
humus1	−0.261	0.108	0.015	−0.472	−0.051
humus2	−0.135				
humus3	0.139	0.086	0.105	−0.029	0.308
humus4	0.135	0.102	0.185	−0.064	0.334
humus5	0.122	0.142	0.391	−0.157	0.400
moisture1	−0.597	0.320	0.061	−1.224	0.029
moisture2	0.185				
moisture3	0.412	0.229	0.071	−0.037	0.862
saturation1	0.430	0.327	0.188	−0.210	1.070
saturation2	−0.397				
saturation3	−0.366	0.344	0.288	−1.040	0.309
saturation4	0.333	0.445	0.455	−0.540	1.205

Similarly, the results for the level of soil moisture may be somewhat unexpected at first glance, showing higher defoliation at plots with better water availability. As shown in Figure 10.2 there was a peak of defoliation in the 1980s, most probably triggered by the extremely dry summer in 1983. Trees growing on dryer site conditions should be more adapted to drought than trees on more moist sites normally not suffering from drought. Thus the drought in 1983 may have been more harmful for beech trees on normally moist sites, leading to the results calculated for the covariate moisture in Table 10.3.

Figures 10.11 and 10.12 display the posterior mode estimates of the nonlinear effects obtained in model (M6) that combines an MRF prior and uncorrelated random effects. Additionally included are results from an autoregressive version of model (M3), where indicators formed of lagged response variables $Y_{i,t-1}$ are included as additional covariates (Figure 10.11) and model (M1) that neglects spatial correlations (Figure 10.12). In model (M6) the age effect is rapidly increasing for young trees, reaching a maximum at an age of 55 years. Afterwards, it is decreasing for a short period followed by a slight increase up to an age of about 210 years. The estimated time trend recovers the empirical trend of category 2 shown in Figure 10.2. Therefore, model (M6) may be well suited to distinguish between healthy trees and defoliated trees but may not discriminate well between slightly and severely defoliated trees. This is due to the assumption of one common time trend in model (10.14). We will investigate this further below. Inclination of slope turns out to have an almost linear effect and is also not significant. In contrast, the effect of canopy density shows significant deviations from a horizontal line. Obviously, an increased canopy density leads to higher probabilities for lower categories and, hence, for a more healthy state of the trees. Note that this conclusion is specific for beeches and does not necessarily hold for other tree species.

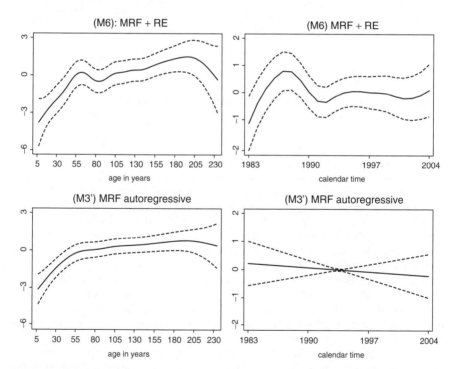

Figure 10.11   Posterior mode estimates for the time trend and the nonlinear age effect.

While the estimated nonparametric effects remained quite stable as long as any spatial effect was included in the model, some important deviations can be found when the spatial effect is excluded or when an autoregressive component is added. In the former case, the nonparametric estimates become more wiggly, as demonstrated in Figure 10.12

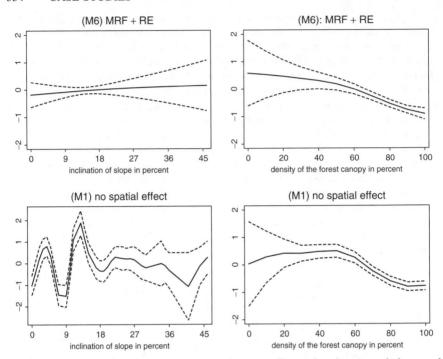

Figure 10.12   Posterior mode estimates: nonlinear effects of inclination of slope and density of the forest canopy.

for the effects of inclination and canopy density. Obviously, the nonparametric effects absorb some of the influences that are otherwise covered by the spatial component. The inclusion of an autoregressive term leads to less pronounced nonparametric effects, as shown for age and the calendar time in Figure 10.11. This is due to the effect that the lagged variable $Y_{i,t-1}$ already contains quite a lot of information on Y_{it} and therefore less variation is left that is to be explained by covariates. Hence direct modelling of spatial and temporal correlations is to be preferred if, as in our application, the emphasis is on covariate effects and not on prediction.

Figure 10.13 shows the estimated interaction effect between calendar time and age from model (M6). Apparently, young trees were in a poorer health state in the 1980s but recovered in the 1990s, unlike older trees which showed the contrary behaviour. A possible interpretation is that it takes longer until older trees are affected by unfavourable environmental circumstances while younger trees are affected more quickly but adapt as they grow older. The interaction effect appeared to be stable with respect to the specification of the spatial effect but disappeared when including lagged realisations of the response. Again the explicit modelling of the dependency structure instead of a conditional autoregressive modelling leads to additional insight into influences of covariates on health status.

For comparison, Figure 10.14 shows the estimated nonlinear and spatial effects from model (M6) obtained with fully Bayesian inference. The nonlinear effects are slightly more pronounced in this case: in particular, the effect of inclination shows somewhat more

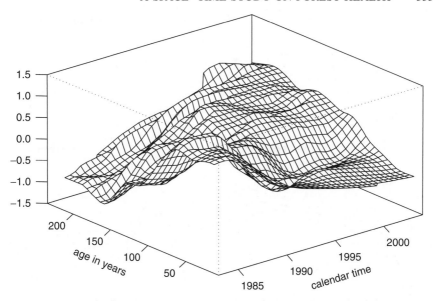

Figure 10.13   Posterior mode estimates: interaction effect in model (M6).

nonlinearity. In addition, the structured spatial effect has a wider range. This corresponds to our experience from many other analyses with simulated and real data: fully Bayesian inference tends to yield higher variance estimates, implying somewhat rougher surfaces.

As a next step, we take a more detailed look at the estimated time trend. Figure 10.2 gives clear evidence that the averaged trend for severe defoliation is different from the trends for the two other categories, while model (10.14) only allows for a common time trend $f_{\text{time}}(t)$. Because the majority of the data are from the two other categories, the estimated trend in Figure 10.11 resembles the corresponding average trends in Figure 10.2 but neglects the different shape of the trend for severe defoliation. We can deal with this issue by introducing category-specific trends $f_{\text{time}}^{(1)}(t)$ and $f_{\text{time}}^{(2)}(t)$ in the probit model, i.e. we assume

$$P(Y_{it} \leq r|.) = \Phi(\theta^{(r)} - f_{\text{time}}^{(r)}(t) - \eta_{it}^*), \quad r = 1, 2,$$

where η_{it}^* is the predictor without the global time trend $f_{\text{time}}(t)$. It follows that

$$P(Y_{it} = 3|.) = \Phi(-\theta^{(2)} + f_{\text{time}}^{(2)}(t) + \eta_{it}^*).$$

While such extensions are conceptually easy to define, complicated constraints on the parameters in the model arise since the inequality

$$\theta^{(1)} - f_{\text{time}}^{(1)}(t) < \theta^{(2)} - f_{\text{time}}^{(2)}(t) \tag{10.15}$$

has to be fulfilled for every value of t. Especially in sparse data situations, numerical problems are likely to be encountered due to the failure of (10.15).

Figure 10.15 shows the posterior mode estimates for the category-specific time trends. While the estimate for $f_{\text{time}}^{(1)}(t)$ is almost identical to the global trend in the previous

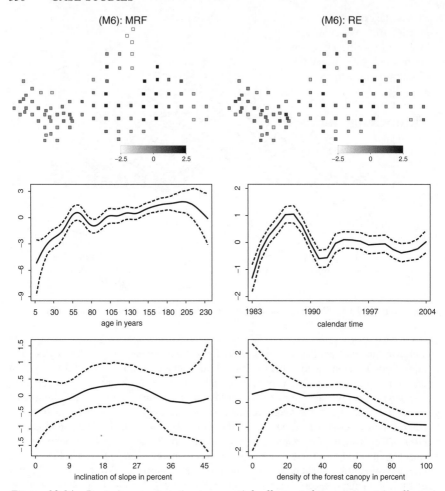

Figure 10.14   Posterior mean estimates: spatial effects and nonparametric effects in model (M6).

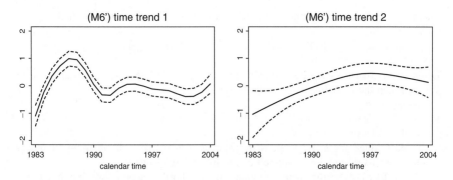

Figure 10.15   Posterior mode estimates: category-specific time trends.

analyses, the estimated trend for severe defoliation reflects the shape of the averaged trend in Figure 10.2. We conclude that it may be generally useful to consider the possibility of category-specific effects in cumulative regression models for categorical responses. With the relatively sparse data at hand, however, we could not model or identify further covariate-specific effects.

Acknowledgements

The authors thank Axel Göttlein for providing the data as well as for helpful comments and discussions. Financial support from the German Science Foundation (DFG), Collaborative Research Center 386 'Statistical Analysis of Discrete Structures' is gratefully acknowledged.

References

Belitz, C., Brezger, A., Kneib, T. and Lang, S. (2009) BayesX–*Software for Bayesian Inference in Structured Additive Regression Models*. Version 2.0.1. Available from http://www.stat.uni-muenchen.de/~bayesx.

Brezger, A. and Lang, S. (2006) Generalized structured additive regression based on Bayesian P-splines. *Computational Statistics and Data Analysis*, **50**, 967–991.

Dierckx, P. (1993) *Curve and Surface Fitting with Splines*. Clarendon Press, Oxford.

Diggle, P. J., Heagerty, P., Liang, K. Y. and Zeger, S. L. (2002) *The Analysis of Longitudinal Data*. Oxford University Press, Oxford.

Eilers, P. H. C. and Marx, B. D. (1996) Flexible smoothing with B-splines and penalties. *Statistical Science*, **11**, 89–121.

Eilers, P. H. C. and Marx, B. D. (2003) Multivariate calibration with temperature interaction using two-dimensional penalized signal regression. *Chemometrics and Intelligent Laboratory Systems*, **66**, 159–174.

Fahrmeir, L., Kneib, T. and Lang, S. (2004) Penalized structured additive regression for space–time data: a Bayesian perspective. *Statistica Sinica*, **14**, 731–761.

Fahrmeir, L. and Lang, S. (2001) Bayesian semiparametric regression analysis of multicategorial time–space data. *Annals of the Institute of Statistical Mathematics*, **53**, 11–30.

Fahrmeir, L. and Tutz, G. (2001) *Multivariate Statistical Modelling Based on Generalized Linear Models*. Springer, New York.

Harville, D. A. (1974) Bayesian inference for variance components using only error contrasts. *Biometrika*, **61**, 383–385.

Kammann, E. E. and Wand, M. P. (2003) Geoadditive models. *Journal of the Royal Statistical Society, Series C*, **52**, 1–18.

Kneib, T. (2006) *Mixed Model Based Inference in Structured Additive Regression*. PhD thesis, Ludwig-Maximilians-University Munich. Available from http://edoc.ub.uni-muenchen.de.

Kneib, T. and Fahrmeir, L. (2006) Structured additive regression for categorical space-time data: a mixed model approach. *Biometrics*, **62**, 109–118.

Liang, K. Y. and Zeger, S. L. (1986) Longitudinal data analysis using generalized linear models. *Biometrika*, **73**, 13–22.

Lin, X. and Zhang, D. (1999) Inference in generalized additive mixed models by using smoothing splines. *Journal of the Royal Statistical Society, Series B*, **61**, 381–400.

Nychka, D. (2000) Spatial-process estimates as smoothers. In *Smoothing and Regression: Approaches, computation and application* (ed. M. Schimek). John Wiley & Sons, Inc., New York.

Rue, H. and Held, L. (2005) *Gaussian Markov Random Fields. Theory and Applications*. CRC/Chapman & Hall, London.

Spiegelhalter, D. J., Best, N. G., Carlin, B. P. and van der Linde, A. (2002) Bayesian measures of model complexity and fit. *Journal of the Royal Statistical Society, Series B*, **65**, 583–639.

Stein, M. L. (1999) *Interpolation of Spatial Data. Some Theory for Kriging*. Springer, New York.

Wild, C. J. and Yee, T. W. (1996) Additive extensions to generalized estimating equation methods. *Journal of the Royal Statistical Society, Series B*, **58**, 711–725.

Zeger, S. L. and Liang, K. Y. (1986) Longitudinal data analysis for discrete and continuous outcomes. *Biometrics*, **42**, 121–130.

Index

Statistical Methods for Trend Detection and Analysis in the Environmental Sciences, First Edition.
Richard E. Chandler and E. Marian Scott.
© 2011 John Wiley & Sons, Ltd. Published 2011 by John Wiley & Sons, Ltd.

Statistics in Practice

Human and Biological Sciences

Berger – Selection Bias and Covariate Imbalances in Randomized Clinical Trials

Berger and Wong – An Introduction to Optimal Designs for Social and Biomedical Research

Brown and Prescott – Applied Mixed Models in Medicine, Second Edition

Carstensen – Comparing Clinical Measurement Methods

Chevret (Ed) – Statistical Methods for Dose-Finding Experiments

Ellenberg, Fleming and DeMets – Data Monitoring Committees in Clinical Trials: A Practical Perspective

Hauschke, Steinijans & Pigeot – Bioequivalence Studies in Drug Development: Methods and Applications

Lawson, Browne and Vidal Rodeiro – Disease Mapping with WinBUGS and MLwiN

Lesaffre, Feine, Leroux & Declerck – Statistical and Methodological Aspects of Oral Health Research

Lui – Statistical Estimation of Epidemiological Risk

Marubini and Valsecchi – Analysing Survival Data from Clinical Trials and Observation Studies

Molenberghs and Kenward – Missing Data in Clinical Studies

O'Hagan, Buck, Daneshkhah, Eiser, Garthwaite, Jenkinson, Oakley & Rakow – Uncertain Judgements: Eliciting Expert's Probabilities

Parmigiani – Modeling in Medical Decision Making: A Bayesian Approach

Pintilie – Competing Risks: A Practical Perspective

Senn – Cross-over Trials in Clinical Research, Second Edition

Senn – Statistical Issues in Drug Development, Second Edition

Spiegelhalter, Abrams and Myles – Bayesian Approaches to Clinical Trials and Health-Care Evaluation

Walters – Quality of Life Outcomes in Clinical Trials and Health-Care Evaluation

Whitehead – Design and Analysis of Sequential Clinical Trials, Revised Second Edition

Whitehead – Meta-Analysis of Controlled Clinical Trials
Willan and Briggs – Statistical Analysis of Cost Effectiveness Data
Winkel and Zhang – Statistical Development of Quality in Medicine

Earth and Environmental Sciences

Buck, Cavanagh and Litton – Bayesian Approach to Interpreting Archaeological Data
Chandler and Scott – Statistical Methods for Trend Detection and Analysis in the Environmental Statistics
Glasbey and Horgan – Image Analysis in the Biological Sciences
Haas – Improving Natural Resource Management: Ecological and Political Models
Helsel – Nondetects and Data Analysis: Statistics for Censored Environmental Data Illian, Penttinen, Stoyan, H and Stoyan D–Statistical Analysis and Modelling of Spatial Point Patterns
McBride – Using Statistical Methods for Water Quality Management
Webster and Oliver – Geostatistics for Environmental Scientists, Second Edition
Wymer (Ed) – Statistical Framework for Recreational Water Quality Criteria and Monitoring

Industry, Commerce and Finance

Aitken – Statistics and the Evaluation of Evidence for Forensic Scientists, Second Edition
Balding – Weight-of-evidence for Forensic DNA Profiles
Brandimarte – Numerical Methods in Finance and Economics: A MATLAB-Based Introduction, Second Edition
Brandimarte and Zotteri – Introduction to Distribution Logistics
Chan – Simulation Techniques in Financial Risk Management
Coleman, Greenfield, Stewardson and Montgomery (Eds) – Statistical Practice in Business and Industry
Frisen (Ed) – Financial Surveillance
Fung and Hu – Statistical DNA Forensics
Gusti Ngurah Agung – Time Series Data Analysis Using EViews
Kenett (Eds) – Operational Risk Management: A Practical Approach to Intelligent Data Analysis
Jank and Shmueli (Ed.) – Statistical Methods in e-Commerce Research

Lehtonen and Pahkinen – Practical Methods for Design and Analysis of Complex Surveys, Second Edition

Ohser and Mücklich – Statistical Analysis of Microstructures in Materials Science

Pourret, Naim & Marcot (Eds) – Bayesian Networks: A Practical Guide to Applications

Taroni, Aitken, Garbolino and Biedermann – Bayesian Networks and Probabilistic Inference in Forensic Science

Taroni, Bozza, Biedermann, Garbolino and Aitken – Data Analysis in Forensic Science